Acts of Occupation

Canada and Arctic Sovereignty, 1918-25

Janice Cavell and Jeff Noakes

UBCPress · Vancouver · Toronto

21 20 19 18 17 16 15 14 13 12 11 10   5 4 3 2 1

Printed in Canada on paper that is processed chlorine- and acid-free.

Library and Archives Canada Cataloguing in Publication

Cavell, Janice
Acts of occupation : Canada and Arctic sovereignty, 1918-25 / Janice Cavell and Jeff Noakes.

Includes bibliographical references and index.
ISBN 978-0-7748-1867-4 (bound) / ISBN 978-0-7748-1868-1 (pbk.)

1. Canada, Northern – Discovery and exploration. 2. Arctic regions – Discovery and exploration. 3. Canada – Boundaries – Arctic regions. 4. Jurisdiction, Territorial – Canada. 5. Canada – Foreign relations – 1918-1945. I. Noakes, Jeffrey David, 1970- II. Title.

FC3963.C38 2010   917.1904'2   C2010-903472-4

e-book ISBNs: 978-0-7748-1869-8 (pdf); 978-0-7748-1870-4 (epub)

Canadä

UBC Press gratefully acknowledges the financial support for our publishing program of the Government of Canada through the Book Publishing Industry Development Program (BPIDP), and of the Canada Council for the Arts, and the British Columbia Arts Council.

This book has been published with the help of a grant from the Canadian Federation for the Humanities and Social Sciences, through the Aid to Scholarly Publications Programme, using funds provided by the Social Sciences and Humanities Research Council of Canada.

Printed and bound in Canada by Friesens
Set in Galliard and New Baskerville by Artegraphica Design Co. Ltd.
Copy editor: Lesley Erickson

UBC Press
The University of British Columbia
2029 West Mall
Vancouver, BC V6T 1Z2
604-822-5959 / Fax: 604-822-6083
www.ubcpress.ca

Contents

Maps and Figures

Maps

Figures

Acknowledgments

We would like first of all to thank Norman Hillmer, an incomparable teacher during our years of doctoral study at Carleton University and an unfailing source of wisdom and advice since then. Our colleagues at Foreign Affairs and International Trade Canada, the Canadian War Museum, and the Canadian Museum of Civilization assisted and encouraged us in many ways; we would like to thank Glenn Ogden in particular for alerting us to the Eaton papers in the Archives of Ontario.

The staff at Library and Archives Canada, Barb Krieger at the Dartmouth College Library, Sarah Strong at the Royal Geographical Society Archives, Caroline Herbert at the Churchill College Archives, Janice Millard at the Trent University Archives, Naomi Boneham and Lucy Martin at the Scott Polar Research Institute, the Special Collections staff at the University of British Columbia Library, and Martin Legault at the Natural Resources Canada Library all provided invaluable assistance. Heather Dichter, Meaghan Beaton, and John Richthammer copied documents for us in London, Peterborough, and Winnipeg, while Jennifer Ellison carried out research at the Archives of Ontario.

Thanks also to Ian Stone and Karen McCullough, the editors of *Polar Record* and *Arctic*, and to the Scott Polar Research Institute and the Arctic Institute of North America for permission to reprint material that first appeared in these journals. The anonymous reviewers of the articles and of the book manuscript made a number of extremely helpful suggestions. At UBC Press, Melissa Pitts, Ann Macklem, and Emily Andrew were enthusiastic about the project from the beginning and provided expert guidance.

Finally, we owe an unusual but very large debt of gratitude to Belle and Rudolph Anderson, who preserved a remarkable trove of Arctic documents. Without the material in the Anderson fonds at Library and Archives Canada, our research would have been far more difficult and far less productive. The Andersons wanted a book like this one to be written; we can only hope they would be pleased with it.

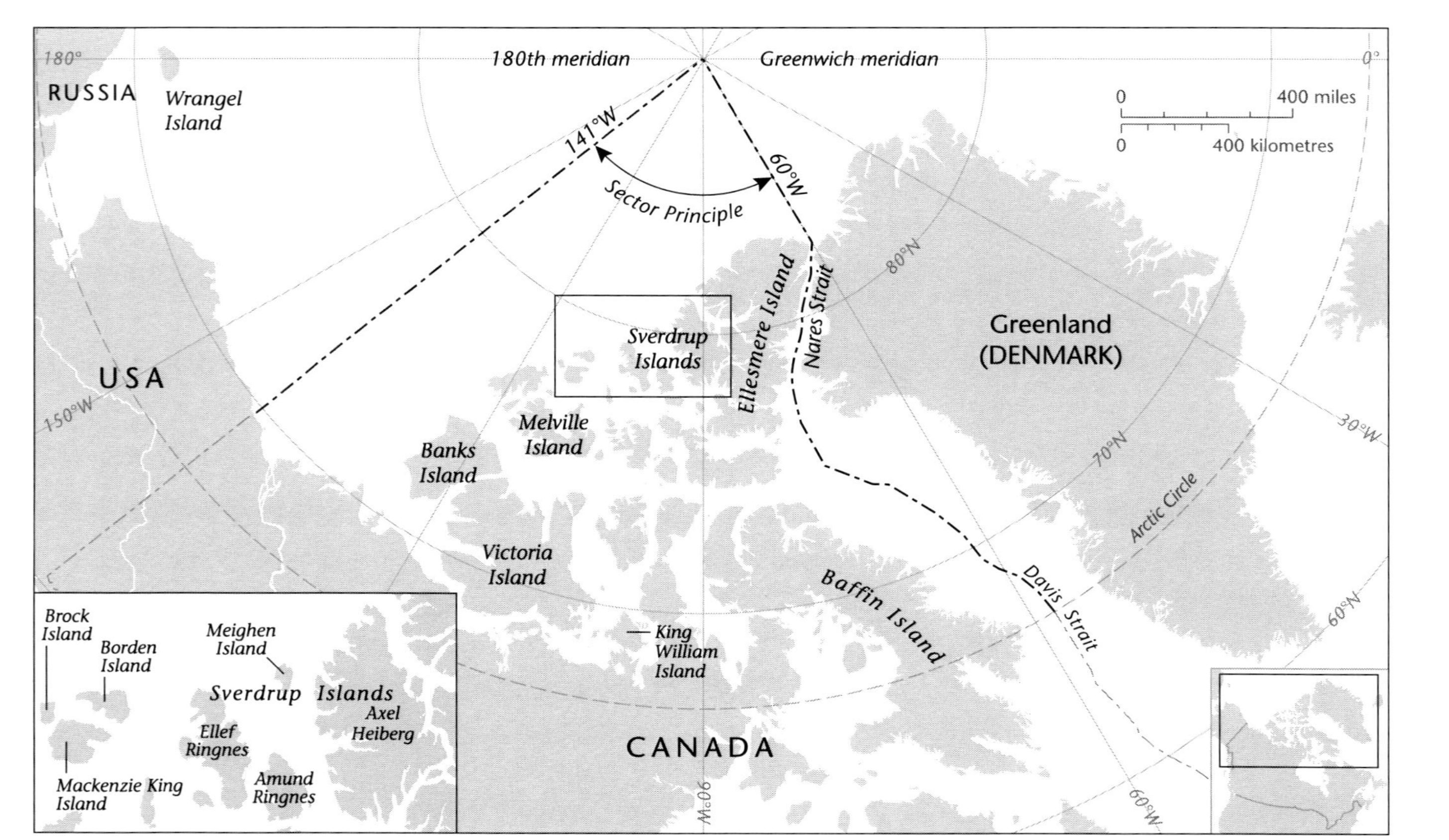

MAP 1 The Arctic (western hemisphere) | Cartographer: Eric Leinberger

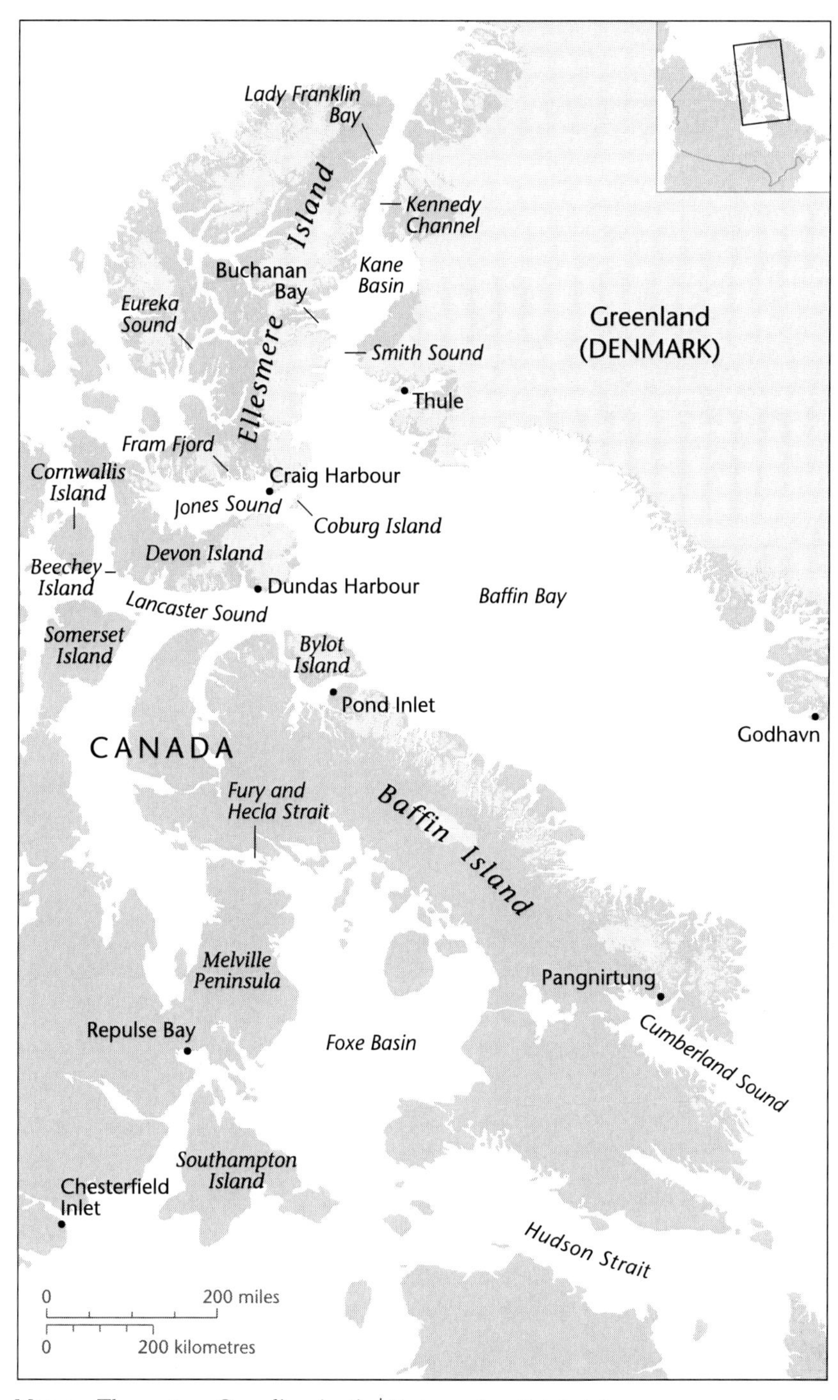

Map 2 The eastern Canadian Arctic | Cartographer: Eric Leinberger

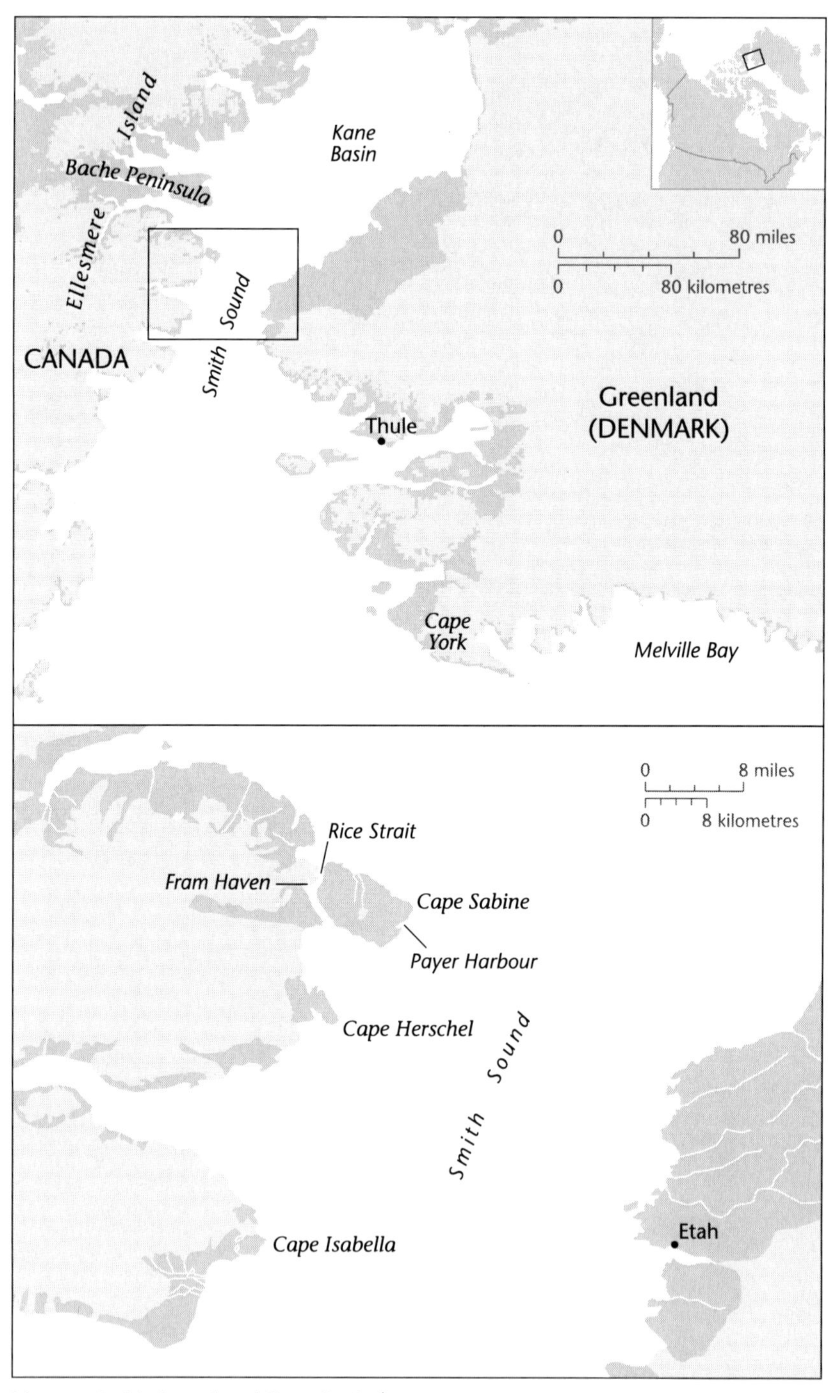

Map 3 Smith Sound and Kane Basin | Cartographer: Eric Leinberger

Acts of Occupation

Introduction
A Policy of Secrecy

I feel positive that the national interest w[oul]d be best served by a continuance of the policy of secrecy.

– J.B. Harkin, 1950

The old man's hands may have trembled slightly, more from age and indignation than from fear, as he read the letters. Certainly, the handwriting of the draft reply he promptly began is more erratic than the firm, backward-slanting script of earlier years. There seem to be signs of mental agitation in the repetitive wording of the draft, which fills eleven large manuscript pages. But that had always been J.B. Harkin's way. Even at the height of his extremely successful civil service career in the 1920s, his interminable memos repeated details, arguments, and recommendations, with Harkin apparently being determined to impose his will through the sheer volume of paper he placed on the desks of his hapless superiors.

Now, on the first day of March 1950, the seventy-five-year-old Harkin sat in his comfortable home on Clemow Avenue in a quiet, affluent Ottawa residential neighbourhood. Since retiring from the civil service, he had occupied his time with volunteer work for the Rotary Club and the Boy Scouts and with plans to write a book about the history of Canada's national parks.[1] This pleasant existence had just been interrupted by a letter from Hugh Keenleyside, the deputy minister of resources and development. The letter was a request for information, prompted by a personal message to Keenleyside from Arctic explorer Vilhjalmur Stefansson. A copy of Stefansson's letter was enclosed for Harkin to read. Stefansson wanted to know more about the secret history of Canada's policy on Arctic sovereignty after 1918, and in particular about the reasons why a planned expedition, to be led by Stefansson himself, was cancelled in the spring of 1921. It was not that Harkin was ashamed of the role he had played in those years; on the contrary, he viewed it as the high point of his career. But he was convinced that the history of what he and others had done must remain shrouded in secrecy for many years yet, with full knowledge kept even from other civil servants. Harkin was immediately suspicious of the motives that had led Stefansson – an inveterate publicity-seeker – to inquire about the documents Harkin had in his keeping.

'J.B. Harkin is, according to my belief and also according to what he has told me, in possession of a good deal of information that has a bearing on the history of sovereignty proposals and acts in Canada which is not available in any records, even secret ones,' Stefansson had written to Keenleyside in 1944. Stefansson hinted that Harkin was a little deranged on the subject, holding 'strongly to an idea that has baffled me, that it is his duty to let certain secrets die with him.' It would, Stefansson argued, be 'a disservice ... to history and maybe to Canada, in relation to its sovereignty problems, if Harkin does not place on record (in secret archives if you like) everything that he has in his memory or can dig up from memoranda which he perhaps intends to destroy.' In late 1949 Stefansson renewed the subject, noting: 'There were so many ramifications ... that I feel sure a study of the entire collection of documents will throw an interesting and perhaps an important light on the development of Canadian policy with regard to the Arctic.'[2] Keenleyside had confirmed that there was relatively little on the Arctic sovereignty questions of the early 1920s in the files of the Department of External Affairs. Now he was asking whether Harkin would send whatever papers he had to External Affairs, 'so that that department could supplement its records and at the same time consider whether any part of the information should be made available to Dr. Stefansson.'[3]

Although the papers Harkin held were government documents, which he had had no right to remove from the official files, his answer was an uncompromising no. His long draft reply suggests a number of reasons for the refusal. First, Harkin considered Stefansson untrustworthy and self-serving. 'There is only one issue & that is whether certain information should be made available to Dr Stef,' Harkin noted. The plea for general information on the development of Canadian Arctic policy 'winds up in a plea for information re an incident concerning himself.' Stefansson was clearly on 'a fishing expedition.' Harkin denied that his secrecy was unreasonable or that he intended his knowledge to die with him. When everyone concerned was dead, the information would be made available in the Public Archives. For unspecified reasons, this delay was in the national interest. 'I am convinced that it w[oul]d not be in the best interests of Canada for me to concur in your suggestions,' Harkin wrote at the top of the fifth page, which is filled with variations on this statement: 'I feel positive that it w[oul]d not be in the national interest for any action to be taken on the lines suggested'; 'The national interest must be the dominant consideration; and I feel positive that the national interest w[oul]d best be served by a continuance of the policy of secrecy.' Besides, unless a minute had been made of the cabinet's decision, Harkin did not think anyone had a record of exactly why Stefansson's expedition was cancelled. On the last page, Harkin revealed that he had 'fought vigorously against the cancellation.' Then, at the very end, he obstinately and almost defiantly asserted once again: 'It is information that Stefansson wants & only he. Let him say what he wants to establish.'[4]

Harkin was right to suspect that Stefansson was on a fishing expedition and that self-interest rather than the cause of historical knowledge was his motive. The explorer's quest for access to Harkin's documents had started just over a decade earlier. In March 1939 Stefansson wrote to former prime minister Arthur Meighen (who had been closely involved in Arctic sovereignty matters during his years in power), observing that there were 'civil servants still at Ottawa' who had invaluable information locked away in their memories, and perhaps in their personal files as well. 'If you think the secrecy is no longer needed, can you take some steps that the prohibitions shall be lifted?' Stefansson asked.[5]

Meighen's reply was prompt, brief, and chilly. The question of confidentiality was 'entirely a matter for those in charge of Canadian affairs today'; therefore, Stefansson should write to the prime minister, William Lyon Mackenzie King, 'stating to him what the confidential matters were and asking, if you so desire, permission for their publication.'[6] This suggestion was an extremely effective brush-off, since relations between King and Stefansson had been distant at best for many years. Stefansson, who was careful never to seem to feel a snub, replied in amiable terms. He presented himself as disinterested and altruistic, concerned only that the material should be 'thrown open for study by anyone who is interested.' He insisted that he 'had no plan of benefiting personally' from the opening of the records.[7]

The claim was only partly true. A few years later, in 1943, Stefansson published a revised edition of his book *The Friendly Arctic,* with a new chapter containing his own account of events in 1918-21. He would no doubt have been pleased to be able to cite official documents on the subject, thus adding credibility to what was in fact a highly misleading narrative. But, failing that, Stefansson had another plan on which to fall back. After his rebuff by Meighen, he promptly forwarded their correspondence to his journalist friend Richard Finnie (the son of one of Harkin's civil service colleagues), along with the suggestion that 'someone like you might want to get after this formerly secret correspondence. You could make out of it, I feel sure, a saleable article – and I think one of some value to the history of Canada.'[8] Stefansson had already sounded Finnie on the subject of Harkin's private papers. Finnie mistakenly thought that Harkin had died, and he promised to make inquiries about the files. He also suggested other possible contacts in Ottawa. Stefansson, in return, continued to push Finnie towards research on 'the history of the secret expedition,' which would make a 'most salable [sic]' article and perhaps even 'a chapter in an eventual history of Canadian exploration.' It seems that Stefansson also had his sights on the private papers of Finnie's father, Oswald Sterling Finnie, who though not 'in on the plotting,' might well have 'heard the inside story' or seen confidential documents.[9]

All Stefansson's efforts were in vain. Oswald Finnie's papers were destroyed in a fire.[10] Harkin died in 1955 and, as he had promised, his collection of documents was placed in the Public Archives of Canada (now Library and Archives Canada), but not until the year of Stefansson's own death, 1962. To anyone who has read them and the other archival records relating to the so-called secret expedition in full, it is at first extremely difficult to see either why Harkin was so determined to protect the documents or why Stefansson was bent on having them released. By that time, the secrets they contained were in no way detrimental to Canada's Arctic sovereignty claims. Harkin's concern therefore appears exaggerated and even irrational.

As for Stefansson, during the 1930s and the war years he was preoccupied with demonstrating his own prescience about the great future of northern air transport, and he accordingly claimed in the 1943 edition of *The Friendly Arctic* that the development of aviation had been among the most important aims of the cancelled expedition. In his first March 1939 letter to Meighen, Stefansson brazenly wrote that the expedition's purposes were 'to discover any lands that might be undiscovered in the Arctic, strengthen Canadian rights to the lands already discovered, and take other steps against the time when trans-Arctic commerce by air would develop.' Stefansson suggested that Meighen had an interest in furthering the release of 'documents relating to those plans which were made under your direction and which have been shown farsighted' by the events of recent years. But, as Harkin's papers would have demonstrated if Stefansson had succeeded in having them released, aviation played no role in the 1920-21 plans.[11] The documents did, however, show with painful clarity just how little the Canadian government trusted Stefansson: Harkin and others, including the deputy minister of the interior, William Wallace Cory, were firmly of the opinion that the explorer was motivated above all by selfishness and vanity. Had Stefansson known exactly what the papers contained, he would surely have been as eager as Harkin to avoid their publication.

The struggle between the two men over the records is of great interest for what it reveals about their stubborn, complex characters, with Harkin's tending to paranoia and Stefansson's to relentless self-aggrandizement and publicity-seeking. Both men seem to have been incapable, not merely of admitting to others, but even of privately believing, that they had ever been wrong. Over the decades each had evolved his own idiosyncratic narrative of past events. Harkin cast himself as the beleaguered defender of Canada's northern claims against foreign threats. To Harkin, Stefansson was an adventurer who, at any time during the early 1920s, might have disclosed confidential information to either Denmark or his adopted country, the United States, with disastrous consequences for Canada. Stefansson, in contrast, saw himself as a prophet thwarted by circumstances and an unimaginative bureaucracy. He had returned from his 1913-18 Canadian Arctic Expedition determined to bring about the creation of a new, greater Canada, wealthy from the development of its far northern resources. For reasons Stefansson never fully understood, this vision was ultimately rejected by Ottawa. To Stefansson, Harkin was an enigmatic figure who could have vindicated him to the world, but refused to do so out of a misguided sense of duty.

Both men deceived themselves. The Harkin papers and other related documents in Canadian and British government files do indeed preserve important events in the history of Canada's Arctic policy from oblivion. However, neither Harkin nor Stefansson emerges as an unsung hero. The most admirable figure in the saga is Danish explorer Knud Rasmussen, who was the object of Harkin's profound suspicion. Unwarranted suspicion of Rasmussen as the possible agent of Danish territorial ambitions in the North was, indeed, the unlikely stimulus for Ottawa's heightened interest in the Arctic archipelago after the First World War. In 1919 Harkin initiated a series of diplomatic exchanges with the Danish government about hunting on Ellesmere Island by Native people from northwestern Greenland. A letter from Rasmussen suggested that the Native hunters should be allowed to continue the practice for the time being.[12] This reply was construed by Harkin as an assertion that the archipelago was a no man's land or *terra nullius,* not subject to the authority of Canada or any other state.

Stefansson, eager to go north again with government sponsorship, did all he could to rouse and heighten Canadian fears. He informed Harkin that Rasmussen planned to lead a Danish invasion of the archipelago and to colonize the islands with Greenlanders. To meet this challenge, Stefansson suggested that a government ship should be sent to patrol the islands and to establish Royal Canadian Mounted Police posts. There should also be a five-year exploring expedition to the archipelago, under his own command. Harkin's apprehension was initially shared by many others, including Prime Minister Meighen and Loring Christie of the Department of External Affairs. It seemed that Stefansson's hopes of leading a new Canadian expedition would soon be fulfilled. However, the famous Antarctic explorer Sir Ernest Shackleton proposed an alternate plan that quickly won over many Ottawa officials. Then came news from London that cast doubt on the existence of a threat from Denmark. Meighen and Christie adopted a more realistic view of the matter, cancelling both the planned 1921 patrol and Stefansson's expedition, but Harkin was unyielding in his conviction that Canada's northern sovereignty was under imminent threat. Indeed, he apparently believed in the Danish plan until his dying day.

After the defeat of Meighen's government in December 1921, Harkin and his allies were able to persuade the new prime minister, Mackenzie King, to send the CGS *Arctic* north in the summer of 1922. This voyage was the first of the annual Eastern Arctic Patrols. The patrols established police posts, customs houses, post offices, hospitals, and other visible signs of Canadian authority on Ellesmere and other northern islands. They were of enormous significance because they brought permanent, continuous government administration to the archipelago. The perception of a threat from Denmark, then, played a key role in the transformation of Canada's earlier Arctic policy – in which proclamations and other purely formal 'acts of possession' were deemed sufficient – into a more active and sustained postwar program that emphasized the need for 'acts of occupation' even on remote and uninhabited northern islands like Ellesmere.

Before the first patrol set out in the summer of 1922, the government sought to keep its plans secret, fearing that early publicity might provoke pre-emptive action by the Danish or American governments. From 1922 until 1925, the patrols had a fairly low public profile, apparently because the civil servants involved believed that too much public attention might lead to awkward questions in Parliament about the amounts being spent on an empty and apparently useless region. As the *Ottawa Journal* shrewdly observed, there were few votes to be gained in the Arctic.[13] After 1925, however, stories and photographs in newspapers and magazines were used to establish the fact of Canada's northern sovereignty in the public mind. But because of Harkin and his policy of secrecy, the general movement towards public awareness did not include an accurate account of events behind the scenes. Historians have long recognized the Rasmussen episode as a catalyst for change, but the full story behind the first patrol remained obscure and poorly understood. A belief still prevails that there was in fact a threat from Denmark.

The first explicit published reference to the supposed threat came as early as 1925, in a *Foreign Affairs* article by American legal expert David Hunter Miller. Perhaps through contacts in Ottawa, Miller had gained the erroneous impression that in 1921 the Canadian government had 'formally notified' both Rasmussen and the Danish government that any discoveries made by Rasmussen during his Fifth Thule Expedition (1921-24) 'would not affect Canadian claims.'[14] Eight years later, further details were provided by V. Kenneth Johnston in the *Canadian Historical Review*. Johnston had seen a few documents from 1920 – that is, from the time when the fears roused by Stefansson were shared by almost everyone concerned in Ottawa. Other documents relating to later developments were not available to him.[15] The Harkin papers, which have been consulted by many Arctic historians over the years, reflect Harkin's own stubborn conviction that he had helped to forestall a Danish invasion of the archipelago. The 1919-25 volume of the series *Documents on Canadian External Relations* (published in 1970) also contributes to the impression that the Danish threat was at least potentially a serious one. It contains only one document on the subject, an October 1920 memo by Loring Christie.[16] Like the papers cited by Johnston, Christie's memo on its own gives an incomplete and misleading picture of the situation. The material required for a full understanding of events is widely scattered in government files and private papers. Historians have therefore almost unanimously accepted the idea that the Danes at one time made a 'flat denial' of Canadian sovereignty over Ellesmere and contemplated a claim to parts of the archipelago.[17] If no such claim was ever actually advanced, this, most writers assume, was because the Eastern Arctic Patrols securely established effective occupation by Canada before Denmark could take action. This book sets the record straight on Rasmussen and the 'Danish challenge' and demonstrates that Stefansson deliberately played on Canadian fears in order to advance his own ambitions.

The story that *Acts of Occupation* has to tell, then, is not same as either the story Stefansson sought or the story Harkin guarded. Instead, so far as the history of the 'secret expedition' is concerned, it is an at times almost farcical tale of misunderstandings, confusions, and deceptions. But it is momentous in a number of ways. The changes in Arctic policy between 1918 and 1925 extended beyond the establishment of effective occupation in the islands lying directly north of Canada. The story of the Danish threat and the cancelled expedition merges with the tale of Stefansson's ill-advised attempt to claim Wrangel Island for Canada.[18] This episode is well known in its outlines, having been told by historian Richard Diubaldo in his 1967 article 'Wrangling over Wrangel Island' and in his book *Stefansson and the Canadian Arctic,* published in 1978.

The occupation of Wrangel Island was Stefansson's hidden goal throughout the years after 1918, even while he was ostensibly urging action on Ellesmere and other eastern Arctic islands. Stefansson hoped to turn his 'secret expedition' towards the western Arctic and the vast unexplored region between the Beaufort Sea and the North Pole. He believed that Wrangel Island could be used as a base for extensive explorations in this area, perhaps leading to the discovery of an unknown Arctic continent. When the expedition was cancelled, Stefansson resolved to act on his own. Without the knowledge of anyone in Ottawa, he sent a party to claim Wrangel for Canada and the British Empire. When the truth was made public in the spring of 1922, Stefansson's daring plan initially appealed to the new and relatively inexperienced prime minister, Mackenzie King.

However, Wrangel Island was situated far beyond the 141st meridian, which Canadians had always vaguely considered to be the western boundary of their Arctic possessions. If Canada did not want Denmark or other nations to make incursions into the archipelago, would it be safe to claim an island that lay outside the traditionally accepted boundary and was much closer to the Soviet Union than to the mainland of Canada? All the civil servants involved were certain that the answer to this question should be no. King, too, eventually decided that no such claim ought to be made. To make matters worse for Stefansson, the four men he had sent to Wrangel Island all died. As a result of this tragedy, Stefansson lost whatever official favour he still retained. He played no further part in Canadian decision making.

Nevertheless, his actions had indirectly brought about much that was positive. By 1925 Canada's haphazard Arctic policy had been transformed into something much more clearly thought out. Canadian officials knew exactly what they wished to claim and how, and they had defined the boundaries within which they intended to work. Senator Pascal Poirier's well-known assertion of the sector principle in 1907 was made on his own initiative, in the form of a resolution that was not adopted. In June 1925, however, Minister of the Interior Charles Stewart stated, as the official policy of King's government, that Canada claimed all lands in the triangular sector between its northern continental coastline and the North Pole. The sector theory never won widespread support among theorists of international law but, combined as it was from the beginning with Ottawa's new focus on effective occupation, it provided an extremely useful framework for government action in the North over the ensuing decades.[19] As David Hunter Miller wrote later in 1925, 'With her claim to sovereignty before the world, Canada is gradually extending her actual rule and occupation over the entire area in question ... while it cannot be asserted that Canada's title to *all* these islands is legally perfect under international law, we may say that as to almost all of them it is not now questioned and that it seems in a fair way to become complete and admitted.'[20] By 1930 Canada's sovereignty over the islands within the sector was firmly established in the eyes of the world.

STEFANSSON'S SWEEPING VISION OF Canada as a northern empire had immense appeal, and there have always been claims that his ideas were rejected because the Ottawa bureaucracy was dominated by narrow, uncomprehending men, motivated by petty jealousy of a great explorer. Richard Finnie, for example, was convinced that 'Ottawa was not ready for Stefansson in his heyday.' Finnie remembered Meighen, King, Harkin, and the rest as 'conservative, cautious and relatively inconspicuous,' while Stefansson 'was a star. He outclassed them in almost every way.'[21] A great explorer and a media star Stefansson undoubtedly was, but he was also immensely ambitious, dogmatic, and devious. Stefansson was willing to use whatever deceptions were necessary to get his way. Both his sincere enthusiasm for northern development and his flagrant misrepresentations of the motives and actions of others knew no bounds.

In *Stefansson and the Canadian Arctic,* Richard Diubaldo concludes that Stefansson's tactless approach to the officials who could have helped him was the main cause of his fall from favour in Ottawa, and that his downfall retarded the development of a satisfactory Arctic policy. With Stefansson gone, Canadian plans for the North were 'meagre' and 'stultif[ied].'[22] This book offers an alternate way of understanding Stefansson's impact on official Ottawa. While Canadian politicians and civil servants grew justifiably suspicious of Stefansson's grandiose plans, his love of newspaper publicity, and his smoothly plausible explanations of his own self-serving behaviour, many of them (in part because of his influence) were genuinely interested in the Far North. Throughout the period covered by *Acts of Occupation,* they were groping towards their own, more pragmatic, vision of the Canadian Arctic and its future. Their concerns link together what formerly seemed to be a jumble of only loosely related events.

The connections between the perceived Danish threat, Stefansson's failed attempt to annex Wrangel Island, and the 1925 sector claim have never previously been examined. The strongly negative official response to Rasmussen's letter – which Stefansson did so much to exacerbate – was only the first in a series of developments that transformed Canada's Arctic policy. The story Harkin was so determined to keep secret is interesting and significant not only in itself, but also because it provides the clue to an even more important drama. By following the sequence of events beyond the cancellation of Stefansson's expedition in 1921, and even beyond the Wrangel Island controversy and his final rejection by Ottawa officialdom in 1924, we can see a formerly obscure period in the history of the Canadian north from a new and illuminating angle.

The story also touches on broader issues in Canadian state formation.[23] By the end of the First World War, Canada had achieved a new level of autonomy within the British Empire. However, in the immediate postwar era the Department of External Affairs remained a small body of civil servants, lacking any significant expertise in the area of international law. Canada, in other words, did not yet possess the necessary political machinery to formulate and implement a truly independent foreign policy. These shortcomings were largely to blame for Ottawa's blundering response to the supposed Danish threat. Stefansson found it a comparatively easy matter to dupe Canadian officials, at least for a time; Harkin, a member of the Department of the Interior with no qualifications whatever for meddling in matters of external policy, was able to take advantage of the confusion and poor communication that often prevailed in Ottawa to push forward with his own plans in defiance of his more sophisticated and better informed superiors. Only the intervention of the British Foreign Office finally resolved the situation, permitting Rasmussen's Fifth Thule Expedition to carry on its groundbreaking ethnographic work. During the Wrangel Island episode, the government again turned to the experts in London for help. But by 1925 Canadian civil servants felt ready to meet a potential threat from American explorers Donald MacMillan and Richard Byrd largely on their own. Unlike Rasmussen, the two Americans did in fact hope to seize parts of the archipelago. The firm Canadian response forced them to abandon their plans.

In the Canadian government's relationships with Great Britain, with other states, and with its own northern hinterland, the years from 1918 to 1925 were a time of important changes, many of them hidden at first beneath the surface of established government routines and therefore imperceptible to members of the public. By the late 1920s, these changes would be made manifest in the far stronger Department of External Affairs created by Mackenzie King and his under-secretary of state for external affairs, O.D. Skelton, and in a greatly expanded northern administration, including a program of scientific and other work related to the annual Eastern Arctic Patrols. Besides a narrative of the events behind the first Eastern Arctic Patrol and the Wrangel Island affair, government documents from the period 1918-25 offer a detailed record of why and how the Canadian government evolved the bureaucratic procedures and other mechanisms that would ensure its control of the Far North. The letters and personal papers of Stefansson, Shackleton, and Rasmussen open up another perspective by showing how ambitious, dynamic explorers interacted with civil servants and politicians. *Acts of Occupation,* therefore, combines episodes of high drama from the closing years of what has been called the heroic age of polar exploration with the quieter, but no less fascinating, story of the Ottawa men who secured Canada's title to the northern archipelago.

1
Taking Hold of the North

Surely ... we know that the reserve and strength of character that is in our northern blood has meant victory in this War. We ought to be proud of the fact that we are a northern people, and we ought not to be afraid to breast the wave. We should take hold of the north.

– *Sir Edmund Walker, after a speech by Vilhjalmur Stefansson to the Empire Club of Canada, 11 November 1918*

VILHJALMUR STEFANSSON RETURNED TO the south from his five-year Canadian Arctic Expedition in September 1918. He left the North reluctantly: in 1915 the government had ordered him to return in 1916, but Stefansson evaded the order. In 1917 Inspector J.M. Tupper, the officer in charge of the Royal North West Mounted Police post at Herschel Island, informed him that Ottawa was urgently attempting to communicate with the expedition. Tupper suggested a quick journey to the nearest telegraph station at Fort Yukon. But, as Tupper indignantly reported to his superiors, Stefansson answered that he 'was not going out of his way' to receive unwanted instructions.[1] Rumours about Stefansson's recalcitrant attitude reached southern Canada, where they sparked unfavourable press commentary. 'While war is on and the public are engrossed in the one big thing, Mr. Stefansson apparently thinks the story he will have to tell on his return will under the circumstances not compel universal attention,' the Toronto *Globe* observed acidly in March 1918.[2] Stefansson stayed in the Arctic as long as he remained confident that he could coax funds from the politicians in Ottawa. He was determined to do work that would establish his credentials as one of the most successful polar explorers of the day. This ambition he certainly accomplished (though at the cost of roughly $535,000 instead of his original estimate, $75,000).[3] But Stefansson had also laid the foundation for years, even decades, of acrimonious controversy about his character and actions.

The controversy centred on two dramatic episodes in the history of the Canadian Arctic Expedition: the deaths of eleven men on or near Wrangel Island and a so-called mutiny by members of the scientific staff. The expedition's ship, the *Karluk,* was caught in the pack ice off the coast of Alaska in the late summer of 1913. Stefansson was away from the ship on a hunting expedition when a strong wind blew the ice and the trapped *Karluk* far to the west. Stefansson proceeded to Collinson Point, where the scientific section of the expedition was working under the supervision of zoologist Dr. Rudolph Anderson. Stefansson (whose supplies had gone with the *Karluk*) wanted to appropriate some of the scientific party's resources in order to make a prolonged journey of exploration over the ice of the Beaufort Sea. Anderson was already angry with Stefansson. Before the expedition began, Stefansson had agreed that Anderson could publish newspaper and magazine articles about his experiences. He later went back on his word, informing Anderson in June 1913 that he had sold exclusive press rights to the *New York Times* and the London *Daily Chronicle.* At Stefansson's request, the government forbade the scientists to lecture or publish until one year after the end of the expedition. 'He admitted to me that what he had done was ethically wrong, that he had lied to me repeatedly, but said that he felt justified in using any means to get my help and get the expedition started,' Anderson later recounted.[4]

The expedition's orders, drafted by the Department of the Naval Service, were poorly thought out. Stefansson was described as the leader of the entire expedition, but only the exploring party was placed under his 'personal direction and control.' The scientific work was 'under the direction of Dr. Anderson.' The orders placed great emphasis on the government's wish that the work done by the scientists 'should be of a high order ... and should mark a distinct advance over previous work.'[5] Arguing that Stefansson's authority as leader did not extend to a complete revision of the priorities set by the government, Anderson refused to let Stefansson take everything he wanted. Stefansson saw this, and later publicly described it, as mutiny, but at the time he could do nothing, since Anderson had the wholehearted support of the other scientists.[6]

The *Karluk* was crushed by ice pressure in January 1914 (Stefansson had been warned before the expedition's departure that the ship was not built to withstand heavy pack). The crew escaped to nearby Wrangel Island.

Eight men were lost on the journey over the treacherous pack ice and three more died on the island, two from a mysterious sickness apparently connected to bad food and one by suicide – or possibly murder. Stefansson was confident that explorers could live off the land, but he had chosen mostly men who lacked any experience in Arctic survival techniques. And, despite Stefansson's theories about the bountiful food resources of the 'friendly Arctic,' Wrangel Island proved to be a poor hunting ground. To make matters worse, the expedition's pemmican was deficient in quality, containing an insufficient amount of fat.[7]

To save his crew, the *Karluk*'s commander, Arctic veteran Robert ('Bob') Bartlett, made a daring and dangerous journey over the sea ice to Siberia. Bartlett's story caused headlines around the world in June 1914. By the time the group on Wrangel Island was rescued, war had broken out. One survivor, William Laird McKinlay, joined the army soon after his return to his native Scotland. 'Not all the horrors of the Western Front, not the rubble of Arras, nor the hell of Ypres, nor all the mud of Flanders leading to Passchendale, could blot out the memories of that year in the Arctic,' he later wrote.[8] McKinlay retained a lifelong bitterness against Stefansson, who in his opinion had not taken the elementary precautions that would have given the men on the *Karluk* a better chance of survival.

Rudolph Anderson and the expedition's other scientists returned to Ottawa in 1916 as ordered, full of anger about Stefansson's high-handed attitude, which they felt showed a selfish disregard for the program of work they were expected by the government to carry out. They had also repeatedly been dismayed by evidence that Stefansson 'ever had his eye on his news reading public' and that, under the cloak of science, the expedition 'was really, at the bottom, a newspaper and magazine exploiting scheme' for Stefansson's personal benefit.[9] Anderson, too, would remain Stefansson's enemy for life. He usually referred to his former colleague as 'Windjammer,' a derisive nickname intended to convey Stefansson's unreliability and boastfulness.[10] (Decades later, when Stefansson's young wife, Evelyn, playfully called him Windjammer, Stefansson 'wasn't amused.')[11] Anderson was American-born. He and Stefansson were old acquaintances, having met – if only briefly – when they were both students at the University of Iowa. Anderson did solid scientific work during Stefansson's 1908-12 expedition, and he joined the Canadian Arctic Expedition on Stefansson's urgent invitation. Once the expedition was over, Anderson and his wife, Belle, settled in Canada. For the rest of his long career, Anderson worked at the Victoria Memorial Museum in Ottawa.

Mrs. Anderson was an intelligent, energetic woman: she too had studied at the University of Iowa, and she had then gone on to earn a master's degree from the University of Wisconsin. She took a keen interest in Arctic matters, and her attitude to Stefansson, like her husband's, was full of distaste and resentment. Belle Anderson was present at the June 1913 encounter between her husband and Stefansson. She was deeply shocked to hear Stefansson say 'with the utmost unconcern that he would lie or do anything to achieve an end he had in view and feel himself justified in doing so.'[12] Five months later Mrs. Anderson gave birth to her first child, a son who died within a few days. In her grief and loneliness, she turned to correspondence with the mothers, sisters, and wives of the men on the missing *Karluk.* The experience left her with a lasting sense of the emotional devastation caused to families by Stefansson's ambitious nature and his careless planning.

After the scientists' return, a few critical newspaper articles appeared, alleging that Stefansson was a publicity hound who cared nothing for careful, systematic investigation of the Arctic. The *Ottawa Citizen,* for example, declared that he had sold himself 'body and soul to the American newspapers and magazines for whom he has written, or plans writing, accounts of his explorations.'[13] But, as with the return of the *Karluk* survivors, the world was too busy with far weightier concerns to take much notice. Rudolph Anderson (who denied responsibility for the *Citizen* article and similar press items) wrote resignedly, 'The government let a certain explorer go at large with a little too much rope, and has to stand for the consequences, and loyal servants of the same do not like to stir up official investigations and scandal, to embarrass a wartime government.' Besides, there was 'no use starting a vague newspaper row with the slippery cuss, as that is just what he wants, anything in the papers favorable or unfavorable is merely playing his game.'[14]

Stefansson explored to the north of the Parry Islands from the spring of 1914 until the autumn of 1915. No members of the expedition were willing to accompany him beyond the early stages of the journey, so he set off with two adventurous Norwegians, Storker Stokerson and Ole Andreason, who had come to the Arctic as trappers and traders. With their help, Stefansson discovered Brock, Borden, Lougheed, and Meighen Islands – some of the last remaining unknown bits of land in the Arctic (later irreverently known in Ottawa as the 'Tory Archipelago'). Additional journeys over the sea ice were made in 1916 and 1917. His travels took Stefansson from the continental coastline in 70° north latitude across the shifting sea ice almost to the eighty-first parallel. In 1914-15 he spent so long out of communication with the world that some gave him up for dead. Even the Inuit declared it was impossible to live for so long on the sea ice, where there was no game. But Stefansson found game. Here were dramatic and solid achievements that he could set against the carping of his detractors.

Next, Stefansson planned to make another long journey across the ice towards Wrangel Island and the coast of Siberia. 'I have never been so eager to do anything,' he later wrote.[15] However, a severe bout with typhoid put an end to this plan as far as Stefansson himself was concerned. A party was sent out, but it was led by Storkerson. For four months Stefansson remained in the hospital at Herschel Island. From there, he began the journey south in April 1918. Stefansson arrived in Vancouver on 16 September and told curious reporters that he would 'rather be up in the Arctic. I hope to return north some day.'[16] He then spent a few weeks in Seattle, on the plea that his poor health obliged him to consult specialists there. Stefansson requested a further leave of absence from his official duties for medical treatment at the University of Iowa hospital. The request was granted, but in early October he decided to omit his stay in Iowa and return to Ottawa at once.

Ever since Stefansson had reached Fort Yukon and read the latest news bulletins from Europe, he had been concerned about the impact of the current international situation on his own fortunes. Eager to link his story with the patriotic fervour roused by the war effort (and perhaps fearful that if he was not seen to be doing some sort of war work, public opinion might oblige him to enlist), Stefansson came up with the idea of lecturing for the benefit of the Red Cross. His shattered health, he explained, made him unfit for any other type of work. But from Ottawa came the blunt answer that although he was free to lecture if he wished, he must first submit a written report for publication by the government. In reply, Stefansson wrote to George Desbarats (the deputy minister of the Naval Service, who was the official mainly responsible for the expedition), arguing in injured tones that he could not understand the government's failure to support him in his eagerness to assist the war effort. Besides, he claimed, by the time a report was written and published, popular interest in the expedition would have died away; his lectures could increase the interest and secure a wider audience for the eventual publication. Desbarats replied coldly that the war effort was indeed of paramount importance and that Stefansson could have made a much greater contribution to it if he had followed instructions and returned home in 1916.

As he travelled across the continent by rail, pausing only for a short visit with his elderly mother (then living in Wynyard, Saskatchewan), Stefansson had much to consider: not only how the story of his exploits would be received by the Canadian government and the public, but also what his responsibilities were to the family he had left behind in the Arctic – his common-law Inupiaq wife, Fanny Pannigabluk, and their nine-year-old son, Alex. In the end, Stefansson decided to forget them both. After leaving the North, he never again acknowledged that Alex was his son. During his stay in Seattle, he began a romance with an attractive local woman, Betty Brainerd. Marriage was later discussed between them, but yet another relationship, this time with married novelist Fannie Hurst, intervened.[17] Despite these other love affairs, Stefansson may initially have intended to return to Pannigabluk and Alex. He was almost thirty-nine years old – an age when many polar explorers felt they were ready to retire. Stefansson, however, did not. 'I am afraid that by now my Arctic work is the only work I am good for, or at least I am less fit for any other work. It is my desire and my dream to continue it,' he had written in 1917. 'I have several alternate plans, one or another of which I hope to put through.'[18] But for any of his plans to become reality, he would need fame and the money it could bring.

Stefansson could certainly hope to make a significant amount of money by writing and lecturing in the immediate aftermath of his return. Despite his nasal voice, Stefansson was a memorable lecturer, with a knack for vivid, humourous stories and an imposing physical presence. He was on the tall side of average height, with strong features, a dimpled cleft chin, and a thick mass of unruly fair hair. His bearing seemed confident to his admirers and cocky to his detractors. A modern Viking, some called him. But after a year or so of notoriety, what would he do? And what if even a short period of fame eluded him? At first, the autumn of 1918 seemed to Stefansson like one of the worst possible times for his return.[19] It would be difficult to get the publicity usually accorded to polar exploits in times of peace. Stefansson liked and needed headlines, but now more than ever the headlines were taken up by events in Europe, where, after four years of deadlock and attrition on the western front, the war had at last turned into a war of movement. Beginning with the surprise attack at Amiens on 8 August (later known as the 'Black Day of the German Army'), the Allied forces swept steadily forward. Canadian troops played a prominent role in the victories at the Drocourt-Quéant Line, the Canal du Nord, Bourlon Wood, Cambrai, and Valenciennes. The rapid succession of Allied triumphs was of all-engrossing interest to readers in both Canada and the United States.

Stefansson went first to Ottawa and made his report to the authorities. He arrived early on 29 October and spent the morning of that day in Desbarats' office. According to Stefansson's later account, the two men agreed that the Red Cross lectures would be cancelled. Stefansson assured Desbarats that after a month's leave for medical care in the United States, he would set to work on his report. In the afternoon, Desbarats informed the press that he was 'highly pleased' with Stefansson's results. Stefansson, in turn, told the assembled reporters with a smile: 'Hardships and sickness [are] largely a matter of the mind. If one thinks or imagines that he is suffering terrible privations and worries about imaginary illness, then he will worry himself into the real thing. But the spirit of optimism is the right spirit. I made up my mind that I was comfortable, and that I had a work to accomplish, and with that thought uppermost I got through.'[20]

Stefansson did not linger in Ottawa. Immediately after his meeting with Desbarats, he hurried to New York. From there, he wrote a letter informing Desbarats that the first lectures in the series (to be given in New York, Brooklyn, Philadelphia, Washington, and Toronto) could not be cancelled after all, because seats had already been sold.[21] Clearly, it was important to Stefansson that he have the chance to publicize the expedition, and he was especially eager to appear in the major American cities. Although he had been born in Manitoba, Stefansson regarded the United States as his home. His Icelandic parents had moved from Canada to the Dakota Territory when Stefansson was only a year and a half old. Stefansson initially became an Arctic explorer through contacts in the American academic world. He valued these contacts both for their practical uses and because they represented a real achievement for an unconventional, impecunious outsider from the West.

Like most Icelanders, Stefansson's parents promoted a love of literature and of learning generally. As a young man, Stefansson aspired to earn a university degree, but his progress was hampered by an arrogant conviction that he already knew more than his teachers. After being expelled from the University of North Dakota for poor attendance and a generally insubordinate attitude to authority, he graduated from the University of Iowa in 1903. Stefansson then went on to graduate work in anthropology at Harvard, funded by a scholarship from the Unitarian Church. He first travelled to the Canadian Arctic as a member of the Leffingwell-Mikkelsen Expedition in 1906-7. In 1908-12 Stefansson led his own ethnographic expedition, sponsored by the American Museum of Natural History. The 1913-18 expedition originally had the backing of the American Museum of Natural History and the National Geographic Society, but when Stefansson approached the Canadian government for an additional grant, Prime Minister Robert Borden decided that Canada would take over the entire venture. It was, then, almost inadvertently that Stefansson became the leader of an official Canadian expedition.

It was a condition of Borden's offer that Stefansson must be naturalized as a Canadian. When his father became an American citizen in 1887, young Vilhjalmur had lost the status of British subject conferred by his Canadian birth. Stefansson took some of the steps that were necessary to recover his former nationality, but he did not complete the process. Nevertheless, he frequently claimed to be Canadian.[22] Whether Stefansson privately considered himself Canadian or American in 1913-18, he unquestionably relied on the strong connections he had built up with American institutions and publishers to make his exploits known to the world. When the Canadian government took over the sponsorship of the expedition, it was agreed that the press contracts and other agreements already made by Stefansson would be honoured. Over the years since 1913, news of the Canadian Arctic Expedition had been released mainly through American publications like the *New York Times*, the *Bulletin of the American Geographical Society*, and the *Geographical Review*, and it was in New York, not Ottawa, that Stefansson made the first extensive public statements after his return.

Stefansson gave a press conference on 30 October and a lecture (sponsored by the American Museum of Natural History and the American Geographical Society) at Carnegie Hall on the 31st. According to Stefansson, the lecture went extremely well despite some 'minor heckling.'[23] When it was over, he got a taste of the feminine admiration fame could bring: an attractive Canadian-born writer, Constance Lindsay Skinner, who had found reading the newspaper reports of his expedition a solace during a time of emotional turmoil, came up to introduce herself. She was not disappointed in her hero. Skinner found in Stefansson a 'blend of daring, self-confidence, energy, genius, and imagination' which reminded her of the early Norsemen.[24] Although the love affair she clearly hoped for never occurred, the two remained good friends until Skinner's death in 1939.[25]

The press conference was reported on page 5 of the *New York Times* and the lecture on page 13.[26] In other years, the *Times* had placed news from Stefansson in a prominent position on the front page. 'Stefansson Discovers New Arctic Land' was the banner headline on 18 September 1915. This story was among the items that had caught and held Constance Skinner's attention. But in the last week of October 1918 and the early days of November, the front pages were dedicated to Turkey's surrender and the collapse of Austria.

Despite the optimistic news arriving daily from Europe, Stefansson evidently feared that the Germans might reject the Allies' terms and fight on to the bitter end, with the result that the war would last for many more months, perhaps even for another year. In 1913 he had signed a contract with lecture manager Lee Keedick of New York. Keedick now informed him that, because of the war, the earlier contract could not be carried out. A successful lecture tour was still possible, but only on terms less favourable to Stefansson. Stefansson duly signed a new two-year contract on 5 November before leaving New York to give his lectures in Philadelphia (6 November) and Washington (8 November).[27] He soon had cause to regret his action: on 9 November came the news that Kaiser Wilhelm II had abdicated, making a speedy end to the conflict almost inevitable.

When the news broke, Stefansson was again in Ottawa. In the course of another long conversation, Desbarats insisted that Stefansson must return to Esquimalt, British Columbia, to wind up the expedition's business there.[28] The two men also discussed the grievances put forward by Rudolph Anderson and the other members of the scientific staff. Stefansson took the attitude that only Anderson's behaviour required examination, and not his own. 'Seeing ... that the Expedition has achieved both a scientific and a popular success ... I feel it would be petty of me were I not to discourage any public humiliation of Dr. Anderson ... the affairs of the Expedition are going so well now that it would probably be best to drop all questions of discipline connected with its past ... I would rather forget them, I think, than receive a public vindication on any point that may seem doubtful or discreditable to those not fully informed,' he wrote after the meeting.[29] But according to Anderson, Desbarats was not convinced that Stefansson had been entirely in the right. Although 'minor difficulties and differences of opinion had arisen,' the deputy minister believed that the 'solutions arrived at [were] the best possible under the difficult circumstances.' Desbarats had never so much as considered any 'public humiliation' of the scientists. Instead, he hoped there would be no public controversy to mar the expedition's record and embarrass the government.[30] Anderson was willing to go along with the policy of silence as long as Stefansson did the same.

From Ottawa Stefansson travelled to Toronto, where he was scheduled to speak on 'My Five Years in the Arctic' at Massey Hall on 11 November, under the auspices of the Empire Club of Canada. He later described the lecture as one of the most important of his career.[31] At noon that day, Torontonians were informed by their mayor that the war was finally over. According to Stefansson, the city then 'went off like a skyrocket' and 'delirious excitement' prevailed.[32] A parade already having been arranged in aid of the Victory Loan program, the atmosphere of jubilation was intensified when wounded veterans and men in training marched through the downtown area to the music of nine pipe and brass bands, including the band of the 48th Highlanders of Canada and the United States Navy Band, led by John Philip Sousa himself. '[A] unique feature of the celebrations,' recorded the *Globe,* 'was the presence of a fleet of airplanes in battle formation which accompanied the troops along the line of march, performing "stunts" amid the skyscrapers and church steeples.'[33] The crowds were estimated at between one hundred thousand and two hundred thousand people. Late in the afternoon, there was a thanksgiving service outside the Ontario provincial legislature at Queen's Park. Some who had intended to hear Stefansson's lecture chose instead to take part in the impromptu celebrations that went on well into the night, or were simply unable to make their way to Massey Hall. Others arrived in a mood of patriotic fervour. Stefansson himself was half an hour late because of the crowds that still thronged the streets and blocked traffic. The proceedings began with the hearty singing of *Rule Britannia.*

Speaking over the noise of whistles, automobile horns, and cheers from the street outside, Stefansson offered this elated crowd much more than an account of his Arctic adventures: he offered a new vision of Canada. The expedition itself was, indeed, given rather short shrift. Stefansson devoted the early portion of the lecture to it, but his emphasis was on his method of living off the land rather than on the actual events. Then he turned to another theme. 'But more interesting to me than the things we have done on the Expedition,' he explained, 'are certain things which our experience on it has led me to see ought to be done, and I am going to tell you of some, which I think the Government of Canada ought to do.'[34] The things Stefansson believed ought to be done would ensure the development of the Far North as an integral part of the nation. The only obstacles to such development were ignorance and prejudice.

Stefansson argued that for millennia, people had considered the lands to the north of their own homes as regions of desolation; nevertheless, new civilizations had sprung up there, and these had invariably been of a higher nature than the old cultures in the south. 'Trench by trench,' he declared, in an obvious attempt to link his theme with the military victory being celebrated, 'the ramparts of ignorance have had to be conquered, as civilization has spread north.' Simple logic and the patterns of human history pointed to one conclusion: the final stage in the evolution of the human race could take place in Canada – if Canadians were not afraid of their destiny. Southern lands were 'suited to the beginnings of high culture,' but 'the test of experience shows that the south is not suitable for civilization's highest development.' Stefansson told his audience: 'We have not come to the ultimate northward movement of the centres of civilization when we have come to London or New York, nor have we at length discovered the ultimate northern frontier ... at the Peace River.' A northern empire for Canada was not a visionary ideal. Instead, there were solid, practical ways to develop the North. Chief among them were the domestication of the reindeer and 'an even more valuable animal – the musk-ox.'

A muskox was like a cow with wool, or a huge sheep that provided meat 'identical in taste' with beef and milk 'with difficulty distinguishable from Jersey milk.' It needed 'no barn to shelter it, no hay to feed it for the winter, for in the farthest islands of the north they live untended, and they are fat in any season of the year.' The Canadian government ought to get 'a thousand or so' muskoxen from the northern islands, which would be 'easy to do'; once the scientists had studied them and pronounced on the best way of proceeding with commercial development, 'we can turn the whole northern half of Canada into grazing lands that shall produce to the square mile as much meat, tallow, milk and wool as do the grazing lands of the Argentine and Australia.' Stefansson concluded by asking for the Empire Club's support in his endeavour to convince the government that his plans were viable.[35]

Sir Edmund Walker, the president of the Bank of Commerce, thanked Stefansson for his speech on behalf of the Empire Club, and he warmly endorsed the explorer's plans. Walker, who had been an advocate of northern development for many years, was clearly overjoyed to have found such a persuasive ally.[36] 'It is one of the most difficult things to make Canadians believe in the value of their own country,' he now remarked. Canada could never take its rightful position in the world 'until we take hold of the resources of this country.' The war had proved the value of Canadians' northern blood, so heroic and so far removed from the 'hysteric quality' of the southern races. 'Surely,' Walker argued, 'we know that the reserve and strength of character that is in our northern blood has meant victory in this War. We ought to be proud of the fact that we are a northern people, and we ought not to be afraid to breast the wave. We should take hold of the north.' Canada must stand behind Stefansson. 'On this great day, when we are celebrating the peace of the world, it does not seem to me a minor event that we are also able to celebrate the return of a man who has risked so much and done so much for this country,' Walker concluded. The audience responded with loud applause and cries of 'hear, hear.'[37]

None of Stefansson's audience on this occasion seem to have recorded their response, but an American university student, Earl Hanson, retained a lifelong memory of another Stefansson lecture in the early 1920s. 'There was nothing of the stirring tale we had expected, about heroic men pitting themselves against tremendous hardships and dangers; instead, Stefansson took us on a voyage of discovery of our own. He opened the gates for us of a new North,' Hanson recalled. 'To me it would have been thrilling to have had a glimpse of some terrible land at one of the "ends of the earth" where only the boldest heroes could go; but it was a hundred times more thrilling to see my own world of everyday affairs suddenly enlarged by thousands of square miles where pioneers would some day stake claims, build homes, and carry the banner of ... civilization.'[38] Much of Stefansson's appeal lay in this ability to inspire belief in a future domesticated Arctic. For his enthralled listeners, the Far North was much more than a realm of adventure and excitement. It became all the more appealing when Stefansson proclaimed that it was destined soon to become part of Canadians' everyday life.

In the heady atmosphere of victory and of hope for a reinvigorated, prosperous peacetime Canada, such visions were welcomed even by normally staid civil servants. The new centre block of the parliament buildings, with its lofty Peace Tower, was then rising on the site of the old building gutted by fire in 1916. It formed a fitting symbol for the spirit of the times. Matters at first moved forward rapidly in Ottawa.

Stefansson himself made his home in New York, living in messy bachelor comfort at the Harvard Club and later in a Greenwich Village apartment (he did not marry until 1941, after the end of his long affair with Fannie Hurst). However, for a time Stefansson also maintained a residence at 109 Metcalfe Street in Ottawa, and he was frequently in the Canadian capital. Stefansson enjoyed the full support of Sir Robert Borden. In 1918 Borden had been prime minister for seven years. As the man who successfully led Canada through the war, he retained considerable prestige despite the lingering controversies over some of his decisions (most notably the conscription policy of 1917). Himself the epitome of integrity in both his political career and his personal life, Borden was a powerful protector. The prime minister saw Stefansson as 'a man of remarkable personality and strength of character as well as of unusual ability' who had 'accomplished marvellous results' in the Far North.[39] Anderson and the expedition's other scientists – most of them employed in civil service jobs – were therefore silent, at least in public. Not until after the publication of Stefansson's narrative, *The Friendly Arctic,* in late 1921 did the scientists' resentment explode into print. Between 1918 and 1920, when Borden retired, Stefansson's influence in Canada was at its height.

Stefansson's aims during this period were to foster the economic development of the Far North (and he was not at all averse to gaining financially himself by the process) and to urge a more aggressive stand by the government on sovereignty matters. Where economic goals were concerned, he believed that 'the logical way' was 'to convince a few "captains of industry" and to induce the governments concerned to give these leaders a fair opportunity.' If a thousand individuals went north to make their fortunes, they would all have to be educated to meet the unfamiliar conditions. But if they were the employees of a large company, they would simply follow the directions of their more knowledgeable superiors. 'An ordinary colony may fail through the conservatism of its members, but a commercial enterprise on a large scale will succeed,' Stefansson declared.[40]

The government could not grant the required concessions unless its sovereignty was secure, and in Stefansson's view, that sovereignty should extend over as large an area as possible. Stefansson was particularly interested in the Beaufort Sea, where a million-square-mile expanse stretching north to the pole remained unexplored. He wanted the government to lay claim to Wrangel Island, which could then be used as a base for exploration into the unknown area. If new land were found there – as Stefansson believed was highly likely – it must be claimed for Canada.[41] Stefansson had all the more reason to push these plans in view of the disappointing financial returns from the 1913-18 expedition. His contract to publish photographs and stories in the London *Daily Chronicle* had produced almost nothing in the way of income. As H.V. Claude-Sussex of the *Chronicle*'s staff explained in January 1919, 'When we entered into the agreement ... we were confident of doing very big things for you with the pictures and other material.' At first there had been 'a very lively public interest' in the expedition. But after August 1914, 'nothing but the war mattered.' The *Chronicle* therefore cancelled the agreement and returned all rights to Stefansson, with the friendly warning that editors, 'whilst open now to consider stories of the kind you have for disposal, are not prepared to pay such big prices for them as before the war.'[42] Moreover, Stefansson was quickly coming to realize that the new lecture contract he had unwisely signed just before the armistice would not bring him much profit. A new and dramatic venture, paid for with government money, could change this bleak outlook.

WRANGEL ISLAND WAS LOCATED far from the Canadian archipelago, closer to the coast of Russia than to any other major land mass. It lay well beyond the 141st meridian, which had been set as the boundary between British North America and Alaska by the Anglo-Russian Treaty of 1825, and which was also traditionally considered by Canadians to mark the western limit of their possessions in the Arctic archipelago.[43] The island was first seen by Henry Kellett, an English naval officer, in 1849.[44] The first landing was made in 1866 by a German, Eduard Dallmann, who did not record any territorial claim.[45] Further landings were made from two American ships, *Thomas Corwin* and *Rodgers,* in 1881. The captain of the *Thomas Corwin* raised the flag and took possession in the name of the United States.[46] The British government did not protest; however, Washington took no action to back up the claim. In 1911 a party from the Russian icebreaker *Vaygach* landed and erected a navigational beacon.[47] Five years later, the imperial government informed the other powers that it considered Wrangel Island Russian territory.

The crew of the *Karluk* raised the Canadian flag on 1 July 1914 as part of their Dominion Day celebration (the sort of ceremony never omitted on any polar expedition, no matter how feeble the condition of its members). Stefansson would later insist that their act constituted a Canadian claim, forestalling the Russian statement by two years. He may genuinely have seen it in this way at the time: to his young son, Alex, he explained that he was in the North to protect Canadian territory from the Russians.[48] However, the men on Wrangel certainly did not intend their flag-raising as a Canadian claim. In 1922 William McKinlay angrily called Stefansson's version a 'cock-and-bull story ... in keeping with much other stuff from the same source.'[49] On his return to the south, Stefansson gave alluring descriptions of Wrangel Island as a northern paradise, abounding with game, and he emphasized its potential strategic value in the future, when major air routes would pass over the Arctic.[50] Privately, he is said to have remarked that the chaotic state of Russia in 1919 presented Canadians with the perfect opportunity to seize the island.[51]

Despite Stefansson's strong position in Ottawa, these arguments were not likely to evoke action from the Canadian government. A claim to Wrangel Island was simply too far removed from any of Canada's past nation-building activities. But to develop the northern areas Canada already possessed as a huge grazing range would fit neatly into established paradigms. Many prominent Canadians were willing to believe that just as the Prairie West had been transformed from a wilderness to a garden by settlement and farming, so the 'Arctic prairies' (as Stefansson called them) could become a pastoral, domesticated landscape. The North would be twentieth-century Canada's new West.[52]

Stefansson therefore moved slowly on Wrangel Island, concentrating instead on the commercial potential of the muskox. Proposals on this matter were already beginning to make their way through the bureaucratic maze even before Stefansson's return, and his presence accelerated the process. As early as 1914, Stefansson had warned Ottawa that the killing of muskoxen by Hudson's Bay Company traders and others might result in the extinction of a species which was essential to the survival of the Inuit.[53] As he observed the animals more closely on his travels through the northern islands, the idea of muskox herding for profit took hold of his imagination.[54]

In February 1917 Stefansson wrote to Sir Robert Borden, Sir Edmund Walker, former American president Theodore Roosevelt, and the premier of British Columbia, Sir Richard McBride, enclosing samples of muskox wool. McBride died before the letter reached him, and his copy was referred to C. Gordon Hewitt, a senior zoologist with the federal Department of Agriculture and a member of the Advisory Board on Wildlife Protection. Hewitt, already a keen conservationist, took up the idea with enthusiasm.[55] On 28 November 1918, Hewitt wrote to Desbarats, expressing his eagerness to discuss the muskox project. Desbarats too became an advocate of Stefansson's plan. In January 1919 Stefansson attended a meeting of the Advisory Board on Wildlife Protection. J.B. Harkin was a member of the board, and it was likely on this occasion that the two men met for the first time. At age forty-four, Harkin was five years older than Stefansson, a vigorous, highly successful, and relatively young bureaucrat. Like Hewitt, he responded to Stefansson's ideas with enthusiasm. He promptly sent samples of muskox wool to several Canadian textile firms, asking for opinions as to its possible commercial value.[56]

In the following month, Stefansson returned to Ottawa, accompanied by a senior official of the US Department of Agriculture. It was decided that the two countries would cooperate in scientific research on the muskox and that the Canadian government would obtain animals from Ellesmere Island. Ellesmere was chosen because it was reportedly the home to large herds of muskoxen and because it was easier to reach by ship than the islands of the western Arctic, where (as Stefansson knew from personal experience) the animals were also abundant. The Canadians would establish an experimental station on the shores of Hudson Bay; the Americans would do the same in Alaska. Soon plans were under way to acquire a ship for an expedition to Ellesmere. There was talk of refitting Joseph Bernier's old vessel, the *Arctic*. This was the situation when Arthur Meighen, then the minister of the interior, came into the picture.[57]

Meighen, a lawyer with a degree in mathematics from the University of Toronto, had entered politics in 1908. He quickly attracted Borden's notice by his outstanding skill as a debater. Meighen became solicitor general in 1913 and minister of the interior in 1917. Relentless in his logic and merciless in his ridicule of his political opponents, he showed the warmer side of his personality only to his family and a few close friends. Meighen was certainly interested in the muskox project; as he himself wrote, common sense indicated that 'the proper method of utilizing the northland, is by taking possession of, and cultivating the animals native thereto, rather than by forcing the cultivation there, of animals native to milder climates.'[58] However, his cautious, rational style put a serious crimp in Stefansson's plans. Stefansson had brought this on himself: not content with the support of Desbarats, Hewitt, and Harkin, who could have quietly guided the project to completion, he sent a long memo directly to Meighen.[59] Meighen, described by a journalist who knew him well as a 'masterful authoritarian' with a 'passion for efficiency,' was not about to go ahead blindly with such a potentially expensive enterprise.[60] Because it was 'in essence ... a business project,' Meighen wanted a report 'from a body of business men in whose personnel an element of the imaginative will have a place.'[61] Meighen arranged for Stefansson to address a joint session of the House of Commons and the Senate on 6 May 1919; he then set about establishing a royal commission to inquire fully into the subject.

The chairman of the commission, appointed on 20 May, was John Gunion Rutherford of the Canadian Board of Railway Commissioners. The other members were Stefansson, Harkin, and John Stanley McLean, president of the Harris meat-packing company and a personal friend of Meighen. Stefansson considered Rutherford a poor choice, with little enthusiasm for the project.[62] The hearings did not start until January 1920. Just as it seemed that Stefansson's plans were about to be translated into action, they had been thwarted, or at least seriously delayed, by Meighen's cautious attitude. It was a portent for the future: Meighen – austere, almost puritanical, with an incisive, dry intelligence – simply was not the type of man to be sufficiently impressed by Stefansson's flamboyant self-promotion.

Stefansson, who had never visited the eastern Arctic, began to gather information about Ellesmere Island and its muskox herds. He sent a letter to George Comer, a New England whaling captain who knew Ellesmere and northwestern Greenland well. In 1916 Comer had sailed up Smith Sound to bring relief to American explorer Donald MacMillan's expedition, based at the Native settlement of Etah on the coast of Greenland. Comer had remained at Etah during the winter of 1916-17, observing the local people and their ways.

Ellesmere Island and most of northern Greenland were uninhabited, but the Cape York or Thule district, where Etah was located, was the home of the world's most northerly people, the Inughuit. Like all the Aboriginal inhabitants of the eastern Arctic, the Inughuit were originally migrants from the west. The ancestors of the Thule people left Alaska in about 1,000 CE. Most of the migrants settled on Baffin Island, but within two hundred years of the initial migration some were established on northern Ellesmere Island. They remained there until the fifteenth century, when worsening climactic conditions caused by the Little Ice Age drove them to the south. They settled on the Greenland side of Smith Sound. Their descendants, the Inughuit, rarely visited Ellesmere after about 1700, but its rich hunting grounds remained part of their collective memory.[63] So remote was their location that their very existence remained unknown to Europeans until 1818, when the British explorer John Ross 'discovered' Cape York and named it after King George III's second son, Frederick Augustus, Duke of York. Ross (a Scot) dubbed the local inhabitants the

'Arctic Highlanders.' Despite occasional visits from British and American exploring expeditions and more regular calls by Scottish whalers, the Inughuit retained their traditional way of life almost unchanged until the American explorer Robert Peary arrived at Etah in the 1890s.

Peary was a man driven, even more than most explorers, by a relentless need for success and fame. Between 1891 and 1909, he spent a total of nine winters in the Arctic. His first objective was the exploration of northern Greenland, which he completed in 1900, naming the northernmost point Cape Morris Jesup after one of his wealthy supporters. From 1900 to 1909, Peary's goal was the attainment of the North Pole. Whether he actually did succeed in reaching the pole is now considered very doubtful, but he unquestionably made impressive sledge journeys, first across the interior ice cap of Greenland and then across the polar pack.[64]

The Inughuit were essential to Peary's success. He adopted their diet, clothing, and travel techniques to a greater extent than any previous explorer. More than this, he was able to make use of the Inughuit themselves. 'For eighteen years I had been training them in my methods; or, to put it another way, teaching them how to modify and concentrate their wonderful ice technic and endurance, so as to make them useful for my purposes ... It has been my fortune to utilize the Eskimos for the purposes of discovery to a degree equaled by no other explorer ... no more effective instruments for arctic work could be imagined,' he wrote after his last expedition.[65] Peary usually referred to the Smith Sound tribe as 'my Eskimos.'

In the process of making the Inughuit useful for his purposes, Peary changed their way of life profoundly. 'There was not a rifle in the tribe when I first went there,' he noted. Formerly 'dependent on the most primitive hunting weapons,' the Inughuit were now supplied with 'repeating rifles, breech-loading shotguns, and an abundance of ammunition.' Peary also provided 'the best material for their weapons, their harpoons and lances, the best of wood for their sledges, the best of cutlery, knives, hatchets and saws for their work, and the cooking utensils of civilization.'[66]

With their new weapons, and encouraged by the rewards Peary offered to the most successful hunters, the Inughuit exterminated the caribou, which had formerly been abundant in their traditional hunting grounds between Cape York and the Humboldt Glacier. At the same time, Peary introduced (or re-introduced) them to Ellesmere Island. On his expeditions, the Inughuit sledged or were transported by ship to the Bache

Peninsula (which lay almost directly across Smith Sound from Etah) and to Grant Land, as the northernmost part of Ellesmere was then known. There they zealously hunted muskoxen on Peary's behalf and their own.

Peary recorded that prior to 1898 the Inughuit had 'killed one or two musk oxen' on the Bache Peninsula, but the more northerly parts of Ellesmere had not been visited by them within living memory.[67] In 1918 Donald MacMillan added the information that since the year when Peary first took them to the Bache Peninsula, 'the Eskimos have journeyed almost annually to these musk oxen grounds.' Moreover, they had begun to hunt on the west coast of the island as well, finding their way to Bay Fjord and Eureka Sound, 'from which region hundreds of musk oxen were taken out last year.'[68] Now, in the spring of 1919, Comer confirmed MacMillan's account, and he suggested that these hunting forays might eventually result in the extermination of the muskox on Ellesmere.

Comer informed Stefansson that in 1916-17 the Inughuit had brought back about 150 muskox skins from Ellesmere. 'This they never used to do but now having found out it is not so dangerous crossing Smith Sound they no doubt will continue to do this until the Musk ox will become scarce,' Comer wrote.[69] Extinction was certainly not beyond the bounds of possibility. Muskoxen were exceptionally vulnerable to hunters with firearms. When threatened, they did not flee; instead, a herd would form itself into a circle. In this way, the animals could successfully defend themselves with their massive horns against any natural predators. But men with guns could quickly slaughter an entire herd. Stefansson immediately forwarded Comer's letter to Harkin.

In this act lay the unlikely genesis of the Eastern Arctic Patrols. Any Canadian official with an interest in Stefansson's muskox project would naturally have been somewhat alarmed by the report, but only in Harkin's hands would it have had such far-reaching consequences. It would be easy to characterize Harkin – the author of so many long-winded memos – as an unimaginative, plodding bureaucrat, but the truth perhaps is that he was too imaginative. In his mind, Harkin could easily picture the broad, magnificent stretches of Canada's wilderness territories, and he was equally adept at envisioning threats to that pristine national heritage.

In 1911 Harkin had been appointed commissioner of Dominion parks, a role he filled with energy and distinction. He is remembered today mainly as the father of Canada's national parks system. Harkin himself wrote in the rough notes for a book on the history of the parks that, when offered the job, the prospect 'intrigued and stirred my imagination.' But the responsibility weighed heavily on him: 'Eight thousand square miles of the sublimest scenery in Canada had been placed under my protection and I lay awake at nights thinking of the damage one bad fire might do.' Then there were the threats from rapacious business interests. Harkin was determined that there should be no commercial development or exploitation of natural resources on park land. The Canadian people, he defiantly wrote, would always have 'free access to vast areas ... in which the beauty of the landscape is protected from profanation, the natural wild animals[,] plants and forests preserved, and the peace and solitude of primeval nature retained.'[70]

Of course, these threats were not mere figments of Harkin's imagination. But he seems to have had the type of imagination that dwells on and magnifies any potential encroachment. Added to this tendency towards mild paranoia was a remarkable degree of self-confidence, almost the equal of Stefansson's own. Harkin was to a large extent a self-made and self-taught man, and by 1919 he had enough successes behind him to make him believe that he could quickly master any new field. The grandson of Irish immigrants from Donegal, Harkin was born in the rural eastern Ontario community of Vankleek Hill in 1875. (This pleasant village is known today mainly for the number of red brick Victorian homes that still retain their original gingerbread trim.) He was the youngest of the five children of William and Eliza Harkin. His father was Protestant and his mother Catholic; the children were brought up in their mother's faith, and one of the sons became a priest. At the time of J.B. Harkin's birth, the family was a prosperous one. His father, a talented and ambitious graduate of McGill Medical School, was elected to the Ontario legislature in 1875 and re-elected in 1880.[71] However, the family had not accumulated enough financial resources to comfortably withstand the loss of its breadwinner. Dr. Harkin's early death in 1881 meant that his son would not be able to attend university. Instead, young James Harkin became a journalist at the age of only seventeen.

Harkin spent nine years as a newspaperman, first with the *Montreal Herald* and then with the *Ottawa Journal.* A dutiful son and brother, he brought his widowed mother and his unmarried sister, Minnie, to live with him in Ottawa. (His mother remained with him until her death in September 1920.) Despite his success in journalism, Harkin apparently yearned for a job with greater stability and prestige. In 1901 he joined the federal civil service as a second-class clerk in the Department of Indian Affairs; he was also appointed private secretary to Clifford Sifton in his capacity as superintendent of Indian affairs. (Sifton was then serving in the Liberal cabinet as the minister responsible for both Indian Affairs and the Department of the Interior.) Harkin quickly obtained promotion to first-class clerk, and his salary rose steadily with each passing year. In 1904 he was seconded to the Department of the Interior, where he again worked as Sifton's secretary – a much more demanding and prestigious job, for which he received $600 per year in addition to his clerk's salary rather than the $300 per year he had been paid at Indian Affairs.[72]

Harkin continued in this post under Sifton's successor, Frank Oliver. Oliver evidently thought very highly of his abilities: in 1907 he had Harkin officially transferred to the Department of the Interior and immediately named him chief clerk as well as private secretary. Four years later, Oliver offered him the parks appointment. Harkin's first impulse was to refuse, since he 'knew nothing about the parks or what would be expected of me.' Oliver countered his argument by pointing out that Harkin would not be 'hampered by preconceived ideas' and that he could quickly learn the facts and skills he needed to fulfill his new duties.[73] Harkin accepted this reasoning and took the job.

The Liberal government was defeated not long afterwards, and so Harkin began his new task under discouraging circumstances: viewed with suspicion by the Conservatives because of his close association with Sifton and Oliver, he was initially given very little in the way of resources. One of his subordinates, Mabel Williams, later evoked those early days in her popular book *Guardians of the Wild.* From his office on Sparks Street, a drab room with 'bare distempered walls ... ill-assorted furniture, [and a] cheerless bookcase empty of books,' Harkin could look 'through ... rather grimy window-panes at the rather depressing back premises of the Rideau Club.' He had no sources of information other than a few government bulletins and dreary files full of complaints about the condition of the park roads. Nevertheless, he set to work with what Williams admiringly described as 'creative imagination and executive energy.'[74] After eight years as parks commissioner, Harkin had evidently come to believe that Oliver was right, and that he was more than equal to any challenge. Indeed, his immense self-confidence was among Harkin's most striking traits. To some observers, it seemed that he believed he was infallible. As one acquaintance wryly noted, 'King Henry 8, J.B. Harkin, God and the Pope will have a great time some day when they all meet.'[75] Not content with his success as parks commissioner, Harkin was always in search of new ways to prove himself. In March 1919 he took on the task of persuading the government to preserve the nation's historic sites. As a result of his efforts, the Historic Sites and Monuments Board was formed in October of that year, with Harkin as one of the members.[76]

Harkin may have been spurred on to distinguish himself in new areas by the sense that, for all his success, he was still something of an outsider in Ottawa society. He was not a member of the exclusive Rideau Club. When Mabel Williams described the 'rather depressing' view of the club building's back wall from Harkin's office window, she may have intended to convey his sense of exclusion and frustration to former colleagues who could read between the lines. Instead of socializing with the city's elite, Harkin had to content himself with the less prestigious Laurentian Club. This slight ostracism was most likely caused by his rural Irish Catholic background and his lack of the useful social connections a university education could provide. A photograph of Harkin, taken around 1915, shows a good-looking, well-dressed, and very neatly groomed man who stares into the camera with a supercilious, slightly challenging expression. He appears successful and confident, yet somehow there is a tinge of uneasiness about the picture. In early middle age, Harkin remained unmarried, perhaps from a combination of financial caution, family obligations, and social insecurity. (He finally married in December 1924, at the age of nearly forty-nine. His bride, Jean McCuaig, was a former clerk in the Department of the Interior.)

In one of his areas of official responsibility, Harkin was not yet as successful as he would have liked to be. In 1917 the government had passed the Northwest Game Act, which forbade the killing of muskoxen, buffalo, elk, white pelicans, swans, and eider ducks (with the proviso that Aboriginal hunters could kill muskoxen for food when in extreme need); set seasons for the hunting of other species; and banned the use of poisoned bait. The commercial sale of muskox meat and skins was prohibited. It was not considered worthwhile to establish government administration in the Northwest Territories until 1921, following the discovery of oil near Fort Norman. The responsibility for enforcing the new Act in the NWT was therefore initially given to Harkin and the Dominion Parks Branch.[77] Harkin had no possible way of doing this except to work through the Mounted Police, who operated only in certain parts of Canada's north and were usually preoccupied with other tasks.[78] The failure to carry out the job assigned to him must have weighed heavily on Harkin's mind. Now here was proof that strict enforcement of the Game Act was required in even the remotest areas of the Arctic, and that failure to take action might result in the depletion of an economically important national resource.

To Harkin, the solution to the new problem was simple. Greenland, he assumed, belonged to Denmark, so the Danes must stop the Inughuit from encroaching on Canadian territory. After consulting with the other members of the Advisory Board on Wildlife Protection, he turned to the Department of External Affairs. (It is an indication of Harkin's degree of knowledge about foreign policy that he almost invariably referred to External Affairs as the 'State Department,' a mistake which persisted to the end of his life.) Taking hold of the North was about to become one of Canada's first foreign policy challenges in the postwar world.

2
The Danish Threat

> *Information received by me direct from Denmark makes it clear that the Danes are thinking of colonizing and exploring the islands ... north of Lancaster Sound ... There is still time for us to do something about this, and I should like to give you my ideas on the subject.*
>
> – *Vilhjalmur Stefansson to Arthur Meighen, 16 September 1920*

OTTAWA IN 1919 WAS not a capital with a wide outlook on the world. To Canadian politicians, only London and Washington really mattered. Canada had no legations abroad. The idea of separate representation for Canada at Washington had long been in the air, and would soon be taken up with renewed vigour, but for the time being the Canadian government's communications with its American neighbour were conveyed through the British ambassador. In terms of the wider world, what most Canadians meant by independence was the freedom to choose whether they would or would not be involved in British foreign policy initiatives.

The few dealings Canada did have with foreign states purely on its own account involved a long chain of communications. The minister or deputy minister of a Canadian government department would send a memorandum to the Department of External Affairs. (The memo did not have to go separately to the prime minister for approval, since from 1912 to 1946 Canadian prime ministers acted as their own foreign ministers.) Next, the matter would be communicated to the governor general; he, in turn, would write to the Colonial Office in London. The Colonial Office would inform the Foreign Office. If necessary, the Foreign Office would communicate with the ambassador or minister of the country in question. Any reply would wend its way back to the interested parties in Ottawa by the same route in reverse. The whole process could take months, or even years.

Clearly, this was not a procedure suited to the needs of a country with its own independent foreign policy. It was taken for granted that only comparatively minor questions would be dealt with in this way. If an international issue was of major importance, it would be the responsibility of the Foreign Office, with the officials in London consulting the Canadian prime minister on points that involved Canadian interests. To Sir Joseph Pope, the elderly, dignified under-secretary of state for external affairs, all was as it should be. Pope was described by his son as 'a thrice-dipped Tory and one to whom the very idea of political independence or, indeed, any change in our relations with the United Kingdom, [was] quite repugnant.'[1] 'Canada by herself is not a *nation,* and I hope I may never live to see her one,' Pope wrote in 1913.[2] In 1920 he dismissed the excitement then prevailing in some quarters over Canada's membership in the League of Nations. 'I think it is all absurd,' Pope sniffed, 'and I am convinced that Canada's true policy just now is to develop her resources and to leave European questions ... to our Imperial statesmen and the trained experts of Downing Street.'[3]

Sir Robert Borden envisioned a far more active international role for Canada than did Pope. However, the prime minister had no thought of developing the Department of External Affairs into a true foreign ministry. Instead, he hoped that the statesmen of Downing Street would become increasingly open to close and frequent consultation with the self-governing dominions. During the war Canada had gained a strong voice in London through membership in the Imperial War Cabinet, and Borden wanted to see a permanent arrangement along the same lines in peacetime. So External Affairs remained little more than a glorified post office, whose main task was to ensure that correspondence from abroad reached the appropriate Canadian government department. Pope was an expert in matters of protocol and procedure, but he took relatively little part in the formulation of policy. However, the department's brilliant young legal advisor, Loring Christie, was Borden's right-hand man in his dealings with the imperial government. Pope played a limited, but at times important, role in the Arctic drama of the next few years; Christie was at the centre of events until his resignation in 1923.

AT HARKIN'S PRODDING, THE deputy minister of the interior, William Wallace Cory, dispatched two memos on the first step of the long journey from Ottawa to Copenhagen. The first recommended that a copy of the Northwest Game Act should be sent to the Danish government. Harkin, entirely ignorant of the procedure for communicating with foreign states, asked that External Affairs write directly to the head of the Greenland

Administration in Copenhagen, Mr. Dagaard Fensen (a mistake for Daugaard-Jensen). In the second memo, Harkin added the suggestion that in order to ensure the Canadian regulations were observed, the Danes should permit Canadian game officers to be stationed in Greenland. Both recommendations were approved without comment by Cory.[4] The deputy minister, who appears in photographs taken over the years as an ever more portly dignitary, must have possessed a certain amount of intelligence and energy to have reached his high position in the civil service. However, in Arctic sovereignty matters Cory was little more than a cipher, relying without question on the recommendations made by the more determined and forceful Harkin.

Harkin also distributed George Comer's letter to other government officials in areas related to the North (C. Gordon Hewitt, Rudolph Anderson, Duncan Campbell Scott of the Department of Indian Affairs, and James White of the Commission of Conservation). To the comptroller of the Royal North West Mounted Police, A.A. McLean, he suggested that a police post should be established on Ellesmere. Harkin was promptly informed that the plan would involve too much expense and difficulty. McLean did offer the assurance that the police at Herschel Island and Fullerton Harbour would strictly enforce the Northwest Game Act, but since the two posts were far removed from Ellesmere, this declaration was of little help.[5]

Neither Harkin nor Cory seems to have realized that although Danish rule over the southwestern coast of Greenland dated from 1721, when missionary Hans Egede founded a settlement at Godthåb (present-day Nuuk), there was no Danish administration in the North. In 1905 the northernmost limit of Danish colonization had been extended from 73° to 74.5° north latitude. A large portion of the island lay beyond this boundary: the territory of the Inughuit was located between 76° and 77° north. The Danes certainly wished to extend their sovereignty over the entire island, and they had raised the issue at the Paris Peace Conference. However, their claim to northern Greenland rested on rather precarious foundations.

Before 1902 no Dane had even entered the region except as a member of a British or American expedition. British explorers John Ross (1818) and Edward Inglefield (1852) made the important early discoveries in the North. They were followed by Americans Elisha Kent Kane, Charles Hall, Adolphus Greely, Robert Peary, and Donald MacMillan, and by another British expedition under Sir George Nares. From the 1880s until the beginning of the First World War, Scottish whalers called regularly at the Inughuit settlements. Neither exploration nor whaling in the North was regulated by the Danes in any way. In 1916 the United States had formally renounced any claim it might have to the areas explored by Kane, Hall, Greely, and Peary in favour of Denmark. (The declaration was requested by the Danes during the negotiations for the sale of the Danish West Indies to the Americans.)[6] The long history of British exploration and whaling in the area certainly made a British territorial claim possible, although none had ever been put forward. The Danes were apparently anxious to ensure that their claim would not be contested by the British before they ventured on official action north of the 1905 boundary. The suggestion that Canadian officials should enforce Canadian game laws in northern Greenland was therefore virtually certain to arouse serious concerns in Copenhagen, but of this possibility Harkin and Cory were entirely oblivious.

Nor did Pope, to whom Harkin's recommendations were duly forwarded by Cory, realize the untoward nature of the request. Harkin's request was passed on from Pope to the governor general, the Duke of Devonshire. On 31 July 1919, Devonshire sent a letter to Lord Milner, the colonial secretary. Before the war, there might have been discussion of the matter at the Colonial Office, but Milner's policy was not to interfere in the concerns of the self-governing dominions.[7] On 23 August the Colonial Office informed the Foreign Office of the Canadian request without making any comment on it. There, too, it was treated as a matter which, having been approved in Ottawa, did not require further discussion in London. The Foreign Office in its turn addressed a diplomatic note to the Danish minister, Henrik Grevenkop-Castenskiold. The note, signed by Lord Curzon, the foreign secretary, asked that 'the Authorities in Greenland should advise the Eskimos regarding the provisions of the Canadian North-West Game Act,' and it added: 'in order to facilitate this the Government of Canada offers to station such officers in Greenland as may be necessary.' Grevenkop-Castenskiold replied the next day that he had 'not failed at once to submit' the note to his government.[8]

The Danes did not answer the British note until 12 April 1920, almost eight months after they had received it. An additional month passed before their reply reached Ottawa. In the interval, Captain Comer's testimony before the royal commission reinforced the impression that Inughuit hunting might result in the extermination of the muskox on Ellesmere

Island. Comer appeared before the commission on 24 January 1920. He emphasized the danger of extinction: muskoxen, he said, could be 'easily taken by the women' and even old people 'bragged how they killed them.' Comer commented favourably on the possibility of domestication. The next witness, however, was geologist John J. O'Neill of the Canadian Arctic Expedition. O'Neill was evidently determined to prevent Stefansson from doing as he liked with government funds. He observed that he did not think it was 'feasible to start in with a commercial scheme at once.' Chairman Rutherford immediately agreed, pointing out that western Canada was 'loaded up with the skeletons' of unsuccessful enterprises and that if anything were done in the North, 'it should be done in a very careful, cautious way.'[9]

Stefansson was not present at this session because of illness, but when he read the transcript of the evidence he can hardly have been pleased. He attended the hearings on 4-5 February, then decided that his best course was to direct his efforts elsewhere. Several later witnesses were, like O'Neill, former members of the Canadian Arctic Expedition. Rudolph Anderson attended most of the hearings and gave evidence at the last session. It was therefore clear that Stefansson's opponents in the civil service were keenly interested in the matter, and that he would not easily get his way. Other witnesses also expressed their doubts regarding Stefansson's confident plans. Arctic veteran Joseph Bernier pointed out that, far from being a simple matter as Stefansson had claimed, acquiring enough muskoxen for an experimental station would likely take several years. Even though Stefansson considered reindeer far less suitable for commercial exploitation than muskoxen, he now turned to plans for reindeer grazing on Baffin Island, presumably because reindeer could be acquired in Norway without government assistance, allowing him to take immediate action on his own initiative.[10]

Stefansson decided that it was more important to push forward with his reindeer plans than to continue his lectures. He gave Lee Keedick notice that he could not fulfill the rest of their contract owing to poor health.[11] In early March 1920, Stefansson resigned from the commission, 'feeling,' as he explained in a letter to Harkin, 'that I can almost certainly do more for the good of the cause outside of it than I can in it.'[12] Stefansson then left for London, where he discussed his reindeer project with officials of the Hudson's Bay Company. His aim was 'to get at least one sound company to demonstrate on a large scale the great opportunities for capital if invested in the North.'[13] From Ottawa, Harkin supportd his efforts, in the belief that successful private enterprise could open the way for later government ventures in the North.[14]

Stefansson's trip was inspired mainly by his plans for northern development, but he also took the opportunity to mingle in the social circles surrounding the prestigious Royal Geographical Society (RGS). Founded in 1830, the RGS had long been, as Rudyard Kipling described it in a speech at one of its meetings, 'the supreme Court of ultimate appeal and final revision throughout the geographical world.' Kipling continued, 'I confess there is something, to me as terrible as it is touching, in the thought of the men even now scattered under the shadow of death, from the Poles to the deserts, the crown of whose labours, when, please God, they return, will be your judgement.' Kipling knew many explorers, and each one 'took it for granted all he had done availed little till it had been weighed and passed by you.'[15]

The RGS had not yet shown any great interest in hearing and passing judgment on Stefansson. Stefansson's main associations had always been with the American geographical societies and the American press. In England, news of his expedition was released through the middlebrow *Daily Chronicle.* Stefansson did write a few letters to the RGS during the course of the expedition, which were duly published in the society's organ, the *Geographical Journal.*[16] However, there was the matter of the *Karluk:* one of the survivors, Ernest Chafe (then serving with the Royal Newfoundland Regiment), published an account of the tragedy in the May 1918 issue of the *Geographical Journal.* Chafe spoke highly of Bartlett, but his references to Stefansson were noticeably cool.[17]

The RGS had not turned against Stefansson, but neither was it in any hurry to embrace him. It was the custom for returning heroes to give a lecture sponsored by the society. In 1918 Stefansson had received a vague, almost casual invitation to speak at one of the regular evening meetings whenever he might happen to be in London.[18] This offer fell far short of the treatment accorded to Stefansson's fellow polar explorers Robert Scott, Ernest Shackleton, and Robert Peary, all of whom spoke at special meetings and were awarded the RGS gold medal. Stefansson therefore evaded the invitation, perhaps in the hope of a more enthusiastic reception on some future occasion.[19] To an inquiring journalist, he claimed that he had 'been ordered to rest following a loss of voice,' having given too many lectures in America.[20] Soon after his arrival in London, Stefansson spent an afternoon discussing his expedition with the society's secretary, astronomer Arthur Hinks. Hinks told Frank Debenham (a member of Scott's last expedition) that Stefansson showed no emotion when speaking about the deaths of his men on Wrangel Island, 'which made me feel a little bit that I did not like him.'[21]

At the time of Stefansson's visit, Sir Ernest Shackleton was the geographical lion of the hour. The Antarctic explorer had returned from his famous *Endurance* expedition in 1916, but there was no opportunity then for him to give more than a few lectures. First, Shackleton had to ensure the safety of the expedition's Ross Sea party; that done, he promptly joined in the war effort. After the armistice, Shackleton served until 1919 with the British force sent to aid the White Russian cause in Murmansk. The maverick, self-promoting outsider Shackleton had always received less favour from the RGS than his ill-fated rival, the more conventional and gentlemanly Robert Scott. However, the immense courage and determination shown by Shackleton in bringing his men out of the Antarctic alive after the loss of the *Endurance* won him the admiration of even Scott's widow, who had always disliked him. 'I think it is one of the most wonderful adventures I ever read of, magnificent, Shackleton or no Shackleton,' Kathleen Scott declared in her diary.[22] Shackleton's narrative, *South,* was published in November 1919 to great critical and public acclaim.

From December 1919 to May 1920, Shackleton lectured twice daily at the Philharmonic Hall, accompanying his talks with the brilliant photographs and films of Frank Hurley. According to *The Times,* the lectures were among the most popular entertainments in London.[23] The *Daily Chronicle* called Hurley's film one of 'the most wonderful that ever came from a camera'; the *Daily Mirror*'s writer placed it 'among the wonders of the world.'[24] When Shackleton finally delivered his RGS lecture on the *Endurance* expedition on 22 March, Stefansson was likely in attendance. Whether he went to the lecture or not, Stefansson definitely cultivated Shackleton's acquaintance at this time, and Shackleton was also interested in knowing Stefansson. Stefansson may have been motivated by the hope that Shackleton, who had once been outside the charmed circle but now apparently stood inside it, would be happy to befriend another outsider. This Shackleton certainly appeared to do: he frequently invited Stefansson to lunch or dine with him at the Marlborough Club (of which Shackleton had become a member by the personal invitation of King Edward VII).[25]

Stefansson told Shackleton that he hoped to lead a new northern expedition. In return, Shackleton confided that he too was thinking of another polar journey, this time in the Arctic. He had already approached the officials of the RGS, who approved the plan. Stefansson wrote to his new friend on 15 April, assuring Shackleton that he would 'be very glad to put at your disposal in addition to my published writings and those of other members of my various expeditions, also any unpublished information which I may have that could prove of value to you.' He ended: 'So call on me for whatever I can do for you, whether it be with regard to the Eskimo or any other things that appertain to a far northern voyage.'[26] It was a letter Stefansson would later regret, and even deny that he had ever written.[27] Shackleton replied, 'Your advice based on ten years experience of Arctic conditions in the Beaufort Sea and the adjacent islands will be of the greatest value to me ... I will take full advantage of your kind offer.'[28]

Meanwhile, Stefansson had found the Hudson's Bay Company officials highly receptive to his plans. The result of his discussions with them was the formation of the Hudson's Bay Reindeer Company, with Sir Augustus Nanton and Edward Fitzgerald of the HBC as trustees and Stefansson as one of the directors. In June 1920 the Canadian government granted Stefansson a lease to roughly 114,000 square miles on southwestern Baffin Island, along with the right to operate butchering, canning, and tanning operations. The lease was then transferred to the Hudson's Bay Reindeer Company. Stefansson did not intend to stop there: he envisioned expansion to Wrangel Island, Ellesmere Island, and the Mackenzie Valley.[29] And, of course, there was still the potentially even more impressive and profitable muskox project.

It was through Stefansson's visit to London that Knud Rasmussen first became linked in Harkin's mind with illegal muskox hunting on Canadian territory. From Baffin Island trader Henry Toke Munn, who was also in London at the time, Stefansson heard rumours that Rasmussen's private trading station, Thule (near Etah in northwestern Greenland), was doing a lively business in muskox hides from Ellesmere. (Munn, an upper-class British adventurer who emigrated to Canada in 1886 and spent years roaming the West and the 'barren lands,' had searched fruitlessly for gold on Baffin Island. His company, the Arctic Gold Exploration Syndicate, then turned to fur trading. In 1915 he established a post at Pond Inlet.)[30] Concerned by Munn's report, Stefansson wrote to Rasmussen. The tone of his letter was friendly and courteous. 'I have the greatest admiration for your work in northern Greenland and have read with delight all that you have written, or at least all that I have seen of what you have written,'

Stefansson told Rasmussen. He suggested that he and Rasmussen were the only writers who had described the Inuit with true understanding, since 'to my knowledge no explorer except the two of us is able to converse freely with the Eskimos in their own language.' Stefansson assured Rasmussen that if he wanted to obtain muskoxen from Ellesmere for domestication in Greenland, the Canadian government would likely permit him to do so. However, 'the killing for any commercial purpose of musk oxen in any Canadian territory such as Ellesmere Land or Axel Heiberg Land is forbidden by Canadian law. The Canadians are beginning to take a great deal of interest in the musk ox.' Stefansson emphasized that his letter was 'purely ... personal' and 'written entirely on my own responsibility.' Munn had been ignorant of the law until Stefansson told him about it, and the two of them had 'agreed in conversation that it would probably be of value to you to get the information also.'[31]

Stefansson then wrote to Harkin about Rasmussen's activities. He suggested that Rasmussen had probably acted in simple ignorance of the Canadian law. Nevertheless, Harkin responded with characteristic indignation and fired off another memo to Cory. He had received the impression that Rasmussen was a Norwegian, and he asked that the 'poaching' immediately be brought to the attention of the Norwegian government.[32]

Harkin's mistake about Rasmussen's nationality demonstrates his ignorance of the Arctic and its recent history. Although not nearly so renowned a figure as Stefansson in the English-speaking world, the Danish explorer had a solid record of important geographical and ethnographic work in the Far North. Rasmussen was born in Greenland of mixed Danish and Greenlandic ancestry. From childhood he had been fascinated by tales of the Inughuit who lived so far removed from the other Aboriginal inhabitants of the island.[33] He first visited them in 1902-4, as a member of Ludvig Mylius-Erichsen's Danish Literary Expedition (the first Danish expedition to northern Greenland), and a few years later he published a book about them, *The People of the Polar North.* The Literary Expedition was privately sponsored, and its main objective was ethnographic work. However, both Mylius-Erichsen and Rasmussen had definite political aims as well: they were anxious to secure northern Greenland for Denmark.[34]

But by the early years of the twentieth century, the region was almost the exclusive private preserve of the American, Peary. Peary wintered in the Smith Sound area in 1891-92, 1893-95, and 1898-1902. In 1905-6 and 1908-9 he made his base on the northern coast of Ellesmere, taking with him many of 'his' Inughuit from Etah. Peary reacted with intense hostility and competitiveness to intruders in his world. When the Norwegian explorer Otto Sverdrup appeared in Smith Sound in 1898, Peary refused all friendly overtures from his rival and set out in the dead of winter to establish a more northerly base. In the face of such steely determination, combined with unfavourable ice conditions in the area, Sverdrup abandoned his plans to take his ship up Nares Strait and to explore the northeastern coast of Greenland. Instead, he turned his course away from Peary's domain to the islands west of Ellesmere.[35] The presence of a rival, then, merely spurred Peary on to greater efforts.

Peary had a firm economic and emotional hold over the Inughuit. During the periods when he was in the United States, they eagerly awaited his return because they needed fresh supplies of necessities like ammunition and luxuries like biscuits. Even when they resented his commanding ways, it was hard for the Inughuit to deny his requests. 'People were afraid of him ... really afraid ... You always had the feeling that if you didn't do what he wanted, he would condemn you to death,' one Inughuk recalled decades later.[36] So long as Peary was a regular visitor to northern Greenland, there was no hope of establishing even an unofficial Danish presence there. Danish geographer Hans Peder Steensby sourly described him as the 'uncrowned monarch, who considers himself the legitimate owner of all countries and inhabitants from Cape York and northwards.'[37] Perhaps significantly, the Danish Literary Expedition did its work during one of the intervals when Peary was not in the North.

In 1909, however, Peary at last reached the pole (or claimed to have done so), and he left the Arctic for good. At the same time, Rasmussen discovered that Sverdrup had plans to return to Smith Sound and to establish a station of his own. There was also talk of German plans to trade with the Inughuit. Rasmussen urged the Danish government to take action. In March 1910 his proposal for a Danish station was discussed by a committee that included officials from the Ministry of the Interior and the state-owned Kongelige Grønlandske Handel (Royal Greenland Trade), which since 1774 had held a monopoly in the colonized part of Greenland.[38] As Rasmussen himself recorded, he 'received an answer to the effect that the land being considered No Man's Land[,] the Danish Government ... did not see its way to establish a station there' for fear of provoking explicit American or British claims. Any Danish actions 'would therefore have to be left to a private initiative.'[39] Officials hinted to Rasmussen that although 'foreign policy considerations prevented the government from proceeding ... there might later be a possibility of a takeover, once everything had been put in place by private means.'[40] In other words, Rasmussen was to establish a Danish presence while leaving the government free to disavow any responsibility for his actions should Peary persuade the American government to protest.

Rasmussen accordingly acted on his own, hoping that the existence of a private Danish trading post would forestall claims by other governments. His task was made much harder by the fact that the cautious government, determined to avoid any difficulties with the Americans and the British, flatly refused to provide a declaration that it had nothing against the plan. Without such a declaration, potential backers were wary. Nevertheless, Rasmussen finally obtained sufficient funds with the help of his friend, engineer Ib Nyeboe.[41] Accompanied by another young Dane with a taste for northern adventure, Peter Freuchen, Rasmussen returned to Greenland. Freuchen later wrote that he and Rasmussen were 'thrilled to be on our way, even though a heavy debt rested on our shoulders.' The two young men were determined 'to finance our own [expeditions] by trading with the natives, furnishing them with goods, making a living for ourselves, and at the same time making the northern, as well as the southern, part of Greenland Danish territory.'[42] From their station, which they named Thule, Rasmussen and Freuchen carried out important explorations of the northern coastline and interior in 1912 and 1916-18; the results of these expeditions were reported in international publications such as the British *Geographical Journal* and the American *Geographical Review*.

In 1920, Stefansson was well aware of the basic facts about Rasmussen and Thule (as his letter to Rasmussen proves), while Harkin and other civil servants in Ottawa knew nothing about them. This knowledge would prove to be of considerable value to Stefansson in the coming months. Rasmussen would soon become a key figure in the developing dispute between Canada and Denmark, for it was to him that the Danish government had turned for help in formulating a response to the British note. When the Danes received the note, they were unable to deal directly with the matter through the Grønlands Styrelse or Greenland Administration (a branch of the Danish Ministry of the Interior) because the Inughuit lived outside the area subject to Danish rule. In fact, the Danes were caught in a rather delicate situation: while they did not yet have the authority to act in the matter, neither did they wish to make an explicit admission that their government's control did not extend north of the 1905 boundary. Although the Americans had renounced their claim in 1916 (over Peary's vehement public protests), the British did not recognize Danish sovereignty over all of Greenland until September 1920.[43] In this situation, the existence of Thule was a godsend to the Danes. The Greenland Administration sent the British note to Rasmussen and asked for his comments.

When Rasmussen received Stefansson's letter from London, his response to the note had already been sent to Copenhagen. There it was considered satisfactory, and the Ministry of Foreign Affairs simply forwarded a translation to the British Foreign Office without considering whether the remarks it contained would readily be understood in London and Ottawa. The Danish response was received in the Department of the Interior on 12 May – the day on which the last of the royal commission's hearings was held.[44] Stefansson had recently returned to Ottawa after his stay in London. He was present as a spectator at the final hearing, which featured the testimony of Donald MacMillan. MacMillan confirmed that the Inughuit frequently hunted on Ellesmere, and he stated that they often killed game even when they already had enough to eat. 'The Eskimo, if he sees anything alive, kills it for the fun of the thing, whether he needs it or not,' Macmillan reported. He claimed that he had once seen Native hunters slaughter thirty-five muskoxen in a single day. To this picture of wanton over-hunting by ignorant Natives, MacMillan added some inaccurate information about Danish claims to northern Greenland. He was under the impression that the 1916 agreement between Denmark and the United States had fully opened the way for Danish rule in the North. MacMillan described the Inughuit as Danish subjects, and he remarked that the Thule trading station might soon be taken over by the Danish government.[45]

If Harkin returned to his office after the hearing, he must have found Rasmussen's letter waiting on his desk; if not, then it was likely the first thing he saw the next morning. He read it with a dismay born of utter incomprehension. Rasmussen began by saying that 'the Polar Esquimaux of the Smith Sound District, who at one time migrated to Greenland across Ellesmere land,' had long known that there were 'vast herds of musk oxen' on Ellesmere. However, they had seldom hunted these animals, because caribou were available in abundance much closer to home, and the caribou skins were of greater practical use. The Inughuit had begun to hunt on Ellesmere only after the 'ruthless' slaughter of game instigated by Peary had exterminated the caribou in their traditional hunting grounds. According to Rasmussen, the need of the Inughuit for muskox skins was far greater than their need for the meat, since caribou skins, traditionally used for most of their clothing and bedding, were no longer available.

The Inughuit now made hunting journeys to Ellesmere each spring. Rasmussen confirmed that the hunting was on a scale which might eventually result in the extermination of the muskox, but he believed that many years would be required to accomplish this. In the meantime, he was taking steps to import reindeer skins from Europe. Rasmussen argued that the Inughuit should not be forbidden to hunt muskoxen on Ellesmere until his 'effective counter-balancing measures' had been fully implemented. He alone would have to take responsibility for this program, since, as it was 'well known,' the 'territory of the polar esquimaux' was a no man's land, and there was no official authority in the district 'except that which I exercise through my station – an authority which I have hitherto had no difficulty in maintaining chiefly because the polar esquimaux, when reasonably treated, adopt a very rational attitude toward all decisions which the stations [sic] considers it advisable to take.' Rasmussen was evidently alarmed by the possible sovereignty ramifications of any involvement by Canadian officials, and he concluded with the emphatic assurance that 'in order to carry out the protective measures indicated in this statement, I shall need no assistance whatever from the Canadian Government.'[46]

The covering letter from Grevenkop-Castenskiold explained that Rasmussen had been consulted because the matter related only to 'the small tribe of Esquimaux living near Cape York at the Danish Missionary and Trade Station of Thule,' not to the entire Native population of Greenland. He stated that the authorities in Copenhagen had studied Rasmussen's statement and could 'subscribe to what Mr. Rasmussen says therein.' Like Rasmussen, Grevenkop-Castenskiold emphasized that the 'Esquimaux of the polar region in question' had taken to hunting muskoxen on Ellesmere only out of dire need, indirectly caused by the American expeditions.[47]

Harkin had no way of knowing that Rasmussen often referred to northern Greenland as a no man's land.[48] Rasmussen's purpose in making such comments was evidently to emphasize that neither the British nor the American government had ever made a formal claim to the area. But because the covering letter was careful to avoid any direct statement about the lack of an official Danish presence in northern Greenland, and because Harkin had just heard MacMillan state that the Danes had taken over the area in 1916, it is hardly surprising if he entirely misunderstood Rasmussen's meaning.

Moreover, the reference to the 'territory of the polar esquimaux' was misleading to Canadians. 'Polar esquimaux' was not, as Harkin assumed, a general term for all Inuit; instead, it was the name then used in Denmark for the Thule tribe.[49] By 'the territory of the polar esquimaux,' Rasmussen unquestionably meant northern Greenland, not the Canadian archipelago. (In the terminology he used, the Canadian Arctic was the territory of the central Eskimos.)[50] But Canadians, if they knew about the Inughuit at all, would have been more likely to use Peary's name for them, the 'Smith Sound Eskimos.'[51]

The root of the confusion was thus a matter of terminology, aggravated by Canadian ignorance of the sovereignty situation in Greenland.[52] It is important to remember that Rasmussen's letter was written to the Greenland Administration, not to the British or Canadian authorities. With a Danish audience in mind, Rasmussen could feel sure of being understood. But well known though it might have been in Copenhagen that Thule lay in a no man's land, this fact was anything but common knowledge in Ottawa. MacMillan's misleading testimony had convinced Harkin that Danish rule in northern Greenland was secure. Therefore, Harkin immediately concluded that the Danes were set on further expansion, and that the two letters were intended as a direct and deliberate challenge to Canada's northern sovereignty.

Harkin's first step was to show Rasmussen's letter to Stefansson and to ask for the explorer's advice. In response, Stefansson produced a five-page memo that painted a dramatic picture of a serious and imminent threat from Denmark. He began by suggesting that the Inughuit, having migrated from Ellesmere to Greenland several hundred years before, were about to make a 'reverse migration' back to Canadian territory. Stefansson based this claim on an article in the *Geographical Review* by Elmer Ekblaw, a member of the 1913-17 MacMillan expedition. However, the families involved were not, strictly speaking, Inughuit: they were the descendants of Baffin Islanders who had settled at Etah in the 1860s. Nor is there any hint in Ekblaw's article that the movement was instigated by Rasmussen or other Danes.[53]

Stefansson implied that the migrants would carry the Danish flag with them, even though he evidently knew quite well that the Inughuit were not yet Danish subjects. Stefansson was on good terms with Peary until the latter's death in February 1920, and he was aware of Peary's view that the United States could and should have claimed northern Greenland.[54] It is also highly likely that he knew the Danes had neither formally asserted nor gained international recognition of their sovereignty in the Far North. With reference to Rasmussen's 'no man's land' passage, he wrote: 'I should think our Government might well take the position that though we have nothing to say about the political status of North Greenland, where Mr. Rasmussen's main trading station is located, we do have very definite authority over all the lands to the west of Baffin Bay, Smith Sound and Robeson [C]hannel. However, Stefansson also stated that Canada should take steps to halt the Inughuit hunting expeditions 'if the Danes are not willing through International courtesy to see that they are not undertaken by residents of their colonies.' He thus muddied the waters instead of clarifying them, as he evidently might have done.

Stefansson also raised doubts about Rasmussen's personal honesty. He wrote that Rasmussen was 'deceptive' when he claimed the muskox skins were essential for the survival of the Inughuit. They were too heavy to be used for clothing; therefore, their main use was as bedding, which Stefansson did not consider essential for life. Stefansson put forward the alternate explanation that Rasmussen frequently sent the Inughuit into Canadian territory in quest of valuable white fox and polar bear furs; while on these expeditions, the hunters and their dog teams lived mainly on muskox meat. The muskox skins, although relatively unimportant, were carried back to Thule and sold.[55] Harkin therefore regarded Rasmussen as a man whose word could not be trusted.

In Harkin, as Stefansson probably realized, he had found the ideal mark: intelligent, energetic, and influential, yet ignorant of basic facts about the Arctic, in many ways naive, and exceptionally susceptible to any hint that Canada's interests were under threat. A few days after Stefansson wrote his memo, he gleefully told an acquaintance that he expected 'to make Ottawa his headquarters soon.'[56] Harkin moved easily from a belief that Rasmussen was encroaching on Canadian territory to the conviction that the Danish explorer intended to challenge Canada's very claim to ownership of Ellesmere Island. There is no indication in the records that Sir Robert Borden (then in poor health and increasingly preoccupied with plans for his retirement, which was announced a few months later) took any interest in the matter. The Canadian response was left to Harkin and Cory. Disinterested expert advice might well have produced a more moderate and realistic assessment by Harkin, but instead, he acted on the initial response that had been reinforced by Stefansson.

Once he was sure his deception of Harkin had succeeded, Stefansson may have tried to force the government's hand by making the issue public. On 11 June an article appeared in the Toronto *Globe* under the headline 'Musk-Oxen the Cause of Tiff with Denmark: Danish Eskimos Said to

Invade Canadian Islands and Slay Animals for Their Hides during Closed Season.' The article stated that, according to evidence given before the royal commission, the 'Danish Eskimos' often crossed over to Ellesmere, killed as many muskoxen as they pleased, and then, leaving the carcasses 'strewn about,' took the 'valuable furs' back to Greenland. The anonymous author reported that Canada had protested this behaviour and that 'the Danish official who replied gave no satisfaction, his reply practically being to the effect that the matter was none of Canada's business.' The writer concluded by remarking that the next move was 'up to Canada.'[57] Both the information contained in the article and its final demand for action by Ottawa strongly suggest that it was inspired, or perhaps even written, by Stefansson.

A few days later, on 14 June, Harkin sent the first of many memos on the subject to Cory. He suggested that Inuit from Hudson Bay could be settled on Ellesmere as a way to establish Canadian occupation. Fortunately, this idea was soon dropped. Harkin then attempted to formulate an answer to the Danish note. In one of his characteristic long memos, dated 16 June, he once again demonstrated his profound ignorance of the relevant facts. He described the Inughuit as Danish subjects and observed testily, 'It is not any part of the duty of the Canadian Government to provide for the natives of Greenland. That is a duty properly appertaining to the Danish Government.'[58] Cory forwarded the memo to External Affairs. At Sir Joseph Pope's request, it was sent back to Harkin for revision and shortening. The next version, dated 29 July, was slightly more concise, and certainly clearer in its recommendations. The first memo had merely urged that Canada take 'a very strong stand' regarding its sovereignty without specifying how this might be done. The second put forward a concrete plan: 'Denmark should undertake to prevent its natives from going to Ellesmere land for musk-oxen and ... failure on its part to do so will result in the establishment of Mounted Police posts in Ellesmere land to deal with the situation on behalf of Canada.' (There is no indication that the RCMP were consulted at this point.) 'I consider it very important that the Canadian Government should not subscribe to [the idea that Ellesmere was a no man's land] even by the implication of silence but that it should be made clear that Canada has exclusive authority over all the lands to the west of Baffin Bay, Smith Sound and Robeson Channel,' Harkin wrote, echoing Stefansson's memo to him.[59]

Harkin's second memo was forwarded to the Colonial Office by the governor general on 13 July, with a covering letter to Lord Milner which stated: 'My Government consider if Denmark fails to take steps to remedy the situation complained of, it will be necessary for Canada to establish a mounted police post in Ellesmere Island for the purpose of stopping this slaughter, and of asserting Canadian authority.'[60] Pope noted on his copy of the dispatch that he did not consider it 'sufficiently explicit' about Canadian sovereignty claims. He informed Cory and Harkin of his concerns, but they ignored the warning.[61]

Next, Harkin and the other members of the royal commission invited Rasmussen to come to Ottawa and give evidence, which he agreed to do. Since no other witnesses were called after the 12 May session, it seems that this request was simply a device to set up a personal meeting with Rasmussen. Stefansson then left Ottawa for Vermont, where he had been offered free accommodation by an admirer, Dr. Raymond Pearl of Johns Hopkins University. There he wrote his narrative of the 1913-18 expedition, *The Friendly Arctic* (having successfully evaded Desbarats' request that he give priority to an official account for publication by the government).

ON 10 JULY 1920, THE sixty-six-year-old Borden resigned as prime minister. He was succeeded by Arthur Meighen. For all his intelligence and ability, Meighen lacked the personal qualities and charisma needed for success as a national leader. The economic and social problems caused by the transition from war to peacetime were becoming acute, while bitterness over wartime controversies like the conscription issue still lingered. The Unionist government, a coalition of Conservatives and some members of the Liberal party, was increasingly fragmented, and it would soon face strong challenges from the new Liberal leader, William Lyon Mackenzie King, and from the rise of the Progressive Party in the western provinces and rural Ontario. From this point onwards, Stefansson had to consider the possibility that the government which had sponsored his activities since 1913 might soon fall. Even if Meighen managed to remain in power, economic constraints might hamper Stefansson's Arctic plans – unless he could provide some powerful reason for the government to act.

Meighen was replaced as minister of the interior by Sir James Lougheed. Despite a past featuring a dramatic, Horatio Alger-style rise from poverty to wealth, events would show that Lougheed did not possess the ability or experience needed for the task of heading an organization as large and complex as the Department of the Interior. The son of a carpenter, Lougheed grew up in Toronto's Cabbagetown district. He was of Irish Protestant descent on both sides; his upbringing combined Methodist strictness, Orange bigotry, and the pursuit of education. Lougheed's ambitious mother encouraged him to study law, and he gained admittance to Osgoode Hall in 1875. In the 1880s Lougheed moved west. His Calgary law practice flourished after the Canadian Pacific Railway became his major client; combined with successful investments in real estate, it made him a rich man. Lougheed was a dedicated supporter of Sir John A. Macdonald and extremely active in Conservative party politics. He married into a prominent western family, the Hardistys. When his wife's uncle, Senator Richard Hardisty, died in 1889, Lougheed was chosen as his successor. At the age of only thirty-five, he was the youngest member of the Senate. In 1906 he became the Conservative leader in the upper house and, after Borden's victory in 1911, he served in the cabinet as minister without portfolio. Between 1914 and 1918, Lougheed worked vigorously for the war effort. In recognition of his wartime services, he was knighted in 1916.[62]

Lougheed was not the type of man to be overly concerned about the possible extinction of the muskox in and of itself, but Stefansson's gospel of northern economic development likely appealed to him. He was a dedicated supporter of business interests and viewed the budding conservation movement with disdain. As minister of the interior, Lougheed opened Crown lands in Alberta to exploration for oil. He abolished the Commission of Conservation (established by the Laurier government in 1909) on the grounds that it was not sufficiently subject to government control and spent excessive amounts of money.

At the time of his appointment, the new minister, like Meighen, seems to have been well enough disposed towards Stefansson and his plans. Nevertheless, the departure of Borden was a serious matter for Stefansson, who must have known that *The Friendly Arctic* would cause furious controversy. Borden had agreed to write a preface to the book, thus implicitly endorsing Stefansson's version of the loss of the *Karluk* and the 'mutiny' at Collinson Point. However, this apparent endorsement would have been far more useful if Borden had stayed on as prime minister. While Borden was enthusiastic about Stefansson's plans for further exploration in the North, Meighen's decision to appoint a royal commission before proceeding with the muskox project was a clear indication of the line he was likely to take should Stefansson ask the government to sponsor a new expedition.

The explorer's developing ideas were extremely ambitious and daring. The September 1920 issue of the *Geographical Review* contained an article by Stefansson on 'The Region of Maximum Inaccessibility in the Arctic.' He did not explicitly state that what he called the 'pole of inaccessibility' on the Beaufort Sea side of the North Pole (see Map 4) would be the goal of a new Stefansson expedition, but clearly he hoped to explore – and preferably to find new land in – this last unknown part of the Arctic. Stefansson speculated that ice floes from the area of Wrangel Island were carried by a strong current through the unknown area, to emerge in the North Atlantic near Spitsbergen. Seals and other animals living on the floes would naturally be carried along with the ice. A party moving from floe to floe along this path, travelling light and living on seals and polar bears, could cross the Arctic Ocean by way of the pole.[63] It was a bold plan, and one that Stefansson could hardly expect Meighen to eagerly endorse, let alone willingly pay for. At the same time, he knew there was now little to hope for from his muskox and reindeer plans. The royal commission's report did not appear until the following spring, but Stefansson no doubt anticipated that it would, as in fact it did, advocate a gradual and cautious approach to northern economic development. The commissioners also recommended that no further grazing leases should be granted.[64] In this situation, it was very much to Stefansson's advantage to exacerbate Canadian fears about Arctic sovereignty.

While Stefansson was contemplating ways of getting the government to back a new expedition, he received a personal letter from Rasmussen. It was written on 11 May but, given Stefansson's unsettled way of life, it might not have reached him until the later part of the summer. Stefansson was scheduled to start on a lecture tour of American cities in November. In September he began an aggressive campaign in Ottawa for a new expedition, using Rasmussen's letter as his trump card. In order to do this, he had to significantly misrepresent its contents.

The letter was a completely frank and straightforward document. Rasmussen provided a history of his activities at Thule to show how it had come about that he traded in Canadian muskox skins. If Stefansson had previously been unaware that by 'Polar esquimaux' Rasmussen meant only the Thule tribe, this letter should have enlightened him: in it, Rasmussen referred to 'the Polar-Eskimos, generally known under the name of the Smithsound-Eskimos.' Rasmussen described how he had attempted to persuade the Danish government to make an official claim in 1910, but 'because of the American and [British] interest in this country – the matter was given up.' Again, if Stefansson had not previously realized the unofficial nature of Rasmussen's activities, this statement would have made

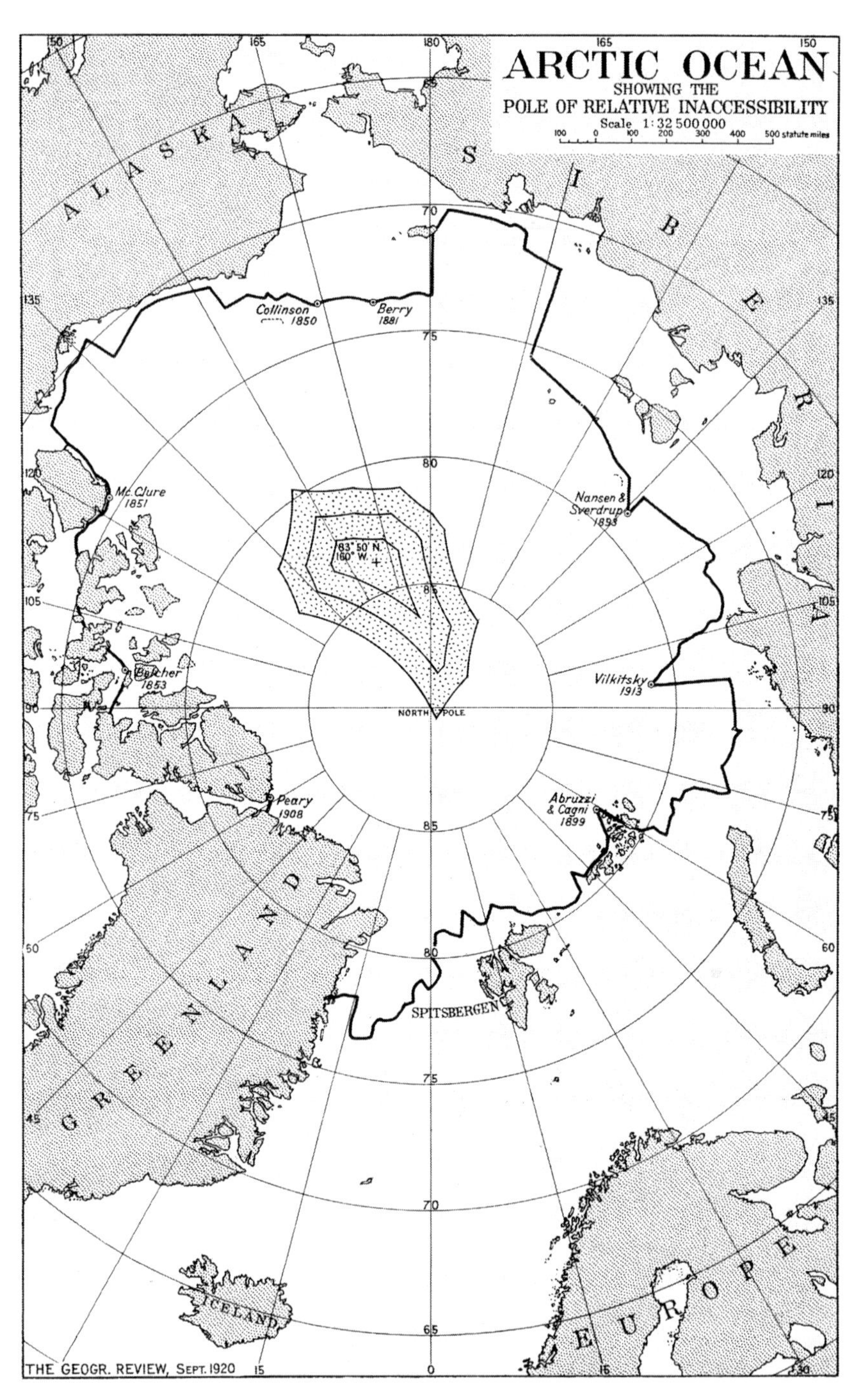

MAP 4 Map accompanying Stefansson's article in the *Geographical Review*, September 1920 | Courtesy of the American Geographical Society

the true situation clear to him. Rasmussen then turned to the muskox issue, assuring Stefansson that he was well acquainted with the Canadian law. He provided a copy of his letter to the Greenland Administration as proof of the assertion. Rasmussen insisted that 'the trading station at Thule has not wished to consider the musk-oxen as an article of merchandise, but on the contrary it has appealed to the Eskimos to chase them as little as possible for their own want.' According to Rasmussen, the station purchased only about ten muskox skins per year. Finally, Rasmussen told Stefansson about his plans for a new expedition, lasting four years, 'during which I shall visit as many as possible of the American Eskimos.'[65] This expedition, intended to retrace in reverse the migration route of the Inuit from the western Arctic to Greenland, was a long-standing dream of Rasmussen's. He had first proposed the project in 1910, but had difficulty funding it until the postwar boom in fur sales.[66] Later known as the Fifth Thule Expedition, Rasmussen's venture would ultimately prove to be one of the most important ethnographic expeditions ever to work in the Canadian Arctic.[67]

For Stefansson, Rasmussen's plans were important because they could be represented as a concrete manifestation of the Danish challenge to Canada's sovereignty. Here, he could claim, was proof that the 'reverse migration' described by Ekblaw was indeed to be undertaken at Denmark's behest for the purpose of occupying the islands. On 16 September Stefansson wrote to Meighen from New York, stating he had received information from Denmark which indicated that the Danes were 'thinking of colonizing and exploring the islands ... north of Lancaster Sound.' 'There is still time for us to do something about this,' Stefansson went on, 'and I should like to give you my ideas on the subject.' Meighen's secretary replied that Stefansson should first meet with Loring Christie.[68]

Stefansson then wrote a long letter to Christie. He did not refer to Rasmussen by name; instead, he told Christie only that 'a private person in Denmark' had written to him. This person was largely responsible for the message sent to Canada. 'I do not know how the communication of the [Danish] Department of State [sic] was actually worded,' Stefansson (who had, of course, been shown the Danish note by Harkin) continued, 'but as formulated by my correspondent in Denmark when he submitted it to them the implication was clear that no Canadian claims were recognized. This appears in a phrase to the effect that the Danes would not interfere with game "south of Lancaster Sound where the Government of Canada has jurisdiction."' Stefansson followed up this entirely mendacious assertion with a warning that 'if we do nothing, I should think that if this case came before an international court it would probably be decided in [Denmark's] favour.'[69] Stefansson had obviously been doing extensive research on the history of the northern islands. He pointed out that Ellesmere had been explored mainly by Americans, while Axel Heiberg Island and the Ringnes group were discovered by the Norwegian Otto Sverdrup. Therefore, Canada could not make any claim to these lands based on the fact of British discovery. Although Canadian sovereignty over them had been proclaimed, there had been no actual occupation to give substance to Canada's declarations.

At the same time, Stefansson wrote another letter to Christie about Wrangel Island – which, with the benefit of hindsight, is not difficult to identify as his true goal. Should a new Canadian expedition led by Stefansson use Wrangel as its jumping-off point, patriotic emotion would be all the more fervent if Wrangel was claimed for Canada in the process. Moreover, if unknown land did in fact exist in the 'region of maximum inaccessibility' north of Wrangel, it would be easier for Stefansson to claim this new territory on Canada's behalf once Wrangel itself was Canadian territory. To Christie, Stefansson argued that Canada was 'the country most logically situated for the development both of lands now known to exist and of others that may be discovered to the north of us.' In his view, it was not 'inevitable that every land north of Alaska shall belong to Alaska ... The countries to the north will belong to whoever appreciates their value and cultivates them.' Unable, at this delicate preliminary stage, to develop his favourite theme as fully as he would no doubt have liked, Stefansson fell back on the island's possible future value for northern aviation and on the fact that it was supposedly 'an excellent base for walrus hunting.' He suggested that because the world was 'gradually approaching a meat shortage, as every food authority concedes,' it was 'certain' that walrus meat would be of commercial value in the future.[70] But even Stefansson must have known that Canada would never claim Wrangel simply to develop a walrus meat industry. At least for the time being, Rasmussen's planned journey through the archipelago was by far the best argument Stefansson had for a new Canadian expedition.

Shortly after these letters were written, Stefansson hurried to Ottawa, where he met with Christie and Borden on 30 September. By Christie's own account, when he first met Stefansson he was impressed by the explorer's high status and accepted his statements about the Danish threat without question.[71] Christie had graduated near the top of his class from Harvard Law School in 1909 and then joined External Affairs in 1913. Like Borden, he was a Nova Scotian, born into a farming family, who had risen through the study of the law. Whether because of these personal similarities or not, he soon became one of the prime minister's most trusted advisors. However, Christie had not studied international law at Harvard, and before 1920 his work had not required him to do any research on sovereignty questions.[72]

Despite Christie's undoubted interest in the matter, he was far too busy with his other duties to make a detailed study of it. Moreover, he was scheduled to leave Ottawa for Europe in early November, in order to attend the first assembly of the League of Nations. Nor did External Affairs have sufficient resources for Christie to pass the task on to a subordinate. It was therefore assigned to the Department of the Interior. The next day, Stefansson attended a hurriedly called meeting of the Advisory Technical Board (ATB), a body created a few months earlier to consider the more complex or technical matters that came before the department.[73] The board was to prepare a report and recommendations for the minister; Lougheed would then advise the cabinet as to how Canada should proceed. Harkin was a member of this body, and he was, of course, present at the meeting.

Stefansson dramatically informed the board members that the purpose of Rasmussen's new expedition was to 'colonize the country with Danish Eskimos from Greenland.' He again claimed that Rasmussen's letter to him included the statement: 'There is no question of our breaking Canadian Game Laws because we are not coming into Canada but a part farther north. It is not under Canadian jurisdiction.' Canada, Stefansson conceded, had 'a great deal of claim on' the archipelago, but 'whether it would be recognized or not is another thing.' He had carefully reviewed Canada's claim to Ellesmere and, as he explained, 'It seems extraordinary, perhaps, that international law says that discovery does not constitute ownership but that the discovery must be followed by exploration and occupation. A mere proclamation that we have taken possession of this island has little significance in law – international law.'[74] Danish exploration and colonization of Ellesmere would give Denmark a better claim than Canada. Moreover, the Norwegians were 'very friendly with Denmark' and might be willing to transfer their rights over Sverdrup's discoveries to the Danes, just as the Americans had transferred their rights over northern Greenland.[75]

Stefansson then attempted to turn the discussion towards Wrangel Island. Somewhat abruptly, he made the provocative statement that there was no reason for the 141st meridian to be accepted as the western limit of Canada's claims. Canada should explore the entire Beaufort Sea and, if islands existed there, it would 'be well for us to own them in view of their becoming valuable at some later time.' There were much larger matters at stake than sovereignty over Ellesmere. 'It is now really a question of territorial jurisdiction over the north in general,' Stefansson declared. First, police posts should be established on Ellesmere. The second step would be exploration. Stefansson then quickly backtracked, arguing in confused words which betrayed the weakness of his position that 'perhaps exploration would be the first thing; to explore the whole area thoroughly to show what it is good for; to show our intentions; and then get commercial companies to establish trading posts there. That amounts to occupation.'

At this point, the discussion almost went badly off the rails from Stefansson's point of view. The board members had received almost no advance notice of the meeting, and they were not initially well informed about its purpose. At first they listened to Stefansson for long stretches in stunned silence, but then a few raised skeptical questions. If Norway's potential claim rested only on discovery, and if discovery was of no value without occupation, why should Canada fear a transfer of Norwegian claims to Denmark? And why did Stefansson advocate another exploring expedition when, according to his own argument, more exploration 'would not get [Canada] any further than she now is?' Harkin, however, made it clear that he stood with Stefansson. 'The whole issue,' he declared, 'seems to me to be: Are the northern islands worth while, or not? That is the first issue: do we want them, or do we not? Apparently if we want them we have to do something to establish our title.' A subcommittee (of which Harkin himself was naturally a member) was formed to report in more detail.[76]

If any doubts had been raised in Harkin's mind by his colleagues' questions, they would quickly have dwindled into insignificance when he heard the latest news from Copenhagen. On the same day as the board's meeting, a letter from Rasmussen's secretary was received in the Department of the Interior. Written in early August, it expressed Rasmussen's regret that he would not be able to come to Ottawa after all. Instead, he was obliged to make a trip to northern Greenland.[77] The letter had not been mailed until September because the secretary was not sure of the correct address.[78] Harkin immediately decided that a dark conspiracy was being played out. He conjectured that his protest had been received in Copenhagen in August; immediately, the Danes had decided to send Rasmussen north, in order to lead an invasion of the archipelago before the Canadian police posts could be established. Rasmussen's letter had been kept unmailed so that he could get away before Canadian suspicions were aroused.[79]

For a time, the panic-stricken Harkin was convinced that, whatever practical difficulties might stand in the way, a Canadian presence must be established at once. On 13 October the members of the Advisory Technical Board's Arctic subcommittee were informed that Meighen wanted their report and recommendations for action 'right away.' As one of the subcommittee members, Otto Klotz, recorded in his diary, he and his colleagues could only state that they were in favour of taking action. They were not yet 'prepared to formulate a definite scheme.' Klotz promptly set about looking up the instructions given to Bernier for his three voyages to the archipelago.[80] On the 14th Lougheed met with Harkin and the surveyor general, Édouard Deville, who was the ATB's chairman. As Deville reported to Klotz, he was ignored, while Harkin 'proceeded to lay before the Minister a memorandum of which Dr. Deville knew nothing.' Armed with Harkin's memo, Lougheed left to attend a cabinet meeting.

Then, at a meeting of the subcommittee on the 15th, Harkin again told the other members that they must immediately produce a set of recommendations regarding the sovereignty expedition. Their proposals would be reviewed by the cabinet the next morning. Klotz noted: 'To most of us I think it looked as if we were being stampeded by something or somebody, altho' Harkin said that it was feared Rasmussen & the Danish government were aiming & taking steps to obtain Ellesmere Land, treating it now as No Man's Land. What truth there is in this we don't know, but the haste with which we were to act appeared unseemly.' Nevertheless, 'we had to carry out the command ... and so there was launched an Arctic expedition ... Our command was – don't think but act, – and we did ... It was after 7 p.m. when I got home, but I was one of the seven that had launched Canada on a great exploration!!' As Klotz sardonically recorded, the ATB's memo contained no specific proposals about a ship, a commander, or 'what was really to be done.' 'And this is fathered by the Advisory Technical Board!!' he wrote in despair.[81]

At the cabinet meeting on 16 October, the assembled ministers decided that instead of sending an expedition by ship in the spring, it would be better to dispatch an immediate overland expedition by the Mounted

Police. There is no record of the reasoning behind this dramatic change in plan, but it is difficult not to suspect that a second memo by Harkin had reached the cabinet along with his colleagues' more reasonable but vague proposals. After the meeting, at a time when Deville and the other subcommittee members remained in ignorance of the cabinet's decision, Harkin met with Cortlandt Starnes, the assistant commissioner of the RCMP, who in turn telegraphed the commissioner, A. Bowen Perry, in Edmonton: 'Reported Danish Government endeavouring get possession Ellesmere Island. [Harkin] [s]tates Government anxious Police Patrol over land take possession first. [A]fraid expedition by water next season too late. I consider such undertaking started at this late season hazardous ... Please wire views.' Perry confirmed that the plan was 'impracticable.'[82] At the next subcommittee meeting, Harkin's idea was greeted with ridicule by his colleagues. Klotz wrote in his diary that they had simply smiled at the thought of the police 'charging' two thousand miles from Winnipeg to Ellesmere, 'establishing posts, law and order among the musk ox and above all establishing sovereignty against the wily Danes until the exploring ship arrives next year.' The subcommittee authorized Deville 'to discuss with the Deputy Minister the matter rationally instead of plunging blindly into a wild ill-considered scheme.' Observed Klotz, 'We all believe in obtaining the sovereignty for Canada, but use reason & common sense to attain the end.'[83]

At the next ATB meeting, Harkin conceded defeat on the idea of an overland expedition, but he argued that a ship could still get as far north as Bylot Island. From there, it should be possible to reach Ellesmere over the ice. Harkin suggested contacting Captain Harris C. Pickels of Mahone Bay, Nova Scotia, a seasoned Arctic navigator who had been as far north as Etah. This proposal too was received with some cynicism: according to Klotz, Deville was 'very dubious about the scare that Harkin is trying to create,' while he himself did not 'take much stock in' the idea of a Danish threat.[84] Nevertheless, a resolution was passed in favour of telegraphing to Pickels for advice.

At yet another meeting held the next day, Harkin put forward the even more impracticable idea that a British dirigible could transport the police to Ellesmere. The Danes, he argued, were probably at that very moment establishing themselves 'on the east coast in the middle of the island.' If the Canadians waited until the next summer, 'the Danes will have had ample time to cover the whole island.' But if a police party was on the spot during the winter, the Mounties could 'establish our rights to at least a portion' of Ellesmere.[85] Klotz, by this time utterly exasperated with Harkin, wondered what the dirigible's crew would do on Ellesmere if they arrived there in the middle of winter, other than 'plant the British flag – and impress the Eskimos if there are any – and the musk ox!' He concluded, 'Our whole business is a comedy.'[86]

Perhaps realizing that opposition from Deville and Klotz might pose a serious threat to his plans, Stefansson had a long and confidential conversation with Klotz that evening. He spoke at length about his project for reindeer grazing on Baffin Island and about his theories of living off the land. As for the Danish threat, Stefansson declared that he too thought it was of little consequence. He suggested that if the HBC were given a grazing lease on Ellesmere, the traders would establish a post on the southern part of the island; this station, along with annual visits by a revenue cutter, would be enough to assert Canadian sovereignty. Skeptical by nature though Klotz was, Stefansson was able to enthrall him as he unfolded his real ambitions. What Stefansson wanted was to lead a five-year expedition west from Ellesmere into the 'region of maximum inaccessibility.' He had brought maps with him, and he outlined possible routes to the North Pole and to Wrangel Island. 'On this latter island he is rather set, as it has no real owner & [he] thinks Canada should possess it,' Klotz recorded. Stefansson suggested that here, too, the HBC would be willing to establish a new post, this time in pursuit of white fox furs. 'He is undoubtedly keen to go North once more & wipe out all the "mare incognitums" from the north polar map ... He is modest & I believe in his Arctic Island scheme & opinions,' Klotz concluded.[87]

Possibly Stefansson had similar conversations with other members of the board, with similar results. At a special meeting held on 27 October, Harkin managed to pass a resolution in favour of an immediate ship expedition to Bylot Island. An airplane was to be carried north in the ship; on arrival at Bylot, both aerial and surface parties could be sent to Ellesmere 'to start occupancy and administration.' If Captain Pickels felt that Baffin Bay could not be navigated at such a late date, then Harkin had his alternate plan: that 'the British Government should be asked to immediately transport a Canadian party to Ellesmere Land by airship [dirigible].'[88]

However, Klotz's more compliant attitude was quickly dispelled by a 7:30 a.m. telephone call from Stefansson on 28 October. Stefansson dramatically announced that, according to a relative of Frits Johansen (a Danish-born member of his 1913-18 expedition), Rasmussen had left Copenhagen in mid-July and returned in early October. Stefansson informed Klotz that he now believed a serious Danish threat did exist after all. Rasmussen had undoubtedly 'gone to establish a post in Ellesmere land.' Klotz, his suspicions fully roused once again, noted: 'This whole Rasmussen business is somewhat nebulous to me, and I am not clear to which quarter to assign the nebulosity.'[89] Deville seemed far more certain about the source of the problem. He informed Cory that he wished to disassociate himself from the resolutions passed on the 27th. 'I am satisfied that the alleged intention of Knud Rasmussen or of the Danish Government to occupy Ellesmere island ... has never existed, otherwise than in Mr. Stefansson's imagination. The wild schemes suggested for the immediate occupation of the island can only result, if they become known, in bringing ridicule over the Department,' he wrote bluntly.[90] Cory and Lougheed decided that it would be 'inadvisable to attempt an expedition until navigation opens up in the Spring.'[91] However, they ignored Deville's pointed observation that the threat from Denmark was in all likelihood a fantasy concocted by Stefansson.

In contrast to Klotz and Deville, Harkin was as alarmed as Stefansson could have wished. Even though Johansen's news proved that Rasmussen had left Denmark before the Canadian protest was received, Harkin's fear that in those few months Rasmussen had taken steps detrimental to Canada's interests remained strong, and he returned to the possibility in memo after memo over the next six months.[92] To Harkin, it was a simple race between two nations, in which the country that first established effective occupation would be declared the winner according to the arbitrary and inescapable rules of international law.

3
An Expedition to Ellesmere Land

I think it would be good policy to defer any public announcement as long as possible. If the Danes are really now in occupation in Ellesmere Land, or intend to take action in that regard next Spring ... a Canadian announcement ... would probably result in special efforts on their part. If they believe that Canada is still asleep they probably will not hurry.

– J.B. Harkin, 6 December 1920

PERHAPS HARKIN SHOULD NOT be blamed for his overreaction. Like most Canadians at the time, he had always taken it for granted that the archipelago was Canada's, and it came as a great shock to him to learn that there could be any uncertainty on this score. 'I had no doubt in my mind when our dispute with the Danes came up with regard to the musk-ox that the land was ours,' he remarked unhappily at the Advisory Technical Board meeting.[1] As Stefansson quite correctly pointed out to both Loring Christie and the ATB, of all the northern islands Ellesmere had by far the most complex history of discovery and exploration. It was indeed with regard to Ellesmere and the islands discovered by Otto Sverdrup that Canada's claims were at their weakest.

In the late nineteenth and early twentieth centuries, sovereignty over polar areas had only begun to be considered by theorists of international law. There was no case law to fall back on: the major decisions that are today considered relevant to Arctic sovereignty – the Clipperton Island case and the Eastern Greenland case – were made in the early 1930s.[2] In 1920 even the broader question of how sovereignty over new territories of any kind could be acquired was not particularly easy to answer. Harkin did not attempt to delve into the complexities of the matter; instead, he considered himself adequately informed after reading a single standard text, Lassa Oppenheim's *International Law*.[3] That Harkin could so easily set off on a mistaken course was due both to the uncertain state of international law on Arctic sovereignty and to his own personality.

Harkin always wrote of 'International Law' rather than 'international law,' and he seems to have conceived of it as a set of fixed, unalterable rules, which governments were bound to obey. The concept of international law as a set of practices evolved over the centuries, sanctioned and hallowed by general acceptance among nations, was obviously unfamiliar to him. But the 'authorities' to whom Harkin was so ready to defer were without any power to issue binding decrees. To write on international law at that time was a matter of seeking to find and elucidate for the benefit of others the patterns and principles underlying the actual practices that governed relations between states. As one such author admitted in 1889, international law as laid down by the theorists was 'often ... quietly ignored or brutally disregarded.' It would, therefore, 'be very unwise of an international lawyer to indulge in the delusion, with which he is often credited, that formulas are stronger than passions ... He only says what ought to be done, given the acquired moral habits of the past, and the rules of conduct which have been founded upon them.'[4] In the last analysis, international law reflected not what nations ought ideally to do, but rather what they were able to do without arousing the active opposition of other states. The history of precedents and actual practices was the bedrock on which theoretical writings had to be founded if they were to be of any value.

There was, however, an interplay between theory and practice: a state would be much more likely to receive the tacit consent of other nations if it could justify its acts by reference to the writings of a well-known legal scholar. Since the sixteenth century, which saw the dawn of both European expansionism and scholarly writing on international law, the relationship between theory and practice on the acquisition of new territory had been especially complicated. Early modern writers seeking to codify international law often turned to Roman law for their conceptual framework. Unfortunately, Roman law was silent on the acquisition by states of previously 'unowned' territory (mainly because statesmen in the ancient world had never envisioned such a process). The theorists therefore had to turn to the Roman law of private property, in which title to things not previously owned by anyone *(res nullius)* could be obtained simply by the act of appropriating them. European scholars argued by analogy that newly discovered land could be taken by the first discoverer in the name of his country. But just how vast tracts of land could be appropriated was not clear. The theorists drew on the Roman practice of *usucapio,* in which legal ownership of property to which an individual had a right but no clear title could be acquired by showing the intention to possess, actual physical possession for a period of time, and use. However, usucapio did not offer sufficient guidelines, because it was intended to apply only to moveable property or to land whose boundaries had already been defined. Considering the great distances involved, the unknown limits of many newly found lands, and the practical difficulties of establishing a colony immediately or even soon after discovery, it was impossible to apply the criteria of physical possession and use stringently. In the sixteenth century, governments could argue that discovery and symbolic acts of possession, such as the reading of a proclamation, were sufficient in cases where their own subjects were the first discoverers, while taking the line that actual settlement was required when the discovery had been made by another nation.[5]

As time passed, theorists and governments alike called for settlement on and actual control of the land, ideally involving acts of political administration, before sovereignty would be recognized. One particularly influential passage was published by the Swiss jurist Emmerich de Vattel in his *The Law of Nations* (1758). According to Vattel, 'All men have an equal right to things which have not yet come into the possession of anyone, and these things belong to the person who first takes possession.' But a nation could not 'appropriate, by the mere act of taking possession, lands which it does not really occupy, and which are more extensive than it can inhabit or cultivate,' for 'such a claim would be absolutely contrary to the natural law, and would conflict with the designs of nature, which destines the earth for the needs of all mankind, and only confers upon individual Nations the right to appropriate territory so far as they can make use of it.' When settled habitation and cultivation were taken as the criteria for possession, nomadic Aboriginal people fell into the class of those who had not established effective occupation. Their lands could, then, be seized by others who would farm or otherwise make use of them. '[T]hese tribes,' Vattel wrote, 'can not take to themselves more land than they have need of or can inhabit or cultivate. Their uncertain occupancy of these vast regions can not be held as a real and lawful taking of possession; and when the Nations of Europe, which are too confined at home, come upon lands which the savages have no special need of and are making no present and continuous use of, they may lawfully take possession of them, and establish colonies in them.'[6] As the era of coastal settlement in the Americas and Australia gave way to the nineteenth-century exploitation of the interior, this was an exceptionally useful idea. Although it certainly was not an uncontested doctrine, the widespread belief that uncultivated land could legitimately be taken by those who would farm it was an important driving force in European expansionism.

But while Europeans might readily agree that they had a moral right to colonize newly discovered lands, they could not so easily settle rival claims among themselves. It was, for example, still an open question in the nineteenth century how far the hinterland of a coastal settlement should extend. Clearly, no nation could quickly establish effective control throughout a large, thinly populated, or as yet unpopulated territory, where the nature of the landscape or the climate made access difficult. In 1885 representatives of the major European nations met at the Berlin Conference to discuss their rival claims in the African interior, and they agreed that none would make a new claim without first notifying the others, or without establishing effective occupation and administration. This was generally considered a precedent for the future; however, even in 1920 its precise impact was still uncertain, since the powers had by no means acted in full conformity with the agreement. Three years after the Berlin Conference, scholars at the Institute of International Law in Lausanne put forward a concise and useful definition of effective occupation. 'Taking possession,' they wrote, 'is accomplished by the establishment of a responsible local power, provided with sufficient means to maintain order and assure the regular exercise of its authority within the limits of the occupied territory.' Formal notification must be provided to other powers, 'either by publication in the form which is customary in each State for the notification of official acts, or through the diplomatic channel.'[7] Occupation, then, was not so much a matter of settlement as of a state's ability to display its uncontested authority over the area in question.

By the early twentieth century, the term *terra nullius* had been coined to describe land that was not yet formally subject to the sovereignty of any nation.[8] There was a broad consensus among theorists that first discovery of a new land gave certain rights, often described as an 'inchoate title.' Such a title was held to prevent settlement by other nations, at least for a time. However, there was no true sovereignty until effective occupation had been established. If many years passed without any action by the discovering country, the inchoate title would lapse, and other nations would be free to establish settlements if they wished. This doctrine was firmly and succinctly stated in one of the leading legal texts of the time,

Oppenheim's weighty two-volume *International Law,* first published in 1905-6. However, this book, to which Harkin turned for guidance, was somewhat cursory in its treatment of the subject.[9]

From Oppenheim, Harkin gained the impression that because Canada had not as yet established effective occupation throughout the northern islands, the inchoate title founded on British discovery had expired, leaving Canada with no better claim than any other nation, and less of a claim than Denmark, should the Danes indeed colonize the islands with Natives from Greenland. He would have done much better to consult another standard text (cited by Oppenheim on the question of inchoate title), William Edward Hall's *A Treatise on International Law.* Hall's analysis was much more extensive and more finely attuned to the realities of practice. He acknowledged that some writers thought full title could never be acquired without continuous occupation, but observed sharply that this doctrine was 'independent of the facts of universal practice.' Hall emphasized that it was impossible to give any precise definition of the time that would have to elapse before an inchoate title could be considered extinct, for this would depend on the characteristics of the land and other circumstances of each individual case. Furthermore, 'when discovery, coupled with the public assertion of ownership, has been followed up from time to time by further exploration or by temporary lodgements in the country, the existence of a continued interest in it is evident, and the extinction of a proprietary claim may be prevented over a long space of time, unless more definitive acts of appropriation by another state are effected without protest or opposition.'[10]

By Hall's criteria, Canada's title was in good standing, though it was not yet perfected. Many expeditions over the years had demonstrated the continued interest of Britain and Canada since the time of the early discoveries; Canada's intention to exercise sovereignty had been publicly proclaimed; and no other nation had ever officially made a rival claim or attempted to establish occupation. Stefansson was correct when he argued that it would not be altogether impossible for another country to do so. But if any threat was to be feared, it would come from the United States or Norway, not from Denmark. And, as matters stood in 1920, neither the Americans nor the Norwegians could have put forward a strong case unless they were able to establish occupation without any protest from Canada.

Unfortunately, none of Harkin's colleagues in the Department of the Interior were able to correct his false impressions. As well as being poorly informed on the finer points of international law, these Ottawa officials knew little about the relevant aspects of Arctic history. They were, of course, well aware that British explorers had made most of the major discoveries in the archipelago, but the very fact that there had also been significant American and Norwegian activity in the region caused them an unwarranted degree of alarm. In addition, they were ignorant of some important and effective steps that had already been taken to ensure international recognition of Canada's sovereignty over Ellesmere Island.

THE FIRST EUROPEAN TO see Ellesmere Island was the English explorer William Baffin. During his 1616 voyage, Baffin discovered and named Smith, Jones, and Lancaster Sounds. He believed that a large, horseshoe-shaped sweep of land joined the Canadian mainland to Greenland; therefore, in Baffin's mind the sounds were merely smaller bays that opened off the large bay which now bears his name. Perhaps for this reason, Baffin gave no name to the coastline of Ellesmere. Until the early nineteenth century, many maps showed Greenland joined to Baffin Island; Baffin Island in turn was joined to the mainland. In 1818 John Ross reached the entrance of Smith Sound, but he too believed that Greenland and Ellesmere were part of the same large land mass. Another Englishman, Edward Inglefield, visited the area in 1852 during the search for Sir John Franklin's missing expedition. Inglefield was the first explorer to enter Smith Sound and the first to realize that the land to the west was separated from that on the east. He called the western shore Ellesmere Land in honour of the Earl of Ellesmere, vice-president of the Royal Geographical Society.[11]

The next year an American, Elisha Kent Kane, passed through Smith Sound to Kane Basin on another fruitless search for Franklin. Kane made his base at Rensselaer Harbour on the Greenland coast near Etah. From there, one of his parties sledged northwards along the coast as far as the entrance to Kennedy Channel. A second party, led by Dr. Isaac Hayes, visited the Ellesmere side. Believing the western coastline to be part of a new island, separated from Inglefield's Ellesmere Land by a strait, the Americans called their discovery Grinnell Land after their expedition's wealthy patron. They had been deceived by the deep inlet later called Buchanan Bay, which appeared on maps as 'Hayes Sound' for the rest of the nineteenth century (see Map 5).[12] In 1860-61 Hayes returned on an expedition of his own. He again sledged along the coast of Grinnell Land, this time travelling as far as Lady Franklin Bay. Hayes believed he had almost reached the northern coast of Grinnell Land, and that the open polar sea was only a short distance away. He raised the American flag, although as a private citizen he was not technically empowered to make territorial claims.

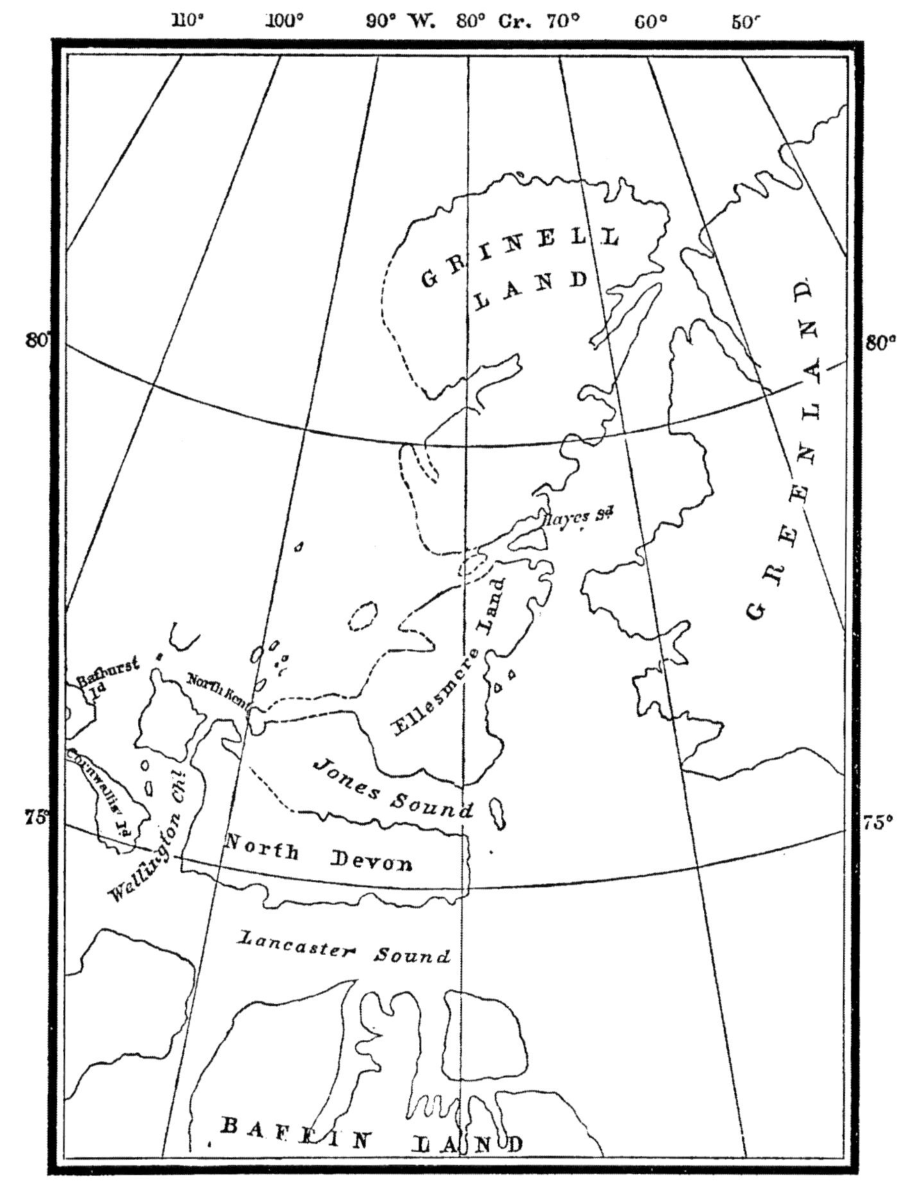

MAP 5 Map of Ellesmere and Grinnell Lands, 1885 | Courtesy of *Science*

Except for Ross's voyage, all the nineteenth-century expeditions up to this point had been privately sponsored. In 1871, however, Charles Francis Hall led an American government expedition. His ship, *Polaris,* passed through Smith Sound, Kane Basin, Kennedy Channel, Hall Basin, and Robeson Channel. He then established a base at Polaris Bay on the coast of northern Greenland. He saw, but did not visit, the northern part of Ellesmere Island. Because Hall believed that Lady Franklin Bay was a strait, he concluded that this was yet another new island. He gave it the name Grant Land in honour of President Ulysses S. Grant. Although Hall made a territorial claim for the United States on the Greenland side, there was no such declaration with regard to Grant Land.[13]

Hall's venture was quickly followed by an official British naval expedition under the command of Sir George Nares. In 1875-76 Nares too found his way through the narrow waters separating Ellesmere Island from Greenland (which are now collectively known as Nares Strait). He made his base on the northern coast of Grant Land, near the present-day settlement of Alert. From there, the British explorers made extensive journeys on both Ellesmere and Greenland. A second party mapped the Lady Franklin Bay area, and in the process proved that Grinnell Land and Grant Land were connected. Nares and his men raised the British flag on several key occasions, most notably at the farthest north achieved by Lieutenant Albert Markham, who led a party from the coast of Grant Land over the frozen polar sea to latitude 83°20'.[14]

Four years later, sovereignty over the archipelago was transferred from Britain to the young Dominion of Canada. In 1879 a draft of proposed legislation was submitted to the Colonial Office, which in turn passed it on to the Admiralty for comment. The draft (based on an address to Queen Victoria from the Canadian parliament) suggested that the eastern boundary of the Canadian Arctic should extend to the north through the waters of Davis Strait, Baffin Bay, Smith Sound, and Nares Strait. All islands between this line and longitude 141° W which belonged to Great Britain (by right of discovery or otherwise) would become Canadian territory. Frederick Evans, the head of the Admiralty's hydrographic department, pointed out that Inglefield had carried British exploration along the proposed eastern boundary only as far north as the entrance to Smith Sound, in latitude 78.5° N. Kane and Hayes had then explored 'to beyond the 82nd parallel,' and they had given the western shore of Kane Basin and Kennedy Channel the name Grinnell Land. Finally, 'our own Arctic Expedition of 1875-6 pushed Northward of the Coasts explored by the

Americans.' Evans suggested that, because of 'these discoveries by American Citizens,' the eastern boundary should extend only as far north as 78.5°; however, this limitation should be 'without prejudice to the rights of this Country established by the discovery of more northern lands made in the late Arctic expedition.'[15] In other words, while Grinnell Land might be claimed by the United States, Grant Land should be Canadian.

The members of the Board of Admiralty concurred with Evans' assessment.[16] It was apparently because of their reservations that the precise extent of Arctic territory transferred to Canada was not defined. The 1880 order-in-council simply stated that 'all British territories and possessions in North America not already included within the Dominion of Canada and all islands adjacent to any such territories and possessions' (excluding Newfoundland) were now part of Canada.[17] British practice, then, was to respect the discoveries made by private American expeditions in Grinnell Land, even though they had not as yet been followed up by official claims or occupation. However, British officials regarded northern Ellesmere – where the Americans had not landed, and where the Nares expedition had done the first exploration and mapping – as their own nation's territory.

In 1881 an official American expedition led by Adolphus Greely established a base at Lady Franklin Bay. The main purpose of this enterprise, which was undertaken as the American contribution to the first International Polar Year, was to make meteorological and other scientific observations. As well, geographical exploration was carried out on both sides of Nares Strait. In the spring of 1882, a party led by Lieutenant J.B. Lockwood reached a new farthest north on the coast of Greenland, where they raised the Stars and Stripes. In 1883 Lockwood explored the interior of Grinnell Land, reaching Greely Fjord on the western coast. Greely bestowed the name Arthur Land (for President Chester A. Arthur) on the unexplored region south of the fjord. The new name did not always appear on maps; instead, the discoveries of both the Hayes and Greely expeditions were usually included under the name Grinnell Land. No territorial claim was made by the Americans on this occasion, nor did they carry out a second flag-raising ceremony.[18]

As the years passed and still no official American claim to Grinnell Land was put forward, Canadians were increasingly inclined to regard it as their own. Nothing of an administrative nature had been done by the Canadian government for many years after the 1880 transfer. In 1895 came the first administrative act, an order-in- council dividing the Canadian Arctic into the districts of Mackenzie, Franklin, Ungava and the Yukon. The archipelago, including Grinnell Land, was contained in the District of Franklin. The eastern boundary of the district ran from the Nares expedition's farthest north (83° 20') 'southerly through Robeson Channel, Kennedy Channel, Smith Sound, Baffin Bay, and Davis Strait.'[19] Some ambiguities in the 1895 order were rectified in order-in-council PC 3388 of 18 December 1897. A map accompanying the 1897 order shows the boundary running from the 85th parallel southward along the 60th meridian, then down the middle of the straits between Grinnell and Ellesmere Lands and Greenland. (The map shows the mythical Hayes Sound separating Ellesmere Land from Grinnell Land.)[20]

Also in 1897, fisheries officer William Wakeham raised the flag on Baffin Island, declaring in the presence of his officers and crew, the agent from a nearby whaling station, and several Inuit that 'Baffin's Land with all the territories, islands and dependencies adjacent to it were now, as they always had been since their first discovery and occupation, under the exclusive sovereignty of Great Britain.'[21] Wakeham had been instructed to ascertain the extent of foreign whaling and trading activities in the North and, if necessary, to 'firmly and openly ... declare and uphold' Canadian jurisdiction.[22] Finding that some whalers believed Baffin Island was an American possession, he considered a flag-raising to be the simplest and best method of affirming Canadian rights. On his return, he was derided by the British press for carrying out such a ceremony on land that had been claimed for England since the sixteenth century. Wakeham hastened to explain to his superiors that there had been 'no attempt to attach or "take possession" of the territory over again – "de novo."' Instead, his aim was to impress the fact of Canadian ownership on the local population.[23]

In 1898-99 Otto Sverdrup wintered on the Ellesmere side of Smith Sound. He made excellent maps of the area, and was the first to realize that there was no strait dividing Inglefield's Ellesmere Land from Kane's Grinnell Land. Sverdrup then sailed through Jones Sound and charted the southwestern coast of Ellesmere, which he named King Oscar Land for the king of Sweden and Norway (the two countries were united under one crown until 1905). Sverdrup also discovered several new islands to the west, including Axel Heiberg, Ellef Ringnes, and Amund Ringnes. Although Sverdrup's was a private expedition, he believed that these new lands now belonged to Norway.[24] His government, however, made no official claim.

The activities of Sverdrup and Robert Peary, combined with the unfavourable and bitterly resented outcome of the Alaska boundary dispute, at last roused the Canadian government to greater activity regarding the Far North. In May 1904 the Geographic Board of Canada gave the entire island the name Ellesmere Land. The move was initiated by the Department of the Interior's geographer, James White, who was concerned about the American names given to different parts of the island and about the fact that the *Century Atlas,* published in the United States, had 'coloured it a different colour as if it did not form part of the Dominion.'[25] White's nomenclature was accepted by other countries.[26] As the *Bulletin of the American Geographical Society* noted, 'In naming the whole Ellesmere Land the Canadian Board acts upon the idea that British sovereignty extends over the entire Arctic archipelago north of our continent.'[27]

White had thus taken an important step forward in securing international recognition of Canada's authority. A man whose contribution to Canadian sovereignty in the North has never adequately been acknowledged, White would remain keenly interested, and from time to time actively involved, in sovereignty issues until his death in 1928. Although he was a graduate of the Royal Military College (RMC) in Kingston, there was nothing military about White's appearance except his straight posture and his large moustache, the ends of which were carefully waxed. He was thin, scholarly-looking, and somewhat cantankerous. After his graduation from RMC in 1883, White joined the Geological Survey of Canada. He did topographical fieldwork in the Rocky Mountains, Ontario, and Quebec, then transferred to the Department of the Interior in 1899. As the department's geographer (later chief geographer), White determinedly built up an efficient central organization for the production of maps. Previously each department had created its own maps as needed; inevitably, the results were of uneven quality. Under White, the geographer's office became the main government agency performing such work.[28] White took particular care that naming should be accurate and consistent. Place names in the Arctic were a special interest of his, and in 1910 he published a detailed list of Arctic names and their origins.[29]

White's sovereignty concerns were shared by others in the government, most notably Clifford Sifton, the minister of the interior. Sifton firmly believed that 'everything north of the Mainland of Canada between Greenland on the one side and the Alaska line on the other' should be Canada's.[30] Prime Minister Wilfrid Laurier resisted all suggestions for a sweeping sovereignty claim, which he believed would 'simply arouse a storm at this juncture,' but he agreed that Canada should 'quietly assume jurisdiction' in the Far North. Only after there were 'men stationed everywhere' would it be time for a proclamation.[31]

As a result of Sifton's efforts, an official Canadian expedition was sent to the islands in the *Neptune,* under the command of a member of the Geological Survey, A.P. Low. Low's instructions were to proceed as far north as Kennedy Channel 'and visit as much territory as the state of the ice will permit.'[32] As it turned out, ice conditions made it impossible for the *Neptune* to get farther north than Smith Sound. Low made a point of landing at Payer Harbour, Peary's winter quarters of 1901-2.[33] He then proceeded to nearby Cape Herschel, where on 11 August 1904 'a document taking formal possession in the name of King Edward VII, for the Dominion, was read, and the Canadian flag was raised and saluted.'[34] The proclamation ran: 'In the Name of King Edward VII, and on behalf of the Government of the Dominion of Canada, I have this Day taken Possession of the Island of Ellesmereland and All the Smaller Islands Adjoining It. And in Token of such formal Possession have caused the Flag of the Dominion of Canada to be hoisted upon the land of Ellesmereland: and have deposited a Copy of this document, sealed in a metal box, in a cairn erected on the conspicuous Headland of Cape Isabella.'[35] Low made similar proclamations on the islands of North Devon and North Somerset. In addition, as he reported to Joseph Pope, 'The Sovereignty of the Dominion was asserted in Hudson Bay, and at Ponds Inlet, Cumberland Gulf and Cape Haven, all on the eastern side of Baffin Island, by the collection of Customs.'[36]

There was no protest over the Geographic Board's action or Low's Ellesmere proclamation from the Americans. American explorer Frederick Cook wrote in 1911 that no nation had yet 'assume[d] the responsibility of claiming or protecting' the island, and he reported with characteristic grandiloquence that he had 'affixed the name of Acpohon as the general designation.'[37] However, since Cook's claim to have reached the North Pole had by then been exposed as a fraud, the name was disregarded even by his compatriots. An official of the US Hydrographic Survey wrote to the Geographic Board in 1916, stating that his department intended to follow Canada's decisions on the Arctic chart it was then preparing.[38]

A few months after Low was sent north, Sifton asked the Dominion astronomer, Dr. W.F. King, to write a report on Canada's title to the islands. King evidently had a keen intelligence, but he was not particularly well versed in matters of international law. He had no legal training: his LLD was an honorary degree, bestowed by the University of Toronto for his work on the Alaska boundary problem.[39] King fretted over the question of whether, in the absence of any effective occupation by Britain, the 1880 transfer to Canada should be considered valid. He was also somewhat alarmed by the vagueness of the order's wording, for which he could not account with the limited documentation at his disposal. However, King noted with relief that no objection to Canada's title had ever been raised by another state, and he seemed fully aware of the weight that such tacit consent carried. Finally, he made the sensible observation that by their nature, Arctic territories could not be occupied in the same way as more temperate lands. Therefore, 'complete occupation by settlement is not to be expected.' But given that foreign, and particularly American, whalers, traders, and explorers were increasingly active in the Arctic, it was essential for Canada to assert its jurisdiction more forcefully.[40] The maps accompanying the report, drawn up under James White's authority, showed the boundary lines established by the 1895 and 1897 orders-in-council extending as far north as the pole. White later wrote that he had prepared the maps 'really, though not avowedly, to offset the claims of Sverdrup.'[41] This action was to hold great significance for Canada's future claims in the archipelago.

Soon afterwards, the Laurier government purchased the sturdy *Gauss*, specially built for Erich von Drygalski's German Antarctic Expedition. The ship was renamed the *Arctic*. Under the command of Captain Joseph Elzéar Bernier, the *Arctic* made voyages to the archipelago in 1906-7, 1908-9, and 1910-11. In August 1907 Bernier made a second Canadian proclamation on Ellesmere – one of many such ceremonies he carried out. Indeed, sovereignty declarations were something of an obsession with Bernier. The captain was well known in government circles for his vanity and his ambition to be remembered as an Arctic hero. According to the reminiscences of Richard Finnie, who grew up as the son of an influential civil servant, Bernier's repeated proclamations caused considerable annoyance in Ottawa. 'He made a great to-do about planting the flag ... I recall that this was an irritant to my father and others who took the view that the islands north of the Canadian mainland already belonged to Canada since they were ceded by Great Britain. Ergo, by planting the flag at every opportunity Bernier was only stirring up an idea in some people's minds that maybe the islands he didn't get around to could be claimed by explorers of other countries,' Finnie wrote.[42] The recollection was accurate: in 1927 Oswald Finnie frostily described Bernier's flag-raisings as 'unfortunate incident[s]' which did nothing to strengthen Canada's claims.[43] Another senior bureaucrat agreed that, although Bernier was an 'able' and 'valorous' navigator who had done good service for his country, his claims to have 'taken legal and valid possession of many hundred thousand square miles' of Arctic territory on behalf of Canada must be regarded as excessive.[44]

Bernier's Ellesmere declaration did not result from any pressing official concerns about Canada's title to the island. He was not specifically instructed to take possession of either Baffin Island or Ellesmere. Bernier nevertheless performed ceremonies on both. His proclamations on a few islands discovered by British explorers but not yet claimed by a representative of Canada (for example, Banks Island) did have the sanction of the Department of Marine and Fisheries, which was placed in charge of his expeditions. However, this directive seems to have been caused by a combination of muddled thinking and Bernier's determination to enhance his public image by formally raising the flag as often as possible. Wakeham had carried out a ceremony on Baffin only because some of the whalers he encountered were not aware of Canada's sovereignty, and he was under no illusion that his act constituted a new taking of possession. Low had not received specific instructions to formally proclaim Canada's sovereignty, but his flag-raising on Ellesmere was desirable because of the potential foreign claims to Grinnell Land and King Oscar Land. No such ceremony was required where Canada's title could be based on the 1880 transfer from Britain. Nevertheless, the officials involved in the planning of Bernier's voyages apparently hoped that declarations on Banks and a few other islands would be a sort of insurance. Bernier was initially given the authority to annex any new islands he might discover. But, not content with this directive, before his departure in the summer of 1906 he asked for and received revised instructions permitting him to 'take formal possession' of 'all lands and islands on your way.' In particular, he wanted to claim Banks and Albert Lands.[45] To Bernier, it must have seemed that Wakeham had inaugurated a new tradition of Canadian nation building in the Far North. Sovereignty declarations provided a kind of stage for Arctic explorers, and Bernier intended to exploit the theatrical possibilities of such acts to the utmost.

In fact, Bernier was determined to do much more than just claim (or reclaim) islands on behalf of Canada. He had a staunch, though not particularly well informed, ally in Conservative senator Pascal Poirier. In February 1907 Poirier proposed a Senate resolution calling for Canada to formally extend its northern boundary as far as the pole. His speech is often described as a milestone in the evolution of Canada's Arctic policy. However, a close examination quickly dispels this impression. Poirier argued that Canada's right to all the Arctic islands was based on four factors: the rights gained by British discovery (which he mistakenly believed had automatically passed to Canada in 1867); the 1763 Treaty of Paris (which ceded any French rights in Hudson Bay to Great Britain but said nothing about the archipelago, most of which had not been discovered by Europeans when the treaty was signed); and the charter of the Hudson's Bay Company (which did not in fact give the HBC any authority over the northern islands). Finally, Poirier put forward a new justification. He recounted that during a visit by Bernier to the Arctic Club in New York, the explorer and his colleagues had 'proposed and agreed ... that in future partition, of northern lands, a country whose possession to-day goes up to the Arctic regions, will have a right, or should have a right, or has a right to all the lands that are to be found in the waters between a line extending from its eastern extremity north, and another line extending from the western extremity north.' Poirier, then, was suggesting that a casual conversation among explorers could somehow provide a legal basis for Canada's claim.

In Poirier's view, it was essential for Canada to make a declaration immediately, and so 'lay the foundations for a glorious and everlasting empire.' Future changes in climate might make the Far North suitable for agriculture. 'Possibly in a thousand years,' Poirier speculated, 'Canada will be a well settled country up to the north pole ... and possibly we shall find out there the place where Eden was, because we are aware that some dreamers, serious dreamers ... think that Paradise Lost was up at the pole, and that those four big rivers spoken of flowed from the pole.' Poirier conceded that these speculations were 'in the field of hypothesis.' However, he considered it 'pretty certain' that some day the Arctic would 'be dotted with cities and villages, and fields tilled by the husbandman.'

The resolution was not passed, or even voted on. Speaking on behalf of the government, Senator Richard Cartwright assured Poirier that, though there was no plan to assert jurisdiction 'quite so far north' as the pole, the problem of Canada's northern boundary was receiving serious consideration. The 'various acts of dominion' carried out by Wakeham,

Low, and Bernier were 'sufficient to maintain our just rights,' and there was no need 'to enter into any formal declaration.'[46] In April 1907 the Laurier government forwarded a copy of Poirier's speech and Cartwright's reply to the Colonial Office, thereby implying that Canada did not yet intend to make a public statement about the northern limit of its possessions.[47] When Bernier returned from the North in the autumn of 1907, he informed Ottawa that he had landed on the southern coast of Ellesmere, at a spot he named King Edward VII Point. There he had proclaimed Canadian ownership over 'all adjacent islands, as far as ninety degrees north.' If Bernier hoped that Poirier's speech had successfully cleared the way for such a claim, he was disappointed. There is an 'x' beside this passage in his report, presumably made by the minister of marine and fisheries, Louis-Philippe Brodeur.[48] Bernier's published narrative said nothing about a claim extending to the pole, no doubt because of strict verbal orders from Laurier and Brodeur.

Undaunted, Bernier prepared to make a similar proclamation during his next voyage and to leave a permanent record of it in the North. On Dominion Day 1909, at Winter Harbour on Melville Island, he erected a tablet recording his claim to the entire archipelago, 'from 60 degrees west longitude to 141 degrees west longitude, and as far north as 90 degrees north latitude' on behalf of Canada.[49] These boundaries included northwestern Greenland as part of the Canadian sector. Bernier had far exceeded his instructions, and he knew that he might face a stern rebuke from Ottawa. But as it turned out, this time circumstances forced the government to tacitly sanction what he had done.

In the month before Bernier reached Quebec at the end of his voyage, the Far North was almost continually in the headlines. On 2 September the news of Frederick Cook's claim to have discovered the North Pole was published; a few days later came Robert Peary's competing claim and his accusation that Cook had perpetrated a hoax. Peary reported that he had raised the American flag at the pole, thereby annexing it and the surrounding region 'for and in the name of the President of the United States.'[50] In a telegram to the Associated Press he announced dramatically: 'Have nailed the Stars and Stripes to the Pole.'[51] On 10 September the colonial secretary, Lord Crewe, informed the governor general that a question about Canada's northern claims would be asked in the British House of Commons. Despite his reluctance to indulge in sweeping rhetoric which might provoke opposition from other nations, Laurier evidently felt that under the circumstances a statement from Canada was required. Crewe was told Canada did in fact lay claim to 'all lands intervening between the American border and the North Pole.'[52] Choosing his words with care, a British official announced on 15 September that 'while the Canadian Government had not made any formal declaration ... it was believed that they considered themselves entitled to claim all the land' between the border and the pole.[53]

When Bernier arrived at Quebec on 5 October, one reporter thought it was 'evident that he was cautious, and kept back something.'[54] Clearly, Bernier did not know how the government would receive the news of his action. The story of his proclamation was released a few days later, and on 16 October the overjoyed captain gave a full account of his proceedings to the Canadian Club of Ottawa. Despite the high level of press and public interest in the Arctic at the time, no foreign government protested. Laurier praised Bernier's achievements in public (for example, in some brief remarks following the Canadian Club speech), but the prime minister did not make an official statement about the sector claim, either in the House of Commons or elsewhere.[55] No matter how strongly he felt about the archipelago, Laurier did not wish to include northwestern Greenland in Canada's claim. There is no evidence that he saw Bernier's proclamation as anything more than a dramatic but essentially meaningless act, which conveniently helped Canada to counter the publicity given to the equally meaningless proclamation made by Peary. The matter could be dropped once press attention shifted away from the Arctic.

Bernier himself obviously thought otherwise. During a visit to New York in January 1910, he announced that the 'principal object' of his next expedition would be 'a division of the Polar sea.' Therefore, he intended 'to ask Sir Wilfrid Laurier ... to request Britain, the United States, Russia, Sweden and Norway and Denmark ... to designate official representatives to accompany him.' When asked to comment by the Opposition, Laurier curtly responded that Bernier 'had better keep to his own deck.' Delighted by the prime minister's discomfiture, a Conservative MP observed: 'I think somebody ought to take charge of this gentleman. He seems to be running loose.'[56] Poirier spoke in support of Bernier in the Senate a few days later, but a fellow Conservative, James Lougheed, observed scathingly that Bernier should never have been permitted to 'emblazon' his intentions to the world by 'indiscreet interviews.' From the government side, Senator Cartwright readily agreed that 'officers ... entrusted with such expeditions, owe it to the country and to themselves, not to speak without due warrant and authority.'[57] The public heard nothing more from Bernier about his embarrassingly grandiose and unrealistic plan.

Bernier's propensity for theatrical gestures obscured humdrum but very significant activities. In his unglamorous capacity as a fisheries officer, he advised the whaling captains that by a 1906 amendment to the Fisheries Act, they were now required to take out Canadian licenses for operations either in Hudson Bay or in Canadian territorial waters north of 55° latitude. Bernier collected the license fees without difficulty from British and American whalers; millionaire American sportsman Harry Whitney was also among those who paid. As well, Bernier was a customs inspector, collecting duty on goods brought into the country for trade with the Inuit.[58] But even after three voyages by the *Arctic*, no permanent administrative apparatus had been established anywhere in the archipelago. By 1911 the decline of the northern whaling industry had removed the main stimulus for official action.

Nevertheless, the Liberal program was continued on a grander scale by Robert Borden's Conservatives after their victory in the 1911 federal election. Taking to heart W.F. King's observation that American activities in the Arctic could gradually erode Canadian sovereignty, Borden made the momentous decision that Stefansson's 1913-18 expedition should be sponsored by the Canadian government alone rather than by the American geographical societies and other US institutions on which Stefansson had previously relied for the major part of his funding. As Borden explained in a letter to Gilbert H. Grosvenor of the National Geographic Society, 'The Government of Canada feels ... with regard to the present exploration, that it would be more suitable if the expenses are borne by the Government more immediately interested and if the expedition sails under the flag of the country which is to be explored.'[59] Both Stefansson and the societies willingly acquiesced.[60]

It was, then, taken for granted both in Ottawa and in American geographical circles that the 1913-18 expedition would significantly enhance Canada's official northern presence. When Stefansson was authorized to annex new lands lying north of Canada, he was also wisely cautioned that it was 'not desirable that any stress should be laid on the fact that a portion of the [archipelago] may not already be British.'[61] If Stefansson heeded this advice, there would be no further embarrassments like the ones caused by Bernier. Indeed, Stefansson's major theme before the summer of 1920 was that Canada, now secure in its dominion over the archipelago, should begin to economically exploit the islands, while at the same time extending its sway to any new land that might be discovered in the Arctic. He did not call Canada's authority over Ellesmere into question until it was clearly in his interest to do so.

Although the potential for foreign claims in the Arctic did at least theoretically exist, both the long period of tacit consent to Canada's title by other nations and the realities of international politics made such claims highly unlikely. Great Britain was still very much a military and diplomatic force to be reckoned with, and in a crisis of this nature the British government would undoubtedly come to Canada's aid. Stefansson was privy to discussions about the sovereignty ramifications of the Canadian Arctic Expedition by Borden and members of the cabinet in 1913, and so he can hardly have been unaware of these factors. Therefore, the conclusion that he deliberately misled Harkin, Meighen, and Christie is inescapable. For the moment at least, Stefansson's fame as an explorer and his confident, persuasive manner blinded them to what should have been obvious facts.

However, it was in London rather than in Ottawa that the response to Denmark's supposed denial of Canadian sovereignty was conveyed to Danish officials. This part of the chain of communications was entirely beyond the reach of Stefansson's influence, since he did not yet have any contacts in the British government. The members of the Foreign Office's Northern Europe Department were evidently puzzled and not a little irritated when Harkin's lengthy memo arrived on their desks at the end of August 1920. One fact was quite clear: the memo did not meet the standards of the British diplomatic service. It was simply not possible to pass such a wordy, obscure document to the Danish minister. Harkin's amateurish effort would have to be condensed, clarified, and polished. 'I do not think we can very well send a copy of this despatch ... so I have prepared a draft letter embodying the salient points,' minuted Fitzhardinge Maxse.[62]

In the process, Harkin's meaning was transformed. This change seems to have occurred partly due to Harkin's lack of clarity and partly because Maxse and his colleagues doubted that any challenge to Canadian sovereignty was intended. Harkin's memo began by quoting Rasmussen's 'no man's land' passage and the explorer's statement that he would require no help from the Canadian government. It also gave extracts from the Danish foreign ministry's covering letter. 'From these quotations,' Harkin wrote, 'it would appear that neither Mr. Rasmussen nor the Danish

Government seem to recognise that Canadian authority is dominant and exclusive in Ellesmere land but that on the other hand there seems to be an inference that Denmark has some authority in the area in question. I consider it very important that the Canadian Government should not subscribe to these ideas even by the implication of silence but that it should be made clear that Canada has exclusive authority over all the lands to the west of Baffin Bay, Smith Sound and Robeson Channel.' It was hardly a succinct and direct statement of Canadian concerns, and in any case, to a professional diplomat Rasmussen's letter would have appeared much less threatening than it did to Harkin. The diplomatic exchanges regarding British recognition of Danish authority were then in their final stages, so those in London were undoubtedly far better informed about Greenland sovereignty issues than were the Canadians. And, as a mere matter of practical politics, it must have seemed highly unlikely that a small power like Denmark would openly challenge the British Empire. The Danes could not expect such an action to go uncontested, and a contested occupation would not create a valid title.

The rest of Harkin's memo was taken up by a rebuttal of Rasmussen's claim that the Inughuit were hunting on Ellesmere only out of need. He repeated Stefansson's theory that valuable furs and skins, not meat, were the real object of their forays into Canadian territory. Harkin emphasized the economic importance to Canada of preserving the muskox from extinction, in the process falsely claiming that the herds on Ellesmere were the last that remained in the Canadian Arctic. The well-being of the Inughuit was the Danish government's concern, not Ottawa's. Maxse conceded that on these questions 'the balance of argument is in favour of the Canadian Government.'[63] The British note consequently focused on the issue of hunting, with no direct reference to Harkin's sovereignty concerns. (In an amusing reflection of the attitudes of the British upper classes, the phrase 'shoot musk oxen' was changed to 'kill musk oxen' at two points in the final note: when gentlemen hunted for sport, it was 'shooting,' but Native people could only 'kill' game.) The note was sent on 7 September, the day after the British government officially recognized Danish sovereignty over all of Greenland. It stated that the Canadian government was 'unable to concur in the opinion expressed by Mr. Rasmussen that the danger of extermination can scarcely yet be described as imminent.' While Ottawa was 'only too anxious to co-operate with the Danish Government in this matter,' unless the Inughuit could be 'restrained' from hunting on Ellesmere, the

Canadians would 'be compelled to establish mounted police posts ... to protect their own interests.'[64]

Harkin and others in Ottawa believed that a clear line had been drawn in the snow, when in fact the Danes remained unaware of the strong Canadian concerns about sovereignty. Acting on the assumption that the news of Canada's stand had in all likelihood spurred the Danes on to greater efforts, Harkin pressed for decisive action at the earliest possible date. Canadian plans, he cautioned, must be kept secret so as not to further alarm the Danes. It was the beginning of Harkin's decades-long obsession with absolute secrecy on the matter.

In October and November 1920, Harkin was not alone in arguing for swift action in the North. Christie, for example, supported Stefansson and Harkin in a memo to the prime minister, written on 28 October. Christie had recently been in Washington, where he found that Stefansson was very well regarded in American geographical and political circles. His respect for the explorer's opinions was consequently high. He wrote, 'The necessity for taking concrete steps to confirm the Canadian assertion of sovereignty over the northern arctic islands has now become more urgent; for information has been received that the Government of Denmark, instead of merely contemplating an expedition next year to settle Ellesmere Island as previously reported, have actually sent their expedition; indeed it is understood that it reached the scene of action in the summer of 1920.' In Christie's opinion, action on Ellesmere was 'urgent,' while action elsewhere was 'necessary but not so urgent.' Christie recommended an expedition led by Stefansson, along with 'police administration at strategically selected points' and the 'quiet, unostentatious settlement of Wrangel Island by some Canadian development company, such as the Hudson's Bay Company.' However, Christie's tone was far less alarmist than Harkin's. He had consulted Hall as well as Oppenheim, and he noted that exactly what constituted effective occupation was 'a question in law to be judged by the light of the circumstances of each case. The peculiar present conditions of arctic communication and habitation would undoubtedly be factors in this question. What might reasonably be required to establish the fact of occupation in a temperate zone country might well be unreasonable in the arctic zone.'[65] Unlike Harkin, then, Christie realized that even the meagre program already undertaken by Canada in the eastern archipelago might be considered enough to satisfy the requirements for occupation in such a remote and inhospitable area, where foreign intruders were few and their activities very limited in scope. Only if Canada did not respond to a major Danish incursion would its claim be extinguished.

Stefansson himself sent Meighen a memo a few days later, recommending several steps that would reinforce Canadian sovereignty in the Far North. They included police posts, mapping, a revenue cutter service, the encouragement of private enterprise, and a five-year exploring expedition in the unknown region north of Wrangel Island.[66] Borden, too, wrote to Meighen on 3 November. He stated that in his opinion 'a good deal of importance should be attached' to Stefansson's arguments about both Ellesmere and Wrangel, and he recommended that 'such steps as are reasonably necessary' should be taken.[67]

However, Pope – who had been attending the Advisory Technical Board meetings on Sir James Lougheed's suggestion – expressed reservations about Wrangel. Stefansson's plan to claim the island was discussed at the 24 November meeting of the board. Otto Klotz recorded that although Stefansson was eager to 'foist' Wrangel Island on Canada, none of the board members supported his project. Édouard Deville remarked that 'a claim made for it, would lessen our moral claim on the island[s] north of & contiguous to Canada,' which Klotz considered 'good reasoning.'[68] Pope agreed that the Wrangel plan belonged to the realm of fantasy, and he promptly wrote a memo to the prime minister on the subject.[69] While he was in favour of an expedition to Ellesmere and the establishment of police posts, the under-secretary pointed out that Wrangel was '[e]ssentially ... an Asiatic island,' being situated on the very border of the western hemisphere, the 180th meridian. Therefore, an attempt to claim Wrangel 'could only result in weakening our legitimate claims to the Arctic islands contiguous to our own territory.'[70]

Lougheed seemed happy to leave all such questions either to other departments or to his subordinates in the Department of the Interior. In early November he had demanded a report on Arctic issues but, as Deville recorded with sly amusement, the minister 'was not quite clear as to who was dealing with the matter or by whom the report should be made. He was evidently aware that a board was dealing with it and he mentioned the Geographic Board: when I told him that it was the Advisory Technical Board, he answered, in effect, that it was immaterial, so long as he got his report.'[71] Harkin obliged with a twenty-seven-page memo.

This lengthy document began with the dramatic statement that Denmark might 'already have taken steps calculated to wrest from Canada the sovereignty with respect to Ellesmere Land.'[72] Harkin gave quotations from Oppenheim which, in his view, demonstrated that Denmark or any other country could acquire sovereignty simply by establishing effective occupation. Harkin had evidently been informed by Pope about Denmark's successful efforts to obtain recognition from Britain of its sovereignty over all of Greenland.[73] His interpretation of this new information was that the Danes, having 'devoted special attention' to sovereignty matters, must be particularly well aware of the weaknesses in Canada's case. With their control over Greenland secure through American and British recognition, they now wanted to extend their authority to Ellesmere, and possibly to other northern islands as well.[74]

Captain Pickels had made it clear that there was no possibility of making an immediate voyage to Bylot Island, and the dirigible plan had been quietly dropped after Deville's memo to Cory (see Chapter 2). Harkin had to fall back on the original plan for a 1921 voyage. Police posts, he wrote, should be established on Ellesmere, Bylot, and Devon Islands, in order to 'effectually close up what might be called the front door of the Arctic Archipelago.' The police should act as customs and immigration officers, and post offices should be established as well, since Pope had pointed out to him that the establishment of a postal service was 'a high act of administration.' The *Arctic* (which needed extensive refitting) could be sent north at a cost of approximately $103,000.[75]

Harkin sent his memo forward to the minister without consulting Deville. He was then asked to present his arguments in a more concise form, so that they could be considered by the cabinet along with Christie's memo. Even before the official decision was made, Harkin began moving ahead with plans for the *Arctic*'s voyage.[76] His proposals were duly approved by the cabinet. However, since neither Harkin nor anyone else on the Advisory Technical Board felt any enthusiasm for a claim to Wrangel Island, Stefansson's real objective was as far from being achieved as ever. On 25 November Lougheed asked the board to hold a special meeting on the subject of Wrangel, and he readily endorsed the resulting decision not to recommend a Canadian claim.

On 1 December Captain Pickels was offered the command of the *Arctic,* with the generous salary of $500 a month. On the same day, Cory asked for the services of John Davidson ('Jack') Craig, an engineer who had worked on the survey of the Alaska-Yukon border from 1909 to 1913. Craig was to be in overall command of the expedition.[77] He was well educated, holding both BA and BSc degrees from Queen's University.

Craig's Scottish ancestry was immediately apparent to everyone who met him: though Canadian-born, he spoke with a strong Scottish burr. He had joined the Department of the Interior in 1900 at the age of twenty-four. Photographs of him taken during his years on the boundary survey show a handsome, fit young man with an exuberant grin, evidently enjoying the outdoor life and full of zest for his work. The boundary party started in the south and worked its way north in successive years. When the group reached the Arctic Ocean in 1912, Craig and another of the Canadian surveyors went swimming in the icy water, 'just to say they had done it.' Craig had recently returned from yet another trip to the Yukon. In the summer of 1920, he led a party sent to mark the boundary more clearly in the Portland Channel area, where rich silver strikes had recently been made. In early middle age, Craig retained his boyish good looks and his cheerful outlook on life. He and his wife, Gertrude, were a devoted couple. Although childless, they were fond of young people and popular with their nieces and nephews.[78]

With the appointments of Craig and Pickels, the involvement of the Advisory Technical Board came to an end (no doubt to the great relief of Harkin and Stefansson).[79] A few weeks later, the RCMP were presented with a *fait accompli:* Cory informed Commissioner Perry that Craig would call on him with 'full authority ... to discuss the matter of establishing posts and extending patrols in the north.' Perry should telegraph instructions to Dawson in order to ensure that the best qualified men with northern experience would be available for an expedition to the eastern Arctic in the spring; other preparations should begin as soon as possible.[80] On Christmas Eve, Perry meekly reported that the telegram had been sent and that he would consult his minister on the necessary arrangements for the new posts.

Even as matters at last seemed to be proceeding smoothly, Harkin found new reasons to fret. On 9 December a letter from Rasmussen to H.G. Henderson of the governor general's staff (erroneously described as the 'Secretary of State for External Affairs, Governor General's Office, Ottawa') was forwarded to the Department of the Interior by Pope. Rasmussen stated that although he had now returned from Greenland, unfortunately he would not be able to travel to Ottawa. He would be happy to answer any inquiries about reindeer or muskoxen by mail. The explorer then announced his intention to start on an expedition to the archipelago in June 1921.[81] The letter, written in complete innocence of the suspicions being harboured in Ottawa, would certainly have seemed to confirm Harkin's deepest fears.

In addition, by this time Harkin had read W.F. King's report. It reinforced his concern that Britain's inchoate title might have expired before the 1880 transfer, rendering all Canadian claims based on the transfer invalid. There were, of course, the proclamations made by Low and Bernier, but perhaps these had not been properly authorized by the imperial government. Harkin wondered 'whether the Dominion of Canada is a member of the Family of Nations as understood by International Law and ... whether it is competent of itself to take any action which will effectively establish sovereignty.' If it was not, and if the earlier proclamations had not in fact been authorized in London, the new commander might have to repeat the acts of possession carried out during earlier Canadian voyages. It was therefore 'all-important that whatever is done re. the new expedition shall be done in absolutely complete harmony with all the complications of International Law.' 'To finally clear up these matters,' Harkin observed apprehensively, 'would involve perhaps weeks of most careful and intensive investigation from every possible viewpoint. Moreover, I imagine many more complications will arise as the case is further gone into.'[82]

Then there was the matter of secrecy. The Danes might be 'really now in occupation in Ellesmere Land, or intend to take action in that regard next Spring,' and any news reports on the Canadian plans 'would probably result in special efforts on their part. If they believe that Canada is still asleep they probably will not hurry.'[83] But the arrival of Pickels and Craig in Quebec to inspect the *Arctic* had not gone unnoticed by the press, and on 22 December a short item appeared in the *Ottawa Citizen*. It speculated that the *Arctic* might be sent through Hudson Strait to the western Arctic; fortunately for Harkin's peace of mind, no connection was made between the *Arctic*'s refit and the June article on Ellesmere in the *Globe*.[84]

Finally, there was Stefansson, who was far from ready to give up his hopes of a new exploring expedition and a claim to Wrangel Island. Although Harkin and the Advisory Technical Board had failed him, Stefansson still had influential supporters in Ottawa. In his 28 October memo to Meighen, Christie had endorsed the idea that Stefansson should lead a new expedition. Borden, too, approved of the project, even though the cost, according to Stefansson's own estimate, was likely to be half a million dollars or more. In the first weeks of 1921, Stefansson began a determined campaign. He sent letters to Meighen, Borden, and Lougheed, stating that he would consider it 'both an honor and a public duty' to accept the command of a new venture. However, he needed to know the government's intentions by the end of January, because he had been offered a lucrative contract for another lecture tour, this time on the Chautauqua circuit. Stefansson had already signed the contract, but with the provision that it could be cancelled if he gave notice on or before 1 February that he was going north again.[85] A few days later, Borden informed Meighen that Stefansson's letter 'was written at my request, for the purpose of embodying certain representations which he made to me regarding the importance of continuing exploration in the North ... Mr. Stefansson's views as expressed to me seem to have a good deal of cogency, and at any convenient time, I shall be glad to discuss them with you.'[86]

Borden's opinion could not easily be ignored by the government he had so recently led. Harkin was accordingly asked to draft a letter to Stefansson, inviting him 'to undertake an exploratory trip to Ellesmere.' Because Harkin had held frequent discussions with the explorer, it was felt that he 'could put the matter up to [Stefansson] in better shape than anyone else.'[87] However, Harkin (perhaps in collusion with Craig and Cory, though the point is not clear) had decided that this was not to be anything like the expedition Stefansson wanted – that is, one which placed him in a position of unquestioned authority, allowing him to claim Wrangel Island for Canada and to make additional claims in the 'region of maximum inaccessibility' between Wrangel and the North Pole. Perhaps owing to the influence of Rudolph Anderson and other members of the 1913-18 expedition, Harkin's deep-seated fears about Rasmussen were now matched by an equally wary view of Stefansson.

Stefansson had proposed to Harkin and the other members of the royal commission that the name 'muskox' should be changed to 'ovibos,' with a view to future marketing of the meat. He argued that the existing name would suggest to consumers that the meat must have an unpleasant, musky taste, whereas in fact 'it would not be distinguished from domestic beef unless by its superior flavor.' Harkin asked for Anderson's opinion on the matter (like Harkin, Anderson was a member of the Advisory Board on Wildlife Protection). Anderson responded with a long and sarcastic memo, which included quotations from many explorers who had found the meat did, at least at certain times of the year, possess a musky flavour which was not likely to appeal to the average Canadian. He hinted broadly that Stefansson's instinct for effective salesmanship was his dominant trait, and that the name change was intended for no other purpose than 'to help along a more or less dubious promotion scheme.'[88] These comments may well have led Harkin to speak privately with Anderson. He would have had a perfect opportunity to do so after the meeting of the Advisory Board on Wildlife Protection on 29 November, when the name change was considered and rejected.

Stefansson clearly regarded Harkin as his firm ally at this time. On 21 November the explorer wrote confidently that he was sure he and Harkin had 'the same ends in view.'[89] But – whether because of doubts raised by Anderson's memo or for some other reason – by early 1921 Harkin had concluded that Stefansson was 'unsuitable' for the command of an expedition intended mainly to assert Canadian sovereignty. However, he was worried that 'a direct refusal might irritate' the ambitious and forceful explorer.[90] (Strangely enough, Harkin's doubts about Stefansson seem never to have included any suspicion that Stefansson might have invented the Danish threat to serve his own purposes. This may, of course, have been because the supposed threat suited Harkin's purposes equally well). Unlike many other Ottawa officials, Harkin seems always to have taken it for granted that Stefansson was an American, and he was convinced that Stefansson had no true loyalty to Canada, which, though the place of his birth, was no longer his country.[91]

The problem was not merely that Stefansson, as a foreigner, might not be able to make claims which would be 'in absolutely complete harmony with all the complications of International Law.' As Harkin pointed out in memo after memo from January 1921 on, Stefansson knew the weaknesses of Canada's northern claims better than anyone else. If his cherished exploration project were rejected outright, he might go to the Danes or the Americans, hoping that they would sponsor his enterprise. As Harkin later summed it up, 'it not being possible to really look upon him as a real Canadian this expedition idea was developed with a view to providing a positive assurance by this sop to Stefansson's pride and selfishness that there would be no chance of his tipping off the actual situation concerning the north to either Denmark or the United States.'[92] If Stefansson was duped into thinking he could get what he wanted and then sent north in the summer of 1921, he would be unable to make any mischief until after the police posts were established. This scheme was certainly unjust to Stefansson. Ambitious and in many ways unscrupulous though he was, in 1920 and 1921 he seems never to have considered going to other nations if Canada refused him. Harkin's tendency towards paranoia pushed him into a deviousness more than equal to Stefansson's own, and his determination to protect Canada from the consequences of Stefansson's 'pride and selfishness' was thus the true driving force behind the long and complex history of the cancelled expedition, about which the explorer was so persistently curious in later decades.

FIGURE 1 A.P. Low's proclamation at Cape Herschel, Ellesmere Island, 1904 | LAC, PA-038265

FIGURE 2 The Canadian flag on Wrangel Island, 1 July 1914 | Photo by John Munro | LAC, PA-074079

FIGURE 3 Before the departure of the Canadian Arctic Expedition, 1913. Left to right: Robert Bartlett, Vilhjalmur Stefansson, Rudolph Anderson. | Photo by G.H. Wilkins | LAC, PA-074063

► *Facing page, top to bottom:*

FIGURE 4 Belle and Rudolph Anderson, 1913 | LAC, PA-105138

FIGURE 5 Vilhjalmur Stefansson during the Canadian Arctic Expedition, 1916 | Photo by G.H. Wilkins | © Canadian Museum of Civilization, 51099

FIGURE 9 Robert Borden (left) and Winston Churchill in London | LAC, C-002082

◀ *Facing page, counterclockwise from top left:*

FIGURE 6 W.W. Cory | LAC, PA-042945

FIGURE 7 J.B. Harkin, circa 1915 | LAC, PA-121371

FIGURE 8 Arthur Meighen | LAC, PA-145060

FIGURE 10 Sir Ernest Shackleton (left) on board *Quest*, 1921 | Scott Polar Research Institute, P56/77/39

FIGURE 11 *Arctic* at Pond Inlet, 1922 | LAC, e010691184

FIGURE 12 *Arctic* in the ice, 1922 | LAC, e010691185

FIGURE 13 Kakto and his family on board *Arctic*, 1922 | Photo by W.H. Grant | LAC, e002282907

FIGURE 14 James White. The only known photo of White, probably taken circa 1908 | Natural Resources Canada, Geological Survey of Canada, 202313A

Figure 15 Loring Christie | Photo by Yousuf Karsh | LAC, PA-174532

Figure 16 Craig Harbour post under construction, 1922 | Photo by W. H. Grant | Library and Archives Canada, PA-173051

FIGURE 17 Inspector Charles Wilcox on board *Arctic*, 1923 | LAC, PA-102582

FIGURE 18 Allan Crawford on Wrangel Island | LAC, C-050166

FIGURE 19 On board *Arctic*, 1923. Left to right: J. Dewey Soper, Leslie Livingstone, Jack Craig, Frank Henderson | LAC, e010752415

4
A Citizen of the British Empire

> *A story of heroism, courage, endurance and infinite resource and such as ranks men immortal, was that unfolded ... last night by Sir Ernest Shackleton, C.V.O. ... Sir Ernest Shackleton in opening said that he happened to be an Irishman whose privilege it was still to be a citizen of the British Empire. He hoped to be closely identified with Canada in the near future ... Premier Meighen ... expressed pride in being of the same race and blood as the men whose pluck had carried them through such an enterprise.*
>
> – *Ottawa Citizen, 24 February 1921*

SIR JAMES LOUGHEED ENDORSED the plan to deceive Stefansson, agreeing that 'it was in Canada's best interests to get him away to the north as soon as possible.'[1] Harkin accordingly informed Stefansson that there was 'an excellent opportunity for providing for additional exploration in the region immediately to the north and the west of the better known islands.' Harkin made it clear that the explorer would have no authority when it came to the police posts or other steps designed to establish effective occupation. However, Stefansson would be placed in charge of extensive exploration work. His party was to be taken north in the *Arctic* along with the police and then set off independently.

Harkin asked that Stefansson come to Ottawa at some point after 8 February. 'My idea,' he explained, 'is that the ship provided for the other work could be used to transport your party to a suitable point from which the expedition could take its departure overland ... I [am] anxious to go into the details with you personally.' Harkin concluded with a strong hint on the matter of secrecy. 'I know that so far as you are personally concerned you recognize that everything in connection with the expeditions is strictly confidential,' he wrote diplomatically. But, he continued, 'it is imperative that you should also see that nothing is said to the lecture bureau which will enable it to indulge in publicity which might call attention to these expeditions. There is always a temptation for a lecture bureau to endeavour to get free publicity when opportunity offers and in writing to you upon this subject I take it for granted that you will take all necessary safeguards in that connection.'[2] Harkin was referring to the bureau in charge of Stefansson's current tour. He assumed that Stefansson would receive his message in time to cancel the summer 1921 Chautauqua lectures. However, by some oversight the letter was sent to Stefansson's New York address. It was forwarded and delivered to him in Atlanta on 2 February – the day after the deadline. Ecstatic at the news, Stefansson wired to Ottawa: 'consider it both privilege and duty to accept command expedition to carry out purposes we have discussed and have at heart shall be glad accept stop lecture manager know[s] nothing [about] our plans there will be no publicity.'[3]

The situation, already complicated enough by the plan to deceive Stefansson, quickly became even more labyrinthine. Before Stefansson could travel to Canada, a powerful rival appeared on the scene. Sir Ernest Shackleton arrived in Ottawa on 8 February. While there, he stayed at the city's best hotel, the Chateau Laurier. An enormous building in the French Renaissance style, the hotel stood between the parliament buildings and the Ottawa railway station. The Edwardian opulence of its furnishings (the Chateau was the brainchild of railway magnate Charles Melville Hays, who went down on the *Titanic*) made it a most appropriate setting for Shackleton, the quintessential Edwardian adventurer. The next day a headline in the *Ottawa Citizen* ran: 'Shackleton May Seek North Pole on Next Journey.' 'There is,' Shackleton told the *Citizen*'s reporter, 'certainly a large field left in the north for explorations.' After crossing the Atlantic on the *Aquitania,* Shackleton had spent a week with friends in Montreal, then travelled on to Ottawa, ostensibly also to visit friends. He intended to return to Montreal and give a lecture there, donating the proceeds to charity. He might, he hinted to the reporter, lecture in Ottawa as well. He would show the famous film of the *Endurance* expedition, which had not yet been seen anywhere in Canada. 'There is something missing in the bridge between the Motherland and this country, when such inspiring pictures are never shown here,' commented the *Citizen*'s writer, who believed that this story of 'British doggedness, and self-imposed discipline under dire circumstances' would 'assuredly bear repeating in Canada ...

Sir Ernest Shackleton's narrative should have a place on every Canadian boy's bookshelf.'[4]

Shackleton had come to Canada with the express intention of seeking financial support for his planned Arctic expedition from the government and from wealthy Canadians. He had already succeeded in making himself popular in Montreal. On 4 February his friends there had arranged a meeting with Meighen, who was also visiting the city. Shackleton took the opportunity to give the prime minister a detailed outline of his proposed expedition. Its aims were to search for unknown land and carry out scientific research in the Beaufort Sea. Shackleton proposed that the Canadian government should pay half the cost of the expedition. As an indication of the backing he had already received in England, he included letters of approbation from the Royal Geographical Society and from the editor of the *Daily Mail*. The latter had undertaken to give the venture favourable publicity. Finally, Shackleton appended a copy of the letter Stefansson had written to him in London the previous spring.[5] According to Shackleton, Stefansson had given up any idea of further exploration.[6] Meighen, caught off guard by this brash claim, could only observe dryly to Shackleton that he 'did not understand Mr. Stefansson's attitude in the same way ... and ... was of the opinion Mr. Stefansson was anxious himself to take charge of such work in the North as the Government desired to have done.'[7] Shackleton insisted that he had a letter from Stefansson proving the truth of his statement.

On his return to Ottawa, Meighen promptly forwarded all the documents Shackleton had given him to Lougheed. He asked Lougheed to evaluate Shackleton's plan and to make a recommendation on what action, if any, the government should take. Whatever reservations Meighen may have felt about Shackleton based on their short personal encounter and the rumours that had long circulated about him, the chance to employ such a world-famous figure could not be dismissed lightly. Even more than Stefansson, Shackleton met the era's romantic expectations of what a great explorer should be like. His exploits, along with those of his former leader Robert Scott, had undoubtedly given Antarctic exploration a glamour and fascination that not even Stefansson's well-publicized adventures in the North could match.

Shackleton's flamboyant personality was exceptionally attractive to most of those who met him. 'His nature,' as one fellow explorer wrote, 'was such that it was easy for people to admire him.'[8] After Shackleton's death the normally staid Scottish librarian of the Royal Geographical Society, Hugh Robert Mill, vividly recalled the explorer's 'great athletic figure and bronzed countenance with eyes flashing happiness and fun.' Mill found it 'easy to picture' Shackleton as a sixteenth-century adventurer like Drake or Raleigh, 'strutting in all the bravery of Elizabethan finery along the miry quaysides of Wapping.' 'No one ever exemplified better the pure romance of exploration,' Mill concluded.[9] Another friend, magazine editor P.W. Everett, recalled how Shackleton's face, 'heavy and stern in repose,' would become 'all alive and lit up' when he began to talk, 'the deep, husky voice rising and falling with the movement of the story, and sometimes raised, by way of illustrating his point, to a rafter-shaking roar.'[10]

Of course, there were other sides to this magnetic, dominating personality. Shackleton became 'very unpleasant,' indeed almost 'brutally truculent,' whenever he was attacked.[11] A subordinate on the 1907-9 *Nimrod* expedition, during which Shackleton allegedly came within a hundred miles of the South Pole, bitterly described him as 'a coward, a cad ... incapable of keeping his word.'[12] Shackleton undoubtedly had the habit of making promises he knew he would not be able to keep. For example, when recruiting men for the *Nimrod* expedition, Shackleton promised five of them places in the small party that would try for the Pole, even though there would be room for only three.[13] He also promised his followers on this expedition huge financial bonuses, which he could never realistically have hoped to pay.[14] Rumours that Shackleton had deliberately falsified his distances on the 1908-9 southern journey began to circulate soon after the *Nimrod* expedition's return. But for most people, Shackleton's extraordinary leadership in times of crisis more than compensated for any failings.

Shackleton summed up his own character in a letter to his wife, Emily, written during the early months of the *Nimrod* expedition. 'I have a curious nature,' he confessed. 'I am just good as an explorer and nothing else, I am hard also, and damnably persistent when I want anything: altogether a generally unpleasing character: I love the fight and when things [are] easy I hate it though when things are wrong I get worried.'[15] Shackleton's motive for joining Scott's *Discovery* expedition in 1901 was to win fame and perhaps fortune, enabling him to marry Emily, a rich man's daughter. But even after fame had been abundantly achieved, fortune eluded him, and in any case he was too restless to settle down as a family man. In 1921 Shackleton was forty-seven years old, yet he could not give up the dream of another expedition. Whatever the obstacles, the Arctic 'region of maximum inaccessibility' held as much allure for him as it did for Stefansson. As he himself had observed, Shackleton was 'damnably persistent' whenever he wanted anything. And, like Stefansson, he was not always entirely scrupulous.

On the day before Shackleton arrived in Ottawa, Harkin, Jack Craig, and Édouard Deville had met to discuss his proposal. They concluded that the strong scientific program made Shackleton's plan worthy of serious consideration by the government. He would certainly deserve official support, if only so much money were not already being spent on the *Arctic*. Sovereignty was an urgent concern, while science was not. But if Shackleton could get most of his funds elsewhere, perhaps the government should contribute whatever could be spared. Some official connection was certainly desirable because it would ensure that Shackleton's discoveries could be claimed by Canada. The three men noted warily that it was not clear whether Shackleton intended to enter the Arctic from the east or from the west. If from the west, he could not begin his work until 1922; but if from the east, his expedition might come into contact, and perhaps into conflict, with Stefansson's. They all knew that Stefansson would inevitably be upset by this, and the more so if Shackleton too had government support.

At the same time, there was a flurry of telegrams between Harkin and Stefansson. First, Stefansson informed Harkin that it would suit him best to come to Ottawa on 14 February, but he could make the trip at once if necessary. Harkin replied nervously that there was 'no great rush working out details.'[16] Perhaps alarmed by this message and by the news of Shackleton's presence in Canada, Stefansson responded with a letter in which he set out a list of demands and conditions. Harkin, he conceded, was right that the exploration and occupation work could not all be done by the same commander, but everything should be planned by someone with polar experience. The acceptance he had already given was 'contingent upon the backing of the Government being of a character which suits me, and the plans in general of a scope which make it worth while to undertake them.' Stefansson strongly recommended a 'quiet, entirely unostentatious' occupation of Wrangel Island, possibly by a fur trader. A public announcement of his own intention to explore the Beaufort Sea should be made at once, in order to keep other explorers out. With an eye to the future, Stefansson told Harkin that parts of the letter were perhaps 'too bluntly stated' for publication. He therefore wanted to have it returned to him when they met. He would provide another version, more suitable for release to the public. As a precaution, Stefansson did not sign the document before he sent it to Ottawa.[17]

When Shackleton met with Lougheed on 8 February, he was told that the government 'could not at the present time consider taking any action along the lines he proposed.' Shackleton appeared to accept the refusal with good grace, but he hinted that if he succeeded in getting other funds in Canada, he might return to ask for a lesser amount. Lougheed intentionally gave Shackleton the impression that such a request was likely to be granted. 'I, of course, did not commit the Government in any sense ... but in all probability he will again approach me,' the minister reported to Meighen.[18] Given the general distrust with which Stefansson was viewed in Ottawa, and the fact that planning for his expedition remained at a very early stage, it was inevitable that some officials would consider it best to drop Stefansson altogether and support Shackleton instead – especially if Shackleton was able to raise substantial funds from private individuals in Canada. 'One of the first suggestions after the entrance of Shackleton into this matter was that for reasons of economy the Stefansson end of the expedition should be cut out,' Harkin later recorded.[19] Shackleton's status as a British subject was another strong point in his favour.[20]

Harkin and Stefansson met several times between 14 and 17 February. During their conversations Stefansson 'manifested a good deal of nervousness about Shackleton.'[21] He must also have felt some alarm when Harkin warned him that Meighen's government might not remain in power long enough to carry out its Arctic plans. Nevertheless, he and Harkin worked out an agreement that must have pleased the explorer. Stefansson was to be 'in supreme command of the exploratory work.' A camp would be established on Axel Heiberg Island; from there, Stefansson would travel north, then west into the Beaufort Sea. He would then be at liberty to set his course towards Alaska or Siberia, or in any other direction as he saw fit. Meanwhile, another party would do scientific work and establish Canada's claim to the Sverdrup Islands.[22]

Because Stefansson had not been able to notify the lecture bureau about his change of plans before 1 February, he could not now accompany the *Arctic* in the summer of 1921. He proposed that the base on Axel Heiberg should be established without him; then he would go north in 1922. Although Stefansson could not know it, this was a fatal mistake. The purpose of the deception proposed by Harkin was to get Stefansson out of the way until the police posts were established. A journey north in 1922 would not accomplish that aim. It was inevitable that questions would be raised about the propriety of spending so much on a plan from which Canada would derive so little benefit, especially when Shackleton stood ready to do much the same work at a lower cost. But Harkin remained adamant in his support of Stefansson, and so, at least for the time being, did Meighen. On 19 February Stefansson met with the prime minister. Once again, he strongly urged the occupation of Wrangel Island. This time Meighen was convinced. After consulting the cabinet, he wrote a letter formally notifying Stefansson that the government would 'assert the right of Canada' to the island.[23]

However, Loring Christie had come to take a very different view of both Stefansson and Stefansson's plans. Christie had returned from Europe on 31 January. Once back in Ottawa, he was immediately plunged into discussions of Arctic matters with Harkin, Craig, A. Bowen Perry, and others. Christie had likely spoken about the Danish threat with officials in London, and he now took a far more moderate line on this subject. It seems that he attempted to reassure Harkin by pointing out that if the Danes did in fact make an attempt at occupation, 'pressure from Britain' could easily force them to desist.[24]

In the course of these discussions, Christie realized that Stefansson had taken advantage of his absence. Stefansson had often found it convenient to give his own version of Christie's opinions, boldly twisting Christie's words to suit his purposes. Christie was appalled and furious. 'The circumstances in which [Stefansson] was introduced to me,' he later recorded, 'were such as to suggest that he was a person with a sense of discretion and responsibility. As the result however of a number of incidents I felt bound to doubt that, and finally, so far as I myself was concerned, I was forced to the conclusion that it would be best to have no relations whatever with him.'[25] In an effort to boost Stefansson's prestige in the capital, Sir Robert Borden had recently proposed him for membership in the Rideau Club. The result was not what Borden had confidently expected. Stefansson was blackballed, and it may well have been Christie who was responsible.[26]

Stefansson proudly told Harkin the news about Wrangel Island, expecting that he would be delighted. Although Harkin successfully concealed his true feelings, he was horrified by the possibility that the planned northern expedition might end up as a filibustering incursion into Russian territory. He immediately warned Christie of the danger (while at the same time taking care to keep Christie in the dark about his plans to secure an official Arctic expedition of some kind for Stefansson).[27]

Christie had already written a long memo to Meighen in which he argued that a firm stand by Canada would quickly put an end to the Danish plans, if any existed. He now followed this document with another memo strongly opposing a claim to Wrangel. He informed the prime minister that, after 'further reflection,' he had realized that such a plan 'would be very unwise.' The extent of the British Empire was already so great that additions to its territory should be made only for 'compelling practical reasons.' In the case of Wrangel Island, there were none. It did not appear to have any strategic value, and the commercial possibilities were merely 'speculative.' It would be foolish and pointless to claim the island for economic reasons. Even if Wrangel were Canadian, other nations would still be free to establish commercial enterprises there. 'Finally,' Christie concluded, 'by wandering outside our own hemisphere and region we would inevitably detract from the strength of our case for the ownership of the islands immediately north of Canada which we really need and desire.'[28]

Both of these memoranda were submitted to the prime minister on 1 March. Meighen immediately rescinded his 19 February letter to Stefansson. A message from Meighen's private secretary instructed Stefansson to 'make no use of' the letter.[29] Stefansson replied on 7 March that, having been away from New York, he unfortunately had not received the message until that day, and in the interim he had written to the Hudson's Bay Company on the matter. 'It appears now,' he told Meighen, 'that the Canadian office of the company had already placed the strongest recommendations before the head office in England, urging that a post be established on Wrangel Island even though the political status of the island was uncertain.' HBC officials, he claimed, had assured him 'that their knowledge of the attitude of the Government [was] an added reason for establishing a post.'[30] Clearly, Stefansson hoped that the prime minister would continue to back his plan if he believed it would foster a successful commercial enterprise.

But as long as Meighen remained in power, Wrangel Island was a lost cause. A rumour began to circulate that behind Stefansson's proposals about it lay a plot masterminded by either the Danes or the Americans. Robert Logan, an employee of the Department of the Interior who later went north with the first Eastern Arctic Patrol, recalled in 1974 that there had been 'a general fear that, either knowingly or unknowingly, Stef was acting as a front for some "unspecified Scandinavians" or for U.S.A.'s MacMillan, so that the minute Canada accepted Stef's claim that Wrangel

Island was Canadian territory the others would claim possession of Ellesmere by occupation.'[31] In other words, it was becoming ever clearer to Ottawa officials that Stefansson's plan would undermine the informal, as yet poorly articulated version of the sector principle that had long shaped Canadian assumptions about sovereignty over the archipelago.[32]

ALTHOUGH CHRISTIE WAS ABLE to put a stop to the Wrangel Island plan with relative ease, his views on the Danish threat were not so quickly accepted in Ottawa. His difficulties did not arise only because of Harkin's stubborn convictions on the matter. A number of other officials had become involved in a complex web of information gathering, all of it based on the assumptions that Rasmussen's aims were political and that he had the backing of the Danish government. As the process gathered momentum, it took on a life of its own.

By the time of Christie's return from Europe at the end of January, several separate lines of inquiry had begun. The first originated with Jack Craig, who took on the task set by Harkin of determining the status of Canada's sovereignty claims under international law. Craig turned first to the Department of Justice. On 7 January he wrote to the deputy minister, W. Stuart Edwards. In reply, Edwards agreed that 'effective action should be taken as quickly as possible' and that 'we should try to be in possession of the whole case before this summer's expedition leaves.' All available facts about the history of Canadian activities in the North should be gathered, and there should be an investigation of Danish and American actions. Having observed in W.F. King's report (a copy of which Craig had forwarded to him) that the *Century Atlas* showed Ellesmere Island as an American possession, Edwards warned that an American claim to the island should be anticipated. He clearly did not believe any agency of the Canadian government would be able to provide the required information quickly enough. Edwards suggested that the Colonial Office should be asked 'immediately to take the most effective action ... to secure and forward here all available data at the earliest possible moment'; in addition, Sir Joseph Pope might 'communicate direct with the officials of the British Embassy at Washington' (with whom the under-secretary had an excellent working relationship). Edwards himself was unable to offer much in the way of useful advice. Among his few concrete suggestions for the forthcoming expedition was that it should take care to establish settlements 'at the mouths of the largest river systems on the island.' A glance at a map would have revealed that there were no river systems on Ellesmere.[33]

A few days later Craig was informed by Roy Gibson, who was acting as deputy minister of the interior in Cory's absence, about a news item in the *Montreal Gazette* of 4 January. Gibson, in turn, had heard about the article from Commissioner Perry. Datelined New York and titled 'Explorers Are Resuming Work of Pre-War Years,' the article was a brief summary of journeys planned in the polar regions, Africa, South America, the Pacific, and Asia. These included a proposed circumnavigation of Baffin Island by Donald MacMillan; an attempted drift through the Arctic basin by Norwegian explorer Roald Amundsen; and Rasmussen's planned expedition to the Canadian Arctic, described simply as 'another expedition of five years' duration.' The article also stated that a second Danish explorer, Lauge Koch, intended to make 'a scientific survey of North Greenland, his main purpose being to establish Danish sovereignty in that territory' (this was the Danish Bicentenary Jubilee Expedition of 1920-23). Apparently, no one in Ottawa noticed the clear implication that northern Greenland was not already under Danish sovereignty.

Although it told Canadian officials no more than they already knew about Rasmussen's plans and contained no hint that the Danish government was behind his enterprise, this seemingly innocuous article caused an extraordinary flurry of activity in Ottawa. Gibson forwarded the clipping to Pope on 14 January, and he suggested that Pope should ask the Home Office in London for further information. Pope replied that he did not see how the Home Office could help, and that the Colonial Office was lamentably slow in dealing with such matters.[34] He thought it would be better for him to write to the Canadian High Commission. This he did on the same day, asking the secretary in London, William Griffith, to make inquiries at the Colonial Office 'and elsewhere' in order to secure more precise information.[35]

Shortly afterwards, Commissioner Perry made a far more dramatically phrased request to Sir Basil Thomson of Scotland Yard. 'I desire,' he began, 'to ask your assistance in a delicate and indeed very secret matter. It concerns the sovereignty of the islands which lie to the North of the mainland of Canada, and the disposition of the Government of Denmark to dispute our claim.' He asked Thomson to ascertain whether Rasmussen's expedition was the only one planned by Denmark; what Rasmussen's 'exact official position' was; and what 'character and strength' his expedition would have. Most important of all, Perry wanted to know about 'the instructions, *secret and otherwise,* which the Danish Government has given to Mr. Rasmussen and his expedition, or to any other party.' Perry explained to his British colleague that he needed the information in order to draw up instructions for the leader of the police contingent. It was, he emphasized, 'urgently necessary for me to know how far my officer should proceed in the event of a denial of his right to exercise jurisdiction – whether, for instance, he would be justified in using force.'[36] In a second letter written a few weeks later, Perry passed along the information that Rasmussen had returned to Copenhagen from Greenland in October 1920, and he noted: 'We suspect that [Rasmussen] took some action in Ellesmere Land or some other of the Arctic Islands calculated to prejudice the claims of Canada.' Perry asked Thomson to 'ascertain what Mr. Rasmussen did in this visit.'[37]

Meanwhile, Craig was busy drawing up his own list of questions to be answered. It was, he told Cory, essential 'that someone on the ground in England should be in a position to get first-hand information which will shed light on these points' so that the government could make plans 'which will be absolutely safe from every view point.'[38] Cory thought Christie would be ideally placed to deal with the matter (he intended to visit London on his way back from Geneva), and Meighen agreed that the task should be given to him. However, the instructions did not arrive at the High Commission until after Christie had sailed for Canada. It was then suggested that the staff of the Public Archives might be able to carry out the necessary research, especially since archivist H.P. Biggar was in London working at the Public Records Office and could forward copies of any relevant documents to Ottawa.

On 21 January Craig wrote to the Dominion archivist, Arthur Doughty, requesting that the archives staff undertake the work. The main questions to be answered were exactly what territory the British government had intended to transfer in 1880; whether the 1880 order-in-council was 'intentionally indefinite,' and if so, why; and finally, 'Can Canada of itself, that is without specific instructions from the Imperial Government, take any effective action regarding the sovereignty of lands which may be regarded by other nations as outside of Canada?'[39] The task of sifting through the documents already available in Ottawa was eventually delegated to Hensley Holmden, the head of the Maps Division, who set to work with admirable diligence and thoroughness.

In London, Griffith had duly made inquiries at the Colonial Office. Both the Colonial Office and Scotland Yard turned first to the Royal

Geographical Society (RGS) for information. The society's secretary, Arthur Hinks, replied to Thomson's query on 18 February and to the colonial secretary's on the 19th. He reported that the plans for the Fifth Thule Expedition dated back to 1910. Rasmussen had made 'no sort of secret' about them, and 'so far as we know his interests are in Eskimos and non-political.' The RGS was unable to clarify Rasmussen's position in relation to the Danish government, but Hinks pointed out that in the past he seemed to have received very little financial support from official sources. So far as Hinks and his colleagues were aware, Rasmussen had no instructions from the government.[40]

When this reassuring information was received in Ottawa, it made little difference to the climate of suspicion. Inquiries about Donald MacMillan in Washington and New York yielded equally little evidence of any foreign threat. Fears about MacMillan never reached the level of concern felt over the Danish expedition; nevertheless, Stefansson did his best to inflame feelings on this issue as well. He sent Harkin a letter marked 'strictly confidential,' which was filled with venomous observations on MacMillan's character, allegedly gleaned from Isaiah Bowman of the American Geographical Society. According to Stefansson, Bowman had described MacMillan as 'both unreliable in a general way and incompetent as a scientific explorer.' Stefansson suggested that MacMillan intended to set up as a trader and to slaughter large numbers of caribou on Baffin Island. Rather ironically, considering the extent to which he had cast doubt on Rasmussen's honesty, Stefansson ended by referring Harkin to a passage in Rasmussen's letter of 11 May 1920, in which Rasmussen strongly criticized MacMillan for his role in covering up the facts about the shooting of an Inughuk by Fitzhugh Green, a member of the 1913-17 American expedition.[41]

SHACKLETON HAD NOT BEEN idle since his interviews with Meighen and Lougheed. On 22 February he lectured to the McGill Canadian Club in Montreal, concluding with a reference to his 'hopes and dreams of a great exploring expedition into the hitherto inaccessible hinterlands of the Canadian North.' He told his enthusiastic audience that 'when the time came, as it might come shortly, he would turn to Canada for men to lead on the long trail.'[42] The *Montreal Herald*'s reporter thought that Shackleton had 'worthily maintained British traditions of courage, endurance and discipline,' and that his story was 'an inspiration' to Canada's youth. The prospect of an expedition led by Shackleton and manned by

Canadians was more than welcome to this journalist, who commented that there was a 'vast field' for exploration in the Far North and that 'no doubt' the Canadian government would be happy to aid 'such an experienced explorer.'[43]

The next day Shackleton returned to Ottawa, where another lecture had been arranged, with the proceeds to be given to the May Court Club for its charitable work. The capacity audience of Ottawa's elite included the Duchess of Devonshire, Lord Richard Nevill, Prime Minister and Mrs. Meighen, and William Lyon Mackenzie King. When he introduced Shackleton, Meighen did not avoid the issue of the proposed northern expedition. Instead, he stated plainly that 'the purpose of the lecturer's visit to Ottawa was that he was looking to a further exploration of the Great North of this country in the hope of adding still further territory to the Dominion of Canada.' The normally reticent prime minister took the opportunity to jokingly suggest that 'if his desired expedition to the Arctic ultimately materialized, and he would promise to stay away long enough,' Shackleton would be welcome to take several members of Parliament with him. Shackleton, for his part, began by emphasizing his status as a citizen of the Empire, by implication repudiating any hint that as an Irishman he might harbour disloyal sentiments. He hoped, he said, 'to be closely identified with Canada in the near future.'

During the lecture Shackleton was at his best. The *Ottawa Citizen* reported that 'many times during the evening' the audience 'bubbled over with laughter at the genial bonhommie [sic] of the lecturer and his many humorous sallies and descriptive touches.' The slides and film of the *Endurance* expedition were deemed 'most excellent,' and there was loud and 'long continued' applause at the end of the talk. In proposing a vote of thanks, King referred to Shackleton's 'high courage' and to the 'chivalry and courtesy' he had shown by donating the proceeds of his lecture to the May Court Club. When Meighen seconded the proposal, he took the opportunity to express 'pride in being of the same race and blood as the men whose pluck had carried them through such an enterprise.'[44] The proceedings closed with 'God Save the King.' Shackleton then had a late supper with a group that included the Meighens, King, and Montreal newspaperman John Bassett.[45]

A few days later, Cory instructed Harkin to '[g]et [Stefansson] on [the] phone and tell him to pay no attention to reports re Shackleton.'[46] There was ample cause for rumour: besides giving his lecture, Shackleton had met privately with the prime minister and other officials, including Harkin.

For the first time, he was told that the government had already made a commitment to fund the voyage of the *Arctic* and the establishment of four police posts (one on Bylot Island, one on North Devon, and two on Ellesmere) and that there might also be a new Stefansson expedition under government auspices. Therefore, despite the great admiration for Shackleton and his work in Canada, no more money could be spared for Arctic exploration. Shackleton responded with the ingenious suggestion that his expedition should establish two of the posts and the *Arctic* the other two. This arrangement would cut down on the overall expenses, and financial assistance to Shackleton could then be justified on sovereignty grounds. No immediate promises were made, but it was clear that Shackleton's proposal had considerable appeal.[47]

Shackleton then set off on his return journey to England by way of New York. Whether by luck or good planning, a fellow passenger on the *Aquitania* was Lady Eaton, the wife of Toronto millionaire and philanthropist Sir John Eaton. Her letters are filled with glowing references to 'Shak,' who was constantly at her side throughout the voyage. Shackleton told her about 'all he had done in Ottawa and Montreal.' Won over, Lady Eaton wrote to her husband: 'The more I think of the expedition the better I like it, and I do think if you could manage it ... it would be the biggest thing you've yet done and not the most costly. As advertising it would have the back page of the Star wiped clean off the earth.' In her opinion, to sponsor a new Arctic expedition led by such a famous hero would be both 'good business' and 'so patriotic scientific and far sighted.'[48]

While Shackleton was still en route, his supporters announced that his expedition would definitely start that summer. The *Citizen* reported: 'When here last week Sir Ernest made application to the government for financial assi[s]tance for his project and while no decision in regard to it has been reached, the matter is known to be under consideration ... Financial conditions may preclude government assistance, but the disposition seems to be favorable.'[49] However, this apparent attempt to force Meighen's hand did not have the effect Shackleton probably expected. Instead, the government issued a statement confirming that Shackleton had met with the prime minister to discuss his plans, but emphasizing that no decision had yet been reached, and that the announcement was therefore 'to say the least, premature.'[50]

Lougheed asked Cory and Harkin for a recommendation on Shackleton's proposal. In a memo to Cory, Harkin repeated his argument that if Shackleton received government help, there was a 'grave probability'

Stefansson would reveal Canada's plans to the United States or Denmark. 'It would ... be unwise to bank on his Canadian loyalty too much. The Canadian expedition has been developed on the line of keeping him with us through self interest,' Harkin pointed out yet again. It was true, Harkin admitted, that Stefansson's current plan did not suit Canadian purposes as well as would a departure in 1921, but the promise of a new expedition would give him a good reason to remain silent. But if the government aided a rival expedition, Stefansson would be 'practically certain to get up in arms.' Harkin thought that Shackleton should at least be asked to wait until 1922; then Stefansson could still be kept in line until the all-important police posts were established.[51]

ON THE SAME DAY THAT he acknowledged the receipt of the letter informing him there would be no occupation of Wrangel Island, Stefansson wrote a second letter to Meighen's secretary in which he challenged Shackleton's claim that he intended to give up Arctic exploration. A few days later, he was again in Ottawa for discussions with Harkin and Cory. Stefansson also met with Christie to discuss the cancellation of the Wrangel Island plan.

As Christie later wrote in a memorandum for the official file on Wrangel, Stefansson 'made to me what I could only regard as a suggestion that I should change a legal opinion I had already given in such a way as to induce an alteration in the plans of one of the other Departments (the Interior Department). It so happened that such an alteration would have been advantageous to Mr. Stefansson's own personal interests.' So irritated did Christie eventually become by Stefansson's persistence that he 'finally put this aspect to Mr. Stefansson.' He did not record what the response was, merely that he found it unsatisfactory and that he and Stefansson had 'never met since.'[52]

Stefansson may have been so frustrated by this setback that he pushed especially hard in his meetings with Harkin and Cory; in any event, he had far better luck with them. On 12 March, Harkin – though fully aware that a decision might be taken at a higher level against the Stefansson plan – made the first definite commitment to carry it out when he verbally authorized Stefansson to place Lorne Knight, a young American member of the 1913-18 expedition, in charge of preparations for the 1921 advance party. On the 14th, the delighted Stefansson sent a fulsome letter to Lougheed. He thanked the minister for his 'part ... [in the] inception' of the work 'which Mr. Cory tells me is definitely decided on.' Stefansson emphasized that he had refused a $60,000 lecture contract for 1922-23, preferring to lead the new expedition without a salary because he felt he was 'helping to build the foundation of Canada's greatness.'[53]

It was of course Harkin, not Cory, who had led Stefansson to believe that the expedition would certainly go forward. When called to account, Harkin claimed that the commitment he made to Stefansson had been 'unavoidable.' There was, Harkin went on, 'more or less of a feeling' in Ottawa 'that once Canada's Police posts have been definitely established ... there might be a movement to drop the Stefansson expedition of 1922.' Harkin himself did not think there was any real need for an exploring expedition led by either Stefansson or Shackleton, since the police posts were the essential thing. But the money involved was not too high a price to pay to ensure that Stefansson would not interfere with the establishment of the posts. Harkin was convinced that 'up to date ... Stefansson is playing the game absolutely loyally to Canada. In this connection it must always be kept in mind that we owe him a debt of gratitude as the man who actually brought the weakness of Canada's case to the Government's attention.' Because Stefansson had acted with integrity so far, Harkin felt 'a keen personal responsibility in connection with the matter of the complete arrangements being fully carried out ... Canada's good faith must be maintained.' He therefore wanted to proceed immediately with the preparations requested by Stefansson and Knight.[54]

Cory passed this document to Lougheed, and there was a subsequent meeting between the minister, the deputy minister, and Harkin. As Harkin ruefully noted in the margin of his memo after it was returned to him, Lougheed stated that as matters stood, he could not recommend the Stefansson expedition to the cabinet.[55] Lougheed, it seems, was now strongly inclined towards the Shackleton plan; it may be that their Irish ancestry and their common devotion to the British Empire had formed a bond between the two men. It is certainly likely that Lougheed preferred a fellow British subject to an American adventurer of non-British descent. Whatever the exact reason for Lougheed's change of heart may have been, Harkin now found himself very far out on a limb: Stefansson expected that preparations for his expedition would begin in short order, while the minister responsible was no longer in favour of the plan going forward. The bureaucratic processes involved in financing the *Arctic*'s voyage were well advanced, but not a penny had yet been earmarked for the Stefansson expedition. Without the approval of the cabinet, no funds would ever be forthcoming.

To complete his discomfort, the unhappy Harkin was given the task of providing an appropriate answer to Stefansson's enthusiastic letter of 14 March. Stefansson had put his conviction that his new venture was now 'definitely decided on' in writing; it was up to Harkin to tell him that it was not. As everyone concerned realized, Stefansson's letter had probably been sent in order to get a clear commitment from Lougheed. 'Unanswered or evaded,' it was 'virtually an admission that we have contracted with him to lead an expedition.'[56] Other men in such a position would have retreated into grovelling apology, but Harkin, as convinced as ever of his own infallibility, maintained a doggedly defiant stance when addressing his superiors. In two more of his long memos, Harkin reminded Cory that both he and Lougheed had approved the various recommendations Harkin had made with regard to Stefansson. 'If this is to be the policy it constitutes a complete change of policy,' Harkin observed stiffly of Lougheed's decision not to support the Stefansson plan.[57]

Harkin maintained that during Stefansson's most recent visit to Ottawa, Cory too had spoken as if the advance party would definitely go north. He told the deputy minister that he could not answer Stefansson's letter without knowing whether there was any hope at all for the planned expedition. It would be better to make a final decision and so inform Stefansson than to 'go on as at present.'[58] But if the decision went against Stefansson, the explorer's subsequent course would undoubtedly 'be decided entirely on personal and selfish grounds.' It might be reasonable to believe that (as Christie had suggested) pressure from the British government would restrain a small country like Denmark, but what about the more formidable United States? Stefansson, with his genius for publicity, could go to American press baron William Randolph Hearst for sponsorship. Unscrupulous, money-grubbing American journalists would gladly stir up jingoistic sentiment by casting the situation as a dramatic race between an American and a Briton to discover and claim new land in the Beaufort Sea. The 'real point' was 'whether Canada shall guard every possible danger or whether we shall gamble.' A discouraging answer to Stefansson's letter could set off a dramatic and undesirable chain of events.[59] The implication was clear: Lougheed must be pressured to reverse his decision.

In response to Harkin's determined bullying, Cory drafted a memo of his own. He reminded Lougheed that in February the minister had 'agreed to this advance party being sent up with a view to guaranteeing Stefansson remaining loyal to us until we had succeeded in establishing our police posts in the North.' Was Lougheed now against the entire expedition, or only the 1922 portion? If the advance party was still to go, there would be much work to do in preparation. Cory asked Lougheed to 'advise me definitely as to your views.'[60] However, Cory lacked Harkin's impenetrable self-assurance, and the memo was never sent. Instead, the deputy minister apparently did nothing at all, leaving Harkin to pursue a reluctant policy of drift and equivocation.

When Stefansson wired Harkin on 23 March to recommend American scientist Elmer Ekblaw as a member of the expedition, Harkin replied evasively that only Canadian personnel could be considered. Three days later, he finally answered Stefansson's 14 March letter. He claimed that it had been referred to him only because Lougheed was temporarily away from Ottawa. Harkin assured Stefansson that the minister would 'appreciate the high ground you take' on the salary issue; nevertheless, 'you should not absolutely turn down the [1922-23] lecture contract you refer to. You must keep in mind the exceedingly unsettled conditions in Canadian affairs and the possibilities I pointed out to you ... of changes which might interfere with the proposition as we discussed it.'[61] Stefansson's reply, a collect telegram from Galveston, Texas, did not come until two weeks later. According to Stefansson, his mail had been forwarded to the wrong address. 'Referring your letter it is now too late to retain lecture contracts they were cancelled months ago,' he cabled. 'I assume absolute good faith on part officials now in power and am willing to take chances on change of administration stop Knight should soon proceed Ottawa have you selected any men to go north ... please wire summary present situation Queen Hotel Beeville Texas.'[62] It is difficult to imagine a request that would have been more unwelcome to the increasingly beleaguered Harkin.

5
Rasmussen in London

Earl Curzon of Kedleston ... would be glad if the Government of Canada could be immediately informed by telegraph of the facts of the case, and requested to furnish an assurance by telegraph that the [Danish] expedition will be allowed to land and pursue its investigations without interference.

– *J.D. Gregory to the Under-secretary of State for the Colonies, 9 June 1921*

The 'present situation' was far more complex than Stefansson could possibly have suspected. Harkin was uncomfortably aware of many developments on both sides of the Atlantic that would have deeply alarmed and angered him. Not only was Shackleton's campaign for Canadian support progressing well, but in mid-March Rasmussen had met with Canadian and British officials in London. As a result, the British government and the Hudson's Bay Company looked favourably on the Danish explorer's plans.

Rasmussen knew that he could not hope to make a journey through the Arctic archipelago without assistance. Therefore, early in February his associate Ib Nyeboe had contacted the British legation in Copenhagen. Nyeboe announced Rasmussen's plans and asked for the addresses of the Hudson's Bay Company and the Royal Canadian Mounted Police.[1] While this request was still wending its way through the bureaucratic labyrinth (it did not arrive in Ottawa until 16 May), the Danish foreign ministry sent a note to Lord Curzon, informing him that the British note of 7 September 1920 had been 'the subject of serious consideration and of various discussions with Mr. Knud Rasmussen.' As a result, Rasmussen had taken steps 'which will in future make it quite unnecessary for the Eskimos to hunt musk oxen on Ellesmere Land.' Rasmussen would be in London shortly, and he wanted to discuss the matter with the British authorities.[2]

An interview was accordingly arranged with Sir Henry Lambert of the Colonial Office, without any notice to or consultation with Ottawa. Indeed, the first news of the planned meeting was sent to the Canadian capital by Scotland Yard rather than by the Colonial Office. Sir Basil Thomson informed RCMP commissioner A. Bowen Perry that Rasmussen had recently left Copenhagen, 'travelling on an official mission ... regarding the question of Hudson Bay esquimos [sic]. No doubt, you will be informed direct by the Canadian authorities as regards the result of these conversations.' There was, Thomson added, no definite indication as yet that Rasmussen's expedition was backed by the Danish government, but inquiries were still proceeding.[3]

The Colonial Office advised Rasmussen that he should approach the Canadian High Commission in London, which he did on 12 March through a letter written at the Hotel Russell in Bloomsbury. On 15 March Rasmussen and Nyeboe met with William Griffith at Canada House. Griffith, who had served as secretary there since 1903, was described by the high commissioner, Sir George Perley, as 'very attentive and capable' but 'occasionally rather arbitrary.' Perley felt that Griffith's occasional failings should be excused, for 'it takes a special lot of patience to talk to all the people who come to this office with all kinds of difficulties and grievances.'[4] Griffith reported by telegram to Ottawa that Rasmusssen said he intended to undertake a new expedition, financed partly by the Danish government and partly by private citizens. His party would travel from northern Greenland by way of Baffin Island and the northern coast of the continent as far as the Mackenzie River, then return by sledge through the archipelago. Rasmussen would donate the specimens and artifacts he collected to universities in Scandinavia and Canada. Griffith asked for a written outline of these plans.[5]

Griffith then obligingly arranged a meeting between the two Danes and Charles Sale, the deputy governor of the Hudson's Bay Company. Sale noted in a record of their conversation that Rasmussen 'promised to address a letter to us within a few days giving references and guarantees which he feels sure will prove his good faith.'[6] It was tentatively agreed that Rasmussen would be given credit at the HBC posts, with the proviso that the Danish government must guarantee that it would pay his debts if he defaulted.[7] The meeting left Sale 'satisfied in my own mind that Mr. Rasmussen has no object in view beyond that of scientific exploration.'[8] It is clear from Sale's record and his letters to Griffith that Rasmussen and Nyeboe had made an extremely favourable impression on him.

On 16 March Rasmussen called on Sir Henry Lambert, who was evidently well informed about Canada's sovereignty fears, perhaps through conversations with Loring Christie. As Lambert reported to Griffith, 'Mr. Rasmussen's attitude this morning certainly used no language contesting Canada's unrestricted Dominion of Ellesmere land. I myself spoke deliberately on the basis that it was entirely Canadian, and neither he nor Mr. Nyborg (if I caught the name rightly) who was with him directly or indirectly demurred.'[9] Rasmussen stated that he had been unable to import reindeer skins during the war, but he could now do so without any difficulty. It would therefore be unnecessary for the Inughuit to kill muskoxen for their skins. As for the meat, Rasmussen felt that killing for food should be allowed in cases of dire necessity. He undertook to notify the Canadian government whenever this happened.[10]

Because the High Commission did not yet officially report to the Department of External Affairs, Griffith's communications on this matter were addressed to the Department of the Interior. His initial account of the 15 March meeting was dispatched to Lougheed by cable on the 16th; Sale's memo was enclosed in a letter to Cory, sent on 30 March. However, Lambert's important letter was not forwarded to Ottawa until 28 April, when Griffith provided a copy in another letter to Cory. Christie, Sir Joseph Pope, and Prime Minister Meighen therefore remained unaware that, in Lambert's expert opinion, Rasmussen posed no threat to Canadian sovereignty.[11]

The response to Griffith's 16 March telegram was drafted by Harkin and approved by Cory. Harkin thought that Griffith had done well to ask Rasmussen for something in writing, and that the explorer should be required to submit 'a formal application setting forth in detail the points he wishes to visit and the approximate dates on which he expects to arrive at such points.' Perley and Griffith were instructed to read the papers on the sovereignty issue previously sent for Christie's use, and to 'get some recognition ... of our authority' from Rasmussen.[12]

At about the same time, a news item was forwarded to Commissioner Perry by Scotland Yard. According to an article circulated by wire from Copenhagen on 21 March, Rasmussen had discussed his plans 'with the Canadian Government's representative and the Hudson['s] Bay Company' and as a result, the government had agreed to assist his expedition.[13] Rasmussen was a man of great personal charm, accustomed to getting his way by a potent combination of modesty, quiet humour, and effective showmanship. As his friend Peter Freuchen affectionately remembered, Rasmussen was 'an artist in many lines' and 'had a way of doing things' that made him 'the most beloved man I have ever known.'[14] He was renowned for his extraordinary 'ability to get the trust of others.'[15] Aware that he had made friends for himself in London, Rasmussen was evidently taking it for granted that the help he required would be forthcoming. He therefore neglected to send a written outline and guarantees as he had promised.

In early April, Harkin's worries were increased by the return of Shackleton. The British explorer arrived in Ottawa on the 3rd and promptly gave a press interview. He informed reporters that his visit was of a purely personal nature, but he also 'hinted at the possibility' of discussions with the government. He would make no statement except that '[s]omething official' might 'come later on.' The length of his stay, he added, was 'uncertain.'[16] On the 5th Shackleton wrote to Meighen, reminding the prime minister of the earlier assurance that he would be given financial help if he could obtain private backing. Shackleton estimated that he needed a total of $250,000. A friend in England, John Quiller Rowett, had promised $25,000 and, as a result of Shackleton's shipboard acquaintance with Lady Eaton, Sir John Eaton had now pledged $100,000. His British supporters had no objection to Shackleton's resolve that the 'whole character of the expedition' would be Canadian. Shackleton intended to hire young Canadian scientists and to turn over his scientific collections to the Canadian government. He asked Meighen for a prompt decision so that he could plan for a June departure.

Finally, Shackleton enclosed a copy of the friendly letter Stefansson had written to him in London the previous year. Earlier, Shackleton had claimed that Stefansson's letter proved he had no further Arctic ambitions; now he stated only that it had been written after Stefansson mentioned in conversation that he would not be going north again. Shackleton presumably hoped that the prime minister would not notice the shift, or, if Meighen did notice it, that he would not question Shackleton's word. Shackleton concluded with the observation that even if Stefansson did decide to lead another expedition, 'the shores and the seas are wide, and there is room for all.'[17]

On 9 April a meeting was held in Commissioner Perry's office. Shackleton, Christie, and Harkin joined Perry in a discussion of the two rival plans. Christie was by now a strong advocate of Shackleton. His aversion to Stefansson was well established, and he may also have been influenced in Shackleton's favour by John Bassett of the *Montreal Gazette*. Bassett and Christie seem to have been on very friendly terms (a letter to Christie from

Bassett begins, 'My dear Loring'); Bassett and Shackleton had also struck up a friendship. Like Shackleton, Bassett was Irish-born. He had arrived in Montreal as an almost penniless immigrant in 1909. He quickly made his way as a reporter, becoming the Conservative *Gazette*'s Ottawa correspondent in 1910. Marriage to the daughter of an Ottawa lumber baron gave him access to the city's social elite and helped to advance his career. In 1918 he was appointed vice-president of the Gazette Publishing Company. Bassett assiduously cultivated friendships with politicians and senior civil servants; he and Meighen would remain on close terms for life.[18]

Harkin must surely have remonstrated, but despite whatever protests he made, the group in Perry's office sketched out what they called an 'alternative plan.' Shackleton would be given a government grant, half to be paid in 1921 and half in 1922. There would be no need to send the *Arctic* at all. Shackleton would buy the stores already ordered for the *Arctic* from the government. He would transport the police, along with their equipment and supplies, to southern Ellesmere, where only one post would be established. Shackleton would bring additional supplies to the police post in 1922. His venture would be known as the Eaton-Shackleton Canadian Arctic Expedition. A memo outlining the plan (likely written by Christie) was sent to the prime minister, who in turn placed it before the cabinet.[19] Shackleton then departed for a weekend visit to the Eatons in Toronto; a few days later, he boarded the *Aquitania* in New York for his return home. Immediately after his arrival in London, Shackleton wrote to Christie on Chateau Laurier notepaper: 'Just a personal line to thank you for all the trouble you are taking[.] I shall anxiously await news ... When the Government cables or writes let them make it clear and definite including the amount.'[20]

As Harkin faced this new development and anxiously awaited Stefansson's reply to the discouraging letter he had sent on 26 March, he received a news clipping, sent by Craig with a brief note: 'Mr. Harkin – Have you seen this?' The clipping contained an announcement made in Copenhagen on 9 April regarding the Fifth Thule Expedition: the Danish government would contribute 100,000 kroner (about Cdn$18,000, and roughly one-third of the estimated total cost of the expedition).[21] Three days later Stefansson's telegram from Galveston arrived.

Harkin did not reply to this unwelcome query. Clearly, he preferred to wait until the ministers' decision had been made before he wrote back to Stefansson. If the new plan was rejected, a report on the 'present situation' would be comparatively easy to write. But to Harkin's dismay, the issue was debated fruitlessly for weeks. (Unfortunately, there are no records of these discussions.) On 18 April he received a letter from Lorne Knight, asking for 'as early notice as you can give me' as to when he should travel to Ottawa.[22] The next day brought another telegram from Stefansson: 'Knight should soon proceed Ottawa have you selected any men to go north.'[23] Harkin sent the two documents to Cory, pointing out that 'a message of some kind should go to Mr. Stefansson but of course in view of the uncertainty of the situation with respect to Northern matters I cannot send any reply without instructions from you.'[24] Cory could only answer, 'So far we have not received the decision of [the cabinet].'[25] On 24 April Stefansson wired directly to Cory, with no result. On 4 May Harkin reminded the deputy minister once again that inquiries from both Stefansson and Knight remained unanswered. '[I]t seems to me imperative that an immediate decision should be reached,' he wrote in evident exasperation. Cory's answer was a single word: 'Wait.'[26]

DURING THIS PERIOD OF frustration and delay, the announcement that the Danish government would contribute financially to Rasmussen's expedition sparked an exchange of letters between Harkin and Christie. In Harkin's eyes, the announcement was clear proof that Rasmussen's venture was now an officially sanctioned one, capable of making valid territorial claims on behalf of Denmark. He expressed his worries in a letter to Cory, and suggested that 'some arrangement should be made by which when Rasmussen first touches Canadian territory he should be compelled to pay duty on his supplies and otherwise submit to acts of administration on the part of Canadian authorities.' Cory seems to have been entirely disinclined to make any decisions about, or take any responsibility for, Arctic matters at this point. Instead, he suggested that Harkin do what ought to have been done long before: 'Take [it] up with Christie.'[27]

Harkin accordingly forwarded the clipping and other related documents to External Affairs. Christie replied that a wire would immediately be sent to London. With Meighen's approval, the governor general's office dispatched a message to the colonial secretary: 'Last year Rasmussen and even Government of Denmark showed a disposition to question our sovereignty ... Canadian government request that telegraphic inquiry be made through British Legation at Copenhagen in order to ascertain from Government of Denmark information concerning this proposed expedition, and what its precise object is.' The message noted that Ottawa had 'no objection to an expedition for purely scientific purposes,' but observed:

'It would have seemed more appropriate if Government of Denmark had approached Canadian Government before embarking upon such a proposal. My Ministers would be grateful if you could procure and telegraph this information at an early date.'[28]

This telegram roused the sluggish Colonial Office to action. In response to the earlier requests for information from the High Commission and the RCMP, dispatches about the Rasmussen expedition had already been sent by Sir Charles Marling in Copenhagen on 15 and 19 April. Marling's first report enclosed a copy of Nyeboe's 3 February letter to the British legation; the second recounted a conversation with explorer Einar Mikkelson, who spoke highly of Rasmussen's 'ability and integrity' and asserted that the planned expedition was 'entirely scientific.' On 28 April Marling forwarded an outline of Rasmussen's plans as announced to the Danish Geographical Society a few days before. On the 29th the new colonial secretary, Winston Churchill, replied to the governor general with copies of Marling's three reports and a clipping from *The Times* about Rasmussen's talk to the Danish Geographical Society. Churchill also gave an account of Lambert's 16 March meeting with Rasmussen and Nyeboe, emphasizing Lambert's conviction that the expedition's aims were scientific and that there would be no challenge to Canadian sovereignty. Unfortunately, this important information was sent by ship rather than by telegram.[29]

Also on 29 April (and probably also on Churchill's orders), Lambert wrote to the Foreign Office with a request for more information from Marling. At 6 p.m. the following day, a cipher telegram from Curzon marked 'No Distribution' was sent to Copenhagen: 'Canadian Government are desirous obtaining advance notice of Mr. Rasmussen's movements, the points to be visited and approximate date of arrival at such points. Please ascertain and report to me by telegraph.'[30] Earlier that day Marling had reported by letter that, although the Danish government was giving 100,000 kroner to the expedition, Rasmussen had 'no official status.' Most of the funds would be provided 'by private persons interested in exploration, who may have formed themselves into a small society but without any commercial object in view.'[31] This letter was received in London on 5 May. Because Curzon's telegram asked only for more information about Rasmussen's route, Marling's response by wire on 1 May was merely a more detailed account of the presentation to the Danish Geographical Society. He did not repeat the assurances about Rasmussen's lack of official status contained in his letter of 30 April. Copies of Marling's telegram and his letter were sent to Ottawa by mail on 6 May.

Christie's efforts had gone a considerable way towards clarifying the situation, but even though the British officials concerned in Copenhagen and London were now fully convinced that Rasmussen posed no threat, some of the most crucial information had not yet reached Canada. In Ottawa, the main focus of official attention in late April and early May was on Shackleton and on the new possibility that the Hudson's Bay Company might be the answer to Canada's sovereignty dilemma. On 20 April Shackleton reported from England that, having inspected his ship for the first time, he now realized he could not carry the men and stores for even one police post. The ship was a small (125 ton) wooden vessel bought in Norway by Shackleton's British supporters at a cost of £11,000 (approximately Cdn$55,000) and renamed the *Quest.* Much as Stefansson might have done, Shackleton tried to bluff his way through this difficulty. He sent a telegram to Bassett, asking him to '[t]ry [to] obtain at least fifty thousand dollars as government contribution without stipulation to carry their party.' Or, if the government would give him $100,000, he would charter a second ship for the police party. A third possibility was that the government should give Shackleton a grant, the supplies already purchased for the *Arctic,* and authority as a magistrate. Shackleton would winter on Ellesmere, with his presence replacing the police post.[32]

Bassett apparently replied on his own initiative, proposing that Shackleton should establish a police post in 1921 and postpone his exploration work until 1922. Shackleton promptly agreed, and he seemed to regard Bassett's suggestion as a definite Canadian commitment. Bassett unwisely forwarded Shackleton's cables to Christie and Meighen, along with the comment that Shackleton 'ought to get a definite reply now.'[33] The cables gave the strong impression that Shackleton was unreliable and self-serving: for all his confidently proclaimed plans, the Antarctic hero had proved incapable of providing adequate transportation to the North. Now it seemed that he was conniving to get as much as possible out of the government. As a result, Meighen turned strongly against him. On 9 May the prime minister coldly informed Shackleton that the Canadian government had 'made no definite commitment whatever ... Our arrangements now do not admit of assistance [to] your expedition.'[34] Shackleton's pleas for a reversal of the decision were firmly rebuffed.

However, Meighen's decisive rejection of Shackleton did not mean that the government was now prepared to embrace Stefansson. On 3 May the prime minister had become aware of a new possibility – that the Canadian flag could be carried north at no expense to the government by the

Hudson's Bay Company, which was in the process of expanding its activities into the archipelago. The previous autumn, Stefansson had spoken to Edward Fitzgerald, the deputy chairman of the HBC's Canadian advisory committee, about the government's sovereignty concerns. On 16 April an employee in the HBC's Winnipeg office noticed a news story about Rasmussen's proposed expedition. Company officials therefore decided that Ottawa might be interested in their plans. On 25 April Harkin received a letter from Angus Brabant, the fur trade commissioner, informing him that 'a reconnaissance visit to Ellesmere Land' might be made that summer, when posts were to be established on Baffin Island. Harkin duly passed this information to Cory and Christie, but the prime minister had no inkling of it until George Allan, the member of Parliament for Winnipeg South, forwarded a similar letter from Fitzgerald.[35] Meighen's response was immediate. 'I am glad indeed to have the information of the intentions of the Hudson's Bay Company and I assure you it is most opportune,' he wrote to Allan on 4 May.[36]

Brabant was promptly summoned to Ottawa. On 12 May he met with Harkin and Cory in the morning and then again with Harkin in the afternoon. Harkin's report was discouraging. The HBC, reasonably enough, was not eager to establish a post on Ellesmere, where there was no one to trade with. 'Besides,' Harkin noted, 'the Company's officers did not seem to care to accept responsibilities in connection with the establishment of stations on Ellesmere Land with possibilities of the Danes being there and inevitable complications.' Harkin recommended that the *Arctic* be sent north in June, with orders to establish police posts on Bylot Island and Ellesmere. In September an HBC officer could take over the Bylot post, and the police could return south in the HBC ship, thus reducing the overall cost of the operation. As for Stefansson, 'To eliminate him now would probably appear to him to be an act of bad faith on the part of the Government ... it is important that an immediate decision be reached because matters of personnel and supplies will necessarily take some time to arrange.'[37] At the next cabinet meeting on 18 May, Lougheed – now, apparently, once again on Stefansson's side – pressed strongly for a decision.[38]

The cabinet decided that the HBC presence would be enough. (The traders were appointed to act as customs inspectors in order to give them official status.) Both Stefansson's exploring party and the voyage of the *Arctic* were cancelled. Stefansson later claimed to have been told that the ministers were evenly divided between his supporters and Shackleton's and therefore preferred to send no expedition at all.[39] However, not only is this explanation improbable in itself, but Shackleton's proposals had been dismissed nine days before the cabinet's decision was made. Despite the elimination of Shackleton, the rejection of Stefansson's plan was nearly inevitable, given Christie's attitude and his influence on Meighen. Even Harkin must have known that he was fighting a losing battle. But the decision not to send the *Arctic* came as a complete surprise to many of those involved. 'There is no use trying to write what I think about it,' was all Craig could say when he instructed Captain Pickels to discontinue preparations for the voyage.[40]

No record was made of the reasons behind the decision, but it appears that Meighen and the majority of his ministers were reassured by the reports from London that Rasmussen did not contest Canadian sovereignty. A copy of Lambert's 16 March letter to Griffith had at last arrived in Ottawa on 14 May; Churchill's 29 April letter to the Duke of Devonshire followed on 16 May. These tardy accounts of the meeting between Lambert and Rasmussen almost two months earlier were likely the factor that tipped the balance. Christie later told Borden that the final decision was taken 'as the result of secret information that made the whole affair seem less urgent than it had originally appeared, and that being so, it was felt that in a time of financial stringency the proposed expenditures might well be avoided.'[41]

Harkin was not to be converted to this view. He remained highly suspicious of Danish motives, in part because of a recent letter from the Danish consul in Montreal to the minister of marine and fisheries, asking for any Canadian maps that might be of use to Rasmussen. The consul, Otto Nobel, innocently mentioned the Danish government's financial contribution to the expedition as proof of its serious nature.[42] 'This undoubtedly means that it is an official expedition and as such under International Law would no doubt take steps which would be of real significance in regard to any territorial dispute,' Harkin concluded.[43]

To further exacerbate his fears, Harkin received a preliminary report from Hensley Holmden that emphasized (as W.F. King had done in his 1905 report) that the 1880 transfer of sovereignty from Great Britain was extremely vague about the territories being transferred (see Chapter 3). Holmden had scoured the records available in Ottawa without finding any reason for the vagueness. He speculated that the British themselves had doubted the validity of a title based only on discovery. As Harkin pointed out to Christie, it seemed that Canada could not rest its sovereignty claims on the transfer alone. Instead, a 'special effort' would have to be made to 'extend and increase' the earlier acts of possession.[44]

On 26 May Harkin wrote a rambling, nine-page memo to Cory. He strongly urged that the cabinet's decision should be reversed and the police posts established. The *Arctic* was nearly ready; $55,000 had already been spent on it and could not be recovered. For only $26,000 more, the expedition could proceed. Harkin admitted that Rasmussen himself might not contest Canada's jurisdiction, but this stance was not 'in any way binding on Denmark.' There was 'no positive evidence yet that Denmark is engaged in operations in that connection,' but 'the attitude of Denmark right up to the present continues to suggest suspicions. There has been no repudiation by Denmark of the "No Man's Land" contention though no re-assertion of it. Denmark seems to have adopted a policy of evasion and delay.' According to Harkin, Rasmussen too might be carrying out evasive manoeuvres; most significantly, he had not yet submitted the promised outline of his plans. Harkin did not want Rasmussen to arrive in the Arctic and find 'that Canada is not administering the north islands. Should Denmark wish to raise the question later on his evidence will be available.' Harkin also forwarded a clipping from the American magazine *The World's Work* about a proposed American expedition involving navy personnel. The article was surrounded with marginal annotations by Stefansson. The explorer declared that the Americans were 'after the lands north of Canada, because they want them as territorial possessions...But I can beat them to it if Canada wants me to. It is optional with us if it is our territory or theirs.'[45]

None of these arguments carried any weight with any of Harkin's superiors except Cory. A last-ditch effort by Harkin and Cory to change Meighen's mind was dropped after an exchange in the House of Commons on 30 May.[46] The members of the Opposition, although not fully informed about the purposes of the *Arctic*'s proposed voyage, had been given some inkling of the matter when Parliament was asked to approve the funds. In addition, many of them had heard rumours that an unspecified foreign power harboured territorial ambitions in the North. Questions were therefore asked by William Lyon Mackenzie King and others. Meighen replied firmly that there was good reason to believe 'that no great danger would ensue if the expedition were deferred' for 'at least another year.' Cabinet had decided that under the circumstances, 'the expense would not be justified for the present.' The prime minister assured his listeners that Canada's Arctic interests were 'not imperilled by any other country.'[47]

Between 30 May and 4 June (when the parliamentary session ended), there was no possibility of approaching either Meighen or Christie, who were both deep in preparations for the upcoming Imperial Conference in London. Harkin conceded defeat on the Stefansson expedition, and he sent the explorer a blunt cable – 'entire expedition abandoned' – on 30 May.[48] But when it came to the police posts, Harkin was determined to fight on. His determination was shared by others in the Department of the Interior, and it was largely due to their unyielding attitude that the *Arctic* went north a year later. And, once Meighen and Christie had withdrawn from active involvement in Arctic matters, Harkin's stubborn suspicion of Danish motives nearly resulted in the cancellation of the Fifth Thule Expedition.

THE DANES HAD, IN fact, made a further reply to the Canadian 'protest' on 13 May, and additional reassuring bulletins from Marling had been received in London. But because the matter was not considered especially urgent there, the news was not cabled to Ottawa. The Colonial Office had noted that Marling's telegram of 1 May could not be considered a full response to Ottawa's request for information. After a slight delay caused by the Foreign Office's assumption that Marling's letter of 30 April would be sufficient to allay Canadian concerns, another telegram was sent to Copenhagen on 9 May: 'Canadian Government, as you are aware, have asserted their sovereignty over the Arctic Islands, but last year Rasmussen and even the Danish Government showed a disposition to question such sovereignty in respect of Ellesmere land. This appears to the Canadian Government to be borne out by the fact that Rasmussen expedition is officially supported by Danish Government and that the latter did not approach them before embarking on such a scheme. May we re-assure Canadian Government?'[49] Marling unhesitatingly based his answer on the realities of international politics rather than on the fine points of 'International Law.' He replied by cable two days later: 'I think desired assurance may be given as it would be a direct contravention of declared policy of present cabinet of friendship towards British Empire for Danish Government to attempt such a step as regards Ellesmere Island.' Marling expressed his willingness to approach the Danish foreign minister and asked for instructions on this point.[50] A copy of his cable was sent to the Colonial Office marked 'Urgent'; however, no reply was forthcoming from Churchill until 25 May.

A note received by the Foreign Office from Henrik Grevenkop-Castenskiold on 13 May provided more assurance of Denmark's good will. Having made a formal declaration of their sovereignty over all of Greenland on 10 May, the Danes now offered a guarantee that the Inughuit would not kill muskoxen in Canadian territory except in cases of dire need. Such cases would be reported to the British government. 'I venture to express the hope on behalf of my Government that these measures will satisfy the wishes of His Britannic Majesty's Government,' Grevenkop-Castenskiold wrote.[51] So far as the Foreign Office was concerned, the question had indeed been resolved, and it was now clearly unnecessary for Marling to speak to Danish officials in Copenhagen.[52]

Unfortunately, Ottawa remained in ignorance of this important new development. Marling's cable of 11 May and the Danish note were not forwarded by the Colonial Office until 25 May. Marling's 30 April letter had already been sent on 10 May. With the lack of concern over timely communication that was characteristic of the Colonial Office, all these documents were sent by ship rather than by telegraph. To Harkin, therefore, the situation appeared as dire as ever. Despite the moderate and realistic stand taken by Christie, Meighen, and the cabinet, Harkin's indirect influence now almost put a stop to Rasmussen's plans.

Angus Brabant had returned to Winnipeg after his conversations with Harkin firmly under the impression that Rasmussen's proposed journey was 'solely [for the] purpose [of] advancing Danish claims' to Ellesmere Island. The HBC's Winnipeg office accordingly cabled to the head office in London that all support for Rasmussen should be withdrawn.[53] Brabant informed Harkin about the cable five days after it was sent.[54] Five more days passed before Harkin addressed a memo on the subject to Cory. 'So long as the Government provides officers who will be on hand to make Rasmussen conform to Canadian laws and regulations and thereby automatically admit that the North is Canadian territory there is no particular reason why anything should be done to prevent Rasmussen taking his contemplated trip,' Harkin wrote, in what may have been a final plea for the *Arctic* to be sent. 'I agree,' Cory noted in the margin.[55] It appears that both men then dismissed the matter from their minds; at least, neither of them took any further action.

Fortunately for Rasmussen, the matter was dealt with in London by Charles Sale, who acted promptly and vigorously on his behalf. 'Have received cables from Canada which very seriously affect your expedition we can only explain details in person therefore suggest you visit London immediately,' Sale telegraphed. Sale also informed the Winnipeg office that, in his opinion, the Canadian authorities were 'unduly alarmed.'[56] Busy with last-minute details, Rasmussen was not able to comply with the request at once. He did not arrive in London until 4 June, almost two weeks after he received Sale's message. He expected to leave again two days later.[57] After the situation was explained to him, a shocked Rasmussen reported to Jens Daugaard-Jensen: 'You were right when you said that we must always expect things to go wrong at the last minute. The reason for my being called to London was nothing less than that the Canadian Government does not want me in the region my expedition is to visit because ... people have got the idea that [it] has as objective to take over lands lying within the area of Canada's interest. So I cannot return home until the matter is settled.'[58]

Without the HBC's support, the Fifth Thule Expedition's chances of success were low. Rasmussen's ship, the *Søkongen,* had already sailed from Copenhagen on 27 May. He and his colleagues Therkel Mathiassen and Kaj Birket-Smith were scheduled to follow on the vessel chartered by the Greenland Administration for the various dignitaries attending the celebrations to mark the bicentenary of Hans Egede's voyage (3 July). The ship would leave on 18 June and could not possibly be delayed to accommodate Rasmussen. Time was therefore of the essence.

An obviously agitated Rasmussen dispatched a note to Griffith at the Canadian High Commission. Not altogether plausibly, he explained: 'I very much regret if I misunderstood you when I was here last. I was brought to understand that you wanted a memorandum on the question regarding to the musk-oxes sent over, and that this memorandum was to be sent to Sir Henry Lambert. This was sent through the Danish foreign office, according to my report, given just after my return.' Rasmussen's English, usually adequate though far from fluent, then descended into near unintelligibility. 'Late different circumstances were the cause of my plan of the expedition first now is definitely drawn up,' he wrote, leading Griffith to suspect for a time that Rasmussen's failure to submit his plans might be part of a calculated deception. However, there was nothing ambiguous about Rasmussen's concluding statement: 'I hope you will understand I am only coming as a friend of Canada and that you will assist me in getting out of these different misunderstanding[s].'[59]

Rasmussen called on Griffith later that day, and the two men proceeded to meet with Lambert at the Colonial Office. There Rasmussen presented the long-delayed outline of his plans, with a covering letter in which he stated that his expedition was 'of a purely scientific character, without taking any interest whatever in political or mercantile matters.'[60] He promised that the Danish Ministry of Foreign Affairs would confirm his statement. According to Rasmussen, Griffith and Lambert then 'declared themselves quite satisfied,' and they seemed certain that everything would proceed smoothly as soon as an official confirmation was received from Copenhagen.[61]

Sale telegraphed Fitzgerald in Winnipeg, who in turn sent a cable to Lougheed, requesting permission to 'render Rasmussen any needed assistance.'[62] The cable was received in Ottawa on Sunday, 5 June, when Lougheed was out of town. Contacted by telephone, the minister reacted with confusion. Unsure whether the directive to withhold help was in fact an official one, he could only reply that Meighen and Christie should be consulted.[63] On Monday, Craig first spoke to Christie and the prime minister's secretary. He then searched through the files and found Brabant's letter to Harkin, 'which probably explains the matter.'[64] Lougheed returned to Ottawa on Wednesday; Meighen and Christie having left for London (they sailed from Quebec on board the *Empress of Britain* on 6 June), the matter was now entirely in his hands. He wired to Winnipeg that there was no objection to the Rasmussen expedition, and recommended that Rasmussen should obtain a letter from the high commissioner 'setting out the purposes for which he is going into the North.'[65] However, Lougheed failed to inform Griffith about this plan.

On 9 June Grevenkop-Castenskiold arranged a meeting with Curzon, at which he presented a note guaranteeing on behalf of the Danish government 'that no acquisition of territory whatsoever is contemplated.' Canadian permission for the expedition to proceed was requested. Curzon (a former president of the Royal Geographical Society with a keen interest in exploration) had shown no great concern over the muskox issue, but now he personally took up Rasmussen's cause in a characteristically decisive and imperious manner. Acting on the belief that the Canadian government had ordered the HBC to refuse assistance, he informed the Colonial Office that Ottawa should 'be immediately informed by telegraph of the facts of the case, and requested to furnish an assurance by telegraph that the expedition will be allowed to land and pursue its investigations without interference.'[66] The result was a 'Clear the Line – Urgent' telegram to Devonshire. 'I think the Canadian Govt will give no more trouble over this expedition,' minuted Sir John Dashwood, one of the third secretaries.[67]

In Ottawa, the telegram was promptly passed to Lougheed – who, however, did not reply to the Colonial Office as requested. Instead, he again cabled Fitzgerald in Winnipeg, repeating his assurance that the HBC could assist Rasmussen as long as the explorer did not dispute Canadian sovereignty.[68] Surprised and irritated by Ottawa's failure to respond promptly, Dashwood wrote: 'The Canadian Govt still seem somewhat anxious about this expedition and without any justification, since the Danish Minister has given a categorical assurance that it has no political or commercial object.'[69]

On Friday, 10 June, Rasmussen appeared at Griffith's office 'in great distress,' since it was imperative that he leave for Copenhagen the next day. Again, Griffith and Rasmussen went together to the Colonial Office, where Lambert was not at first entirely sympathetic. As Griffith recorded, Lambert took the opportunity 'to point out in polite terms to Mr. Rasmussen' that if he had submitted his plans in March, 'the matter would no doubt have been adjusted in time.' It was suggested that Griffith should provide Rasmussen with a letter for the HBC, but in the absence of instructions from Ottawa he did not feel able to do so.[70] Griffith did, however, send a cable asking for authorization, and he wisely chose Pope rather than Lougheed as the recipient.[71] Pope now took charge of the matter. Although not active in the formulation of policy, the under-secretary was an expert in protocol and correct procedure. He tactfully arranged with Cory for cables to be sent to both the High Commission and the Colonial Office.[72]

Immediately after the message from Ottawa arrived, Griffith asked Rasmussen to provide a written undertaking not to contest Canadian sovereignty. Griffith promised that as soon as he had received the document, he would 'communicate with the Hudson's Bay Company in order that you may receive the necessary facilities forthwith.'[73] Rasmussen promptly replied: 'I ... have no hesitation in giving you the assurance required by the Minister of the Interior.'[74] Griffith then sent a letter to Sale, asking him on behalf of the high commissioner to do what he could to help Rasmussen.[75] By the end of the day, the tangle of misdirected communications had at last been straightened, and Rasmussen was able to depart for Copenhagen as planned (just missing the opportunity to meet Meighen and Christie, who arrived in London by train from Liverpool two days later).

Rasmussen carried with him a letter of introduction, which asked HBC officers 'to afford him such facilities as he may require and we may be able to give, and to furnish him with such supplies as the Posts can provide.'[76] 'Have interviewed High Commissioner and Colonial Office everything satisfactorily settled Rasmussen expresses many thanks,' Sale cabled to Winnipeg.[77] To Rasmussen, Sale dispatched a friendly farewell: 'Delighted ... that your visit completely successful ... best wishes for the success of your great adventure.'[78] To Griffith, he expressed his gratitude for 'the very friendly manner in which you negotiated the matter with the various authorities concerned. It has relieved us of the possibility of a very embarrassing situation.'[79] Griffith, for his part, reported to Pope that Rasmussen had 'left a good impression on me. In the first place I thought perhaps his neglect to furnish me with a memorandum as suggested might have been calculated, but in the light of further events I have concluded that he was bona fide throughout, and that he is a man of great intelligence, capacity, and good character.'[80]

WHEN THE DANISH NOTES of 13 May and 8 June were at last received in Ottawa, Harkin grudgingly conceded that 'apparently' Denmark was 'definitely admitting British sovereignty in Ellesmere Land.' However, he was far from ready to accept the Danish guarantees at face value. The experienced diplomats of the Foreign Office were completely satisfied by Grevenkop-Castenskiold's categorical assurances, but Harkin disliked the idea of relying on the Danes to report breaches of the Northwest Game Act. 'It does not seem to me to be good policy for Canada to be dependent upon the actions of a foreign country for the observance of Canadian law upon Canadian territory,' he argued to Cory. '[A]cts equivalent to administrative acts by Denmark in connection with British territory ... possibly would really constitute evidence against us insofar as any other foreign country is concerned.'[81] Pope (who now seems to have been firmly in control of the issue) would have none of these arguments, but as a result of Harkin's lingering suspicions, Canada's answer to the Danish note was rather ungraciously worded.

The Colonial Office was informed that 'in the view of the Commissioner of Dominion Parks, who has given a good deal of study to this whole question, if the programme of the Danish authorities is carried out there will be no reason to fear sudden [depletion] of the musk-ox herds.' It was, however, 'specially important that Mr. Rasmussen should strictly adhere to the policy of refusing to buy or trade in musk-ox skins. It is not considered that there could be any objection to the provision that if any of the Cape York Eskimo are in danger of starvation they may provide food for themselves by killing sufficient Ellesmere Land Musk-oxen to meet the necessity of the case; provided the Danish Government carries out its undertaking to report all such cases to His Majesty's Government.'[82] The Foreign Office shortened this wordy communication, at the same time re-phrasing it in order to 'retain the politeness of language we endeavour to use in these formal communications.'[83] On 15 August, Grevenkop-Castenskiold was handed a note stating that 'His Majesty's Government have received a communication from the Dominion of Canada to the effect that compliance with the measures set forth in [the Danish] note [of 13 May] will meet the requirements of the Canadian Government, who point out that it is especially important that Mr. Rasmussen should strictly adhere to the policy of refusing to buy or trade in musk-ox skins.'[84]

The Danes did indeed observe, and instruct the Inughuit to observe, the Canadian law. Danish explorer Lauge Koch, who had gone north in 1920, hunted muskoxen near Fort Conger on the east coast of Ellesmere in the summer of 1921 with a party of Inughuit, obtaining 'great quantities' of meat. By 1922, he had been informed that such activities were forbidden on Canadian territory, and he obeyed the instructions from Copenhagen despite the relative scarcity of game on the Greenland side of Nares Strait.[85] To all appearances, the issue of sovereignty over Ellesmere and the other northern islands had now been resolved to the satisfaction of everyone concerned in Ottawa and London.

Harkin was nevertheless determined that the *Arctic* should go north in 1922, and he was still firmly convinced that the movements of the Fifth Thule Expedition must be carefully watched. Stefansson had perforce renounced all hope of a new five-year northern expedition sponsored by the government. However, he clung with astonishing tenacity to his plan for the occupation of Wrangel Island. As a result, there was a great deal more Arctic drama still to come. The next few years provided some of the strangest, and one of the most tragic, episodes in the history of Canada's Arctic policy. And, unlike the story of the supposed Danish threat, Stefansson's disastrous attempt to claim Wrangel Island for Canada would receive worldwide newspaper coverage, bringing Arctic sovereignty issues out from behind the closed doors of Ottawa offices and into the realm of public debate.

6
Wrangel Island

In my judgement, no more far-fetched claim could well be imagined, and any attempt to associate Canada with such fantastic pretensions could scarcely fail to prejudice us in the eyes of the world, besides weakening our legitimate claim to certain Arctic islands adjacent to our own territory, in respect of which we have a strong case.

– *Sir Joseph Pope to William Lyon Mackenzie King, 21 March 1922*

Of the three explorers who had hoped to lead expeditions to the Canadian Arctic, only Rasmussen was able to depart for the North as planned in the summer of 1921. He left Copenhagen on 18 June on board the *Bele,* along with archaeologist Therkel Mathiassen and ethnographer Kaj Birket-Smith. The other Danish members of the expedition – Peter Freuchen, Helge Bangsted, and Jacob Olsen – were already in Greenland. The group from Copenhagen arrived at Godthåb on 28 June, after a pleasant voyage in the relatively luxurious ship (it was reportedly the largest ever to sail to Greenland up to that time). Most of the passengers, including Rasmussen, disembarked at Godthåb for the bicentenary celebrations, which were held on 3 July. The ship continued north towards Thule with Mathiassen, Birket-Smith, and most of the Fifth Thule Expedition's equipment on board. On 14 July the *Bele* was wrecked near Upernavik. The *Søkongen,* with some assistance from the SS *Island* (which had carried the Danish royal family to Godthåb), was able to salvage part of the cargo. New equipment for the expedition was ordered from Denmark by telegraph and paid for by the government at a cost of 32,000 kroner.

The *Søkongen* called at Thule for the Inughuit members of the expedition, then sailed south once more to pick up the new supplies. A lengthy stay at Godthåb became necessary when several of the Inughuit came down with the influenza that ravaged Greenland during 1921. Freuchen's wife, Navarana, had already died of the disease at Upernavik, and now another Inughuk member of the expedition, the hunter Iggianguaq, fell victim to the epidemic. Freuchen grimly called this 'the third mortal loss' sustained before the expedition even started.[1] On 7 September the *Søkongen* at last set out for the archipelago. Because of the lateness of the season, it was impossible to make much progress through the ice-congested channels. The party landed on a small, unnamed island in Foxe Basin, which they dubbed 'Danish Island.' There they built a hut and settled down for a winter of scientific observations. The *Søkongen* departed on 24 September and returned to Denmark. It was essential that the expedition quickly make contact with the Hudson's Bay Company posts. Rasmussen led a small sledge party to Repulse Bay in November. There the Danes were very hospitably received by the post manager, a colourful, hard-drinking old whaler named George Washington Cleveland. During this trip Rasmussen had his first encounter with North American Inuit, and he was delighted to find that they could understand the Greenlandic dialect in which he was fluent.

In the spring Rasmussen set off to study the inland Inuit around Baker Lake. To his 'great surprise,' he soon saw a sledge coming towards him, driven by a white man. The stranger proved to be Constable Packett of the Royal Canadian Mounted Police, who was on his way to visit the station at Danish Island. 'It was strange to us to meet with police in these regions; and we were at once impressed by the energy with which Canada seeks to maintain law and order in the northern lands,' Rasmussen dryly commented in his narrative of the expedition.[2] Remembering the events of the previous year, he may well have felt a twinge of apprehension. However, Packett turned out to be a pleasant and reasonable young man. He quickly agreed that there was no need for him to question Rasmussen about the expedition, provided that the Danes would leave a report at Chesterfield Inlet for him to read when he returned. But when they suggested that it was not necessary for him to physically inspect the headquarters at Danish Island, Packett politely demurred. His instructions were to visit Danish Island and make a report on what he found there, so to Danish Island he must go.

This encounter was only the first of many meetings between the Danes and the Canadian police. As with Rasmussen's troubles in early June 1921, the surveillance was the result of Harkin's refusal to accept the Danish assurances of good faith as genuine. Harkin had been outraged by the news (released after the *Søkongen*'s return to Denmark in the fall of 1921) that the expedition had named an island under Canadian sovereignty without first requesting permission from Ottawa. That the name chosen was Danish Island added fuel to the fire. Just after the New Year, Harkin suggested to W.W. Cory that the expedition should be closely watched by the police, in order to ensure that Canadian game laws were observed and that there was no illegal trading with the Inuit. A request was duly made to Commissioner Perry, who promptly replied that the matter was 'receiving immediate attention, and every assistance will be given by this Department.'[3]

SHACKLETON AND HIS MEN had also shown great determination in the face of repeated setbacks. When Shackleton received the news that the Canadian government would make no contribution to his expedition, the *Quest* had already been bought at a cost of £11,000, and one of his men was in Winnipeg negotiating for the purchase of a hundred sledge dogs. At this critical juncture, his friend John Quiller Rowett came forward with an offer to fully finance the expedition. A further donation was made by paper manufacturer Frederick Becker. However, by the time these pledges were made it was too late to do any substantial Arctic work before winter set in. Shackleton could not bear to delay his departure until the next year. 'I feel I am no use to anyone unless I am outfacing the storm in wild lands,' he had written to his wife a few years earlier, and in 1921 this feeling was even stronger.[4]

Shackleton resolved to go south instead of north. His hastily formulated plans, announced in the press on 29 June, included oceanographic work in the South Atlantic, scientific observations in the Weddell Sea, and mapping the coastline of Enderby Land. From Canada, John Bassett's paper, the *Montreal Gazette,* lauded the enterprise as part of the great tradition established by Sir John Franklin and the other heroes of British Arctic exploration. It was just over a hundred years since the nineteenth-century search for the Northwest Passage began, and so 'century is linked to century by the exploits of daring men ... Men such as Sir Ernest Shackleton are needed. They are the salt of the race.'[5] Perhaps Bassett had prepared this piece for the departure of the northern expedition on which Shackleton was now never to embark.

The *Quest* sailed from London on 17 September 1921. It soon became clear that the refitting of the ship had been of poor quality. 'Anxiety has been probing deeply into me ... Engines unreliable; furnace cracked;

water short; heavy gales; all that physically can go wrong,' Shackleton wrote in his diary.[6] On 4 January 1922, the *Quest* reached the harbour of Grytviken on South Georgia, a remote island in the South Atlantic. It was a place Shackleton knew well from the *Endurance* expedition.

South Georgia was the scene of what many considered to be Shackleton's finest moment. After the *Endurance* was crushed by ice in the Weddell Sea, Shackleton and his crew had made their way to Elephant Island, dragging the ship's boats over the ice. From there, Shackleton and five other men had sailed in one of the boats through the world's stormiest seas to South Georgia. On 10 May 1916, they landed on the uninhabited west side of the island. Shackleton and two of his men, Tom Crean and Frank Worsley, crossed the mountainous interior to a whaling station on the east coast – a feat never before accomplished. 'Who are you?' demanded the amazed station manager. He had met the explorers before but did not recognize the bearded, emaciated figures in front of him. Shackleton later wrote of his feeling that providence had guided them on their trek over the mountains. As he recounted, 'it seemed to me often that we were four, not three. I said nothing to my companions on the point, but afterwards Worsley said to me, "Boss, I had a curious feeling on the march that there was another person with us."'[7] T.S. Eliot later immortalized this incident in *The Waste Land:* 'Who is the third who walks always beside you? / When I count, there are only you and I together / But when I look ahead up the white road / There is always another one walking beside you.'[8]

During his first night back on the island in January 1922, Shackleton died of heart failure at the age of forty-seven. His men took the body to Montevideo, but at Lady Shackleton's request Shackleton was returned to South Georgia and buried there. His loyal subordinate Frank Wild then took command and carried out a part of the planned scientific program. The *Quest* returned to England in September 1922. It fell to Wild to recount the expedition's adventures to the Royal Geographical Society. 'I am in the unique position of having served with *all* the British Antarctic Explorers of repute,' he told his audience. 'My opinion is that for qualities of leadership, ability to organize, courage in the face of danger, and resource in the overcoming of difficulties, Shackleton stands foremost and must be ranked as the first explorer of his age.'[9]

In August 1921 the Hudson's Bay Company ship *Baychimo* arrived at Pond Inlet, Baffin Island, where the HBC intended to establish a new post and take over the trade which until then had been dominated by Henry

Toke Munn's Arctic Gold Exploration Syndicate and a few independent traders. On board was a lone representative of the Canadian government, Sergeant Alfred Herbert Joy of the RCMP. Joy was not at Pond Inlet to assert Canadian sovereignty. Canada's claim to Baffin Island was considered beyond doubt, even by the cautious Harkin, because of its proximity to the mainland, the presence of several British traders, and the 1920 lease of a large portion of the island to the Hudson's Bay Reindeer Company.[10] In September 1920 Munn had reported the murder of another trader, Robert Janes, by an Inuk. It seemed reasonable to combine an investigation of the killing with the planned 1921 voyage of the *Arctic,* especially since Janes's father had written to the police in very insistent terms asking what they intended to do about his son's death. When the 1921 expedition was cancelled, Commissioner Perry at first wrote: 'As the sending of detachments in that far distant country is abandoned, [I am] of the opinion that the matter should remain in abeyance.'[11] But the voyage of the *Baychimo* provided 'a good opportunity' to send a police investigator north at once, and so Joy – bearing papers investing him with the powers of justice of the peace, coroner, customs officer, and postmaster – stepped ashore on 30 August.[12]

Joy already had several years of experience in the western Arctic, and he was to become famous for his long sledge patrols in the eastern archipelago. His presence on Baffin Island in 1921 marked a clear step forward in Canadian administration of the Far North, yet it did nothing to resolve the question of sovereignty over Ellesmere and the other islands in the northern part of the archipelago. After leaving Pond Inlet, the *Baychimo* established another new post on Baffin Island (at Pangnirtung on Cumberland Sound), but there was no 'reconnaissance' to Ellesmere. In Ottawa, Harkin and others remained keenly aware of the need for more extensive acts of occupation.

The advocates of a 1922 voyage by the *Arctic* gained a staunch and influential new ally in Oswald Finnie, the head of the recently formed Northwest Territories Branch of the Department of the Interior (renamed the Northwest Territories and Yukon Branch in 1923). Finnie was born in 1876 in Arnprior, Ontario, to a Scottish immigrant family. His father was a successful banker. After graduating from McGill University in 1897, Finnie qualified as a Dominion land surveyor. In 1898 he gained an appointment to the staff of William Ogilvie, the commissioner of the Yukon. He remained in the Yukon for eleven years, then settled in Ottawa as an employee of the Department of the Interior's Mining Lands Branch.

Finnie had a sturdy physique and a determined, tenacious character. Appointed to head the Northwest Territories Branch early in 1921, he had spent most of that year in the North.[13] Once back in Ottawa, he assumed a leading role in discussions of Arctic policy, to a large extent displacing Harkin as the subordinate upon whose opinion Cory chiefly relied in northern matters. Finnie's young son, Richard, had encountered Stefansson in the summer of 1920. The thirteen-year-old quickly became a hero-worshipper. He drew a sketch of the explorer, which Stefansson obligingly autographed; young Finnie kept the sketch for the rest of his life.[14] Richard Finnie would grow up to become one of Stefansson's most devoted adherents. His father, however, maintained a more detached and skeptical attitude.

On 29 December 1921, Arthur Meighen's government went down to defeat at the hands of William Lyon Mackenzie King's Liberals. It must have seemed to Finnie and Harkin that, if properly encouraged by the civil service, King might favour a more active Arctic policy. A few weeks after the election, Finnie wrote to Harkin, noting that there was general agreement in the Department of the Interior about the need for an expedition to the archipelago. He proposed that the matter should be revived once the new minister, Charles Stewart, arrived in Ottawa for the opening of Parliament. Harkin replied that he, too, was strongly in favour of government action. In his opinion, the presence of the Hudson's Bay Company traders was insufficient, even though they were acting in an official capacity as customs inspectors.[15]

Joseph Bernier, now seventy years old but still eager to return to the North, somehow got wind of the matter. Captain Pickels having died of a heart attack in October, Bernier wrote to Mackenzie King with an application to command the *Arctic* 'in the place of Captain Pickerel deceased.'[16] In early February, Cory asked Finnie to consult with Craig and Harkin and to produce a memorandum which the minister could submit to the cabinet.[17] Craig duly set to work on the memo, sending a draft to Harkin for approval. Harkin recommended a longer section on the legal aspect. He observed in a rather patronizing tone, 'I think you may take it for granted that the Minister, like nearly everyone else in Canada, is practically unfamiliar with International Law.'[18]

Craig's final version, dated 16 February, contained many echoes of Harkin's memos from 1920 and 1921. Most notably, Craig repeated Harkin's statement that the Danes had never answered the July 1920 protest.

He wrote that the assurances given by Denmark in June 1921 were 'somewhat reassuring' and 'apparently removed a cause of immediate anxiety,' but they 'did not dispose of the existing doubt as to the validity of Canada's title, and in fact the incident emphasized the weakness of Canada's cause, by drawing attention to the apparent ease with which Denmark, or any other nation, could take steps towards establishing a title in the Northland.' Like Harkin, Craig cited Oppenheim rather than Hall on the duration of an inchoate title. He noted that the total expenditure to that date was $81,683.82 (this sum included the cost of laying up the *Arctic* after the 1921 expedition was cancelled). Only about $6,000 more would be needed. Therefore, Craig strongly recommended that the voyage should proceed.[19] Finnie wrote approvingly that the document 'gives a clear view of the situation and establishes, beyond question, the only possible course the Government should follow.'[20] He forwarded it to the deputy minister's office with the comment: 'We are very anxious, and think it is extremely urgent, that this ship be equipped and that parties be sent to the North this summer.'[21]

Despite these strong recommendations, there the matter rested for the next three months. Stewart did not seem inclined to take the matter up with King and the cabinet, and indeed he did not become keenly interested in Arctic problems until the following year. Stewart, who would remain minister of the interior until 1930 (with a brief hiatus during the short-lived Conservative government of 1926), in many ways resembled his predecessor, Sir James Lougheed. He, too, was a self-made man who rose to wealth after going west. Born in rural Ontario, Stewart took up a homestead near Killam, Alberta in 1905. To keep the farm going, he had to supplement his income. Stewart turned to the real estate business with notable success. In 1909 he entered provincial politics. Between 1912 and 1917, Stewart served as minister of municipal affairs and minister of public works. He then spent four years as premier of Alberta before accepting Mackenzie King's invitation to join the federal cabinet.[22] Stewart was therefore an experienced administrator. During his eight years as minister, he ran the Department of the Interior in an exemplary fashion.

Although he never showed the grasp of northern issues displayed by Meighen (in parliamentary debates he frequently made mistakes on points of detail), Stewart did eventually play a much more significant part in the development of Arctic policy than had Lougheed. However, in the spring and summer of 1922 this contribution still lay in the future. In a large department like Interior there were, inevitably, other matters that seemed more important and pressing to the new minister. That the *Arctic* did go north in 1922 may have been indirectly due to Stefansson, who now dramatically brought Arctic sovereignty to the forefront of the new government's concerns. As in previous years, his aim was a government-backed occupation of Wrangel Island. In pursuit of this goal, he had secretly carried out an exceptionally daring plan.

STEFANSSON WAS IN NEVADA when he received Harkin's 30 May 1921 telegram informing him that his expedition had been cancelled (see Chapter 5). The news, he later wrote, 'broke my heart, at least for the time being.' He recovered from his dejection with remarkable rapidity. By 'a curious accident,' later the same day Stefansson encountered an acquaintance from Vancouver, engineer Alfred J. Taylor.[23] Taylor was a solid citizen and a shrewd, energetic businessman: his company, Taylor Engineering, was involved in the construction of the Lion's Gate Bridge and the development of several successful BC mines. He had first met Stefansson in 1913, before the departure of the Canadian Arctic Expedition. The two men remained in touch thereafter, and Taylor provided advice to Stefansson about his reindeer project. Like so many businessmen of the day, Taylor was entranced by Stefansson's theories about the great potential of Arctic commerce. He was also evidently motivated by strong patriotic feelings. Stefansson later presented him with a copy of *The Friendly Arctic* containing the inscription: 'I hope this book will help to maintain your interest in the Greater Canada which cannot be built on any foundation but that of territories as valued as they are valuable, as broad from North to South as they are from East to West.'[24]

Stefansson confided his problems to Taylor. Believing Taylor to be a very wealthy man, Stefansson proposed that he put up half the money for a private expedition. However, much of Taylor's money was invested in silver mines, and 'through the present condition of the mining industry, he was in financial straits.'[25] Nevertheless, Taylor was anxious to do what he could, and soon Stefansson's 'wrecked hopes' were 'replaced by a plan he and I thought we could carry through.' During his perusal of books on international law, Stefansson had noticed the statement that acts by a private citizen could be relevant to the acquisition of sovereignty, if his government later explicitly endorsed what he had done. He ignored the proviso that such acts had to be of a permanent nature – for example, the founding of a settlement. Stefansson and Taylor became convinced that if they sent a party to Wrangel and the government later issued an official statement supporting them, the island would be securely Canadian. Stefansson felt reasonably confident that he could persuade Meighen (or, in the likely event that an election was held soon and Meighen was defeated, his successor) to issue such a statement.[26]

Taylor returned to Vancouver and consulted his lawyer, who suggested that the party should be sent not by Stefansson personally, but rather by a company incorporated under the laws of Canada.[27] The Stefansson Arctic Exploration and Development Company was accordingly created in Vancouver on 23 June, with Stefansson as president and Taylor as vice-president. The two men announced that the aim of their enterprise was to 'thoroughly explore the resources and country lying to the north of Canada.'[28] Stefansson's plan, then, was to carry out a deception like the one of which Rasmussen had been suspected. As he later explained in a letter to Mackenzie King, 'guarded announcements were made intended to be interpreted by the public as meaning that the purpose of the company was mainly commercial, where in fact my main motive was political.'[29] Although Stefansson did not mention this aspect to his supporters or in any of his later published accounts, the formation of a limited liability company had the added advantage of releasing him from any personal responsibility for debts incurred by the enterprise.

Stefansson now contacted Lorne Knight and another American member of the 1913-18 expedition, Fred Maurer, and asked them if they would become British subjects in order to lead an expedition combining exploration work and commercial profit. Both men answered yes. However, Taylor's lawyer pointed out that the two Americans could not possibly be naturalized before the expedition left. It would be necessary to find someone who was already a British subject.[30]

The task did not prove difficult. In March, with Harkin's knowledge, Stefansson had written to various Canadian university presidents, asking them to recommend young men who would be suitable to do scientific work in the North. As a result, he had received an application from a twenty-year-old geology student, Allan Crawford. The son of a mathematics professor at the University of Toronto, Crawford was intelligent, athletic, and had done summer work in the Algoma District of northern Ontario for the Geological Survey of Canada. His letter exuded a boyish eagerness and enthusiasm. On 24 April, Stefansson replied that Crawford should hear from him in the next two weeks. Then there was silence until mid-June, when Stefansson wrote to inform Crawford that he would be in Ann

Arbor at the end of the month to accept an honorary doctorate from the University of Michigan. If Crawford would meet him there, they could discuss Arctic work.[31] Crawford jumped at the chance.

Stefansson had evidently been reflecting on the possibility that the Canadian government would refuse to endorse an occupation of Wrangel. If this were to happen, he could appeal to Great Britain, since the British were more used to thinking in broad strategic terms. An honorary degree was also being bestowed on the British ambassador to the United States, Sir Auckland Geddes. Stefansson cautiously approached him on the topic of Wrangel. Geddes naturally drew back from making any direct statement about the likely attitude of his government but, according to Stefansson's account of the conversation, the ambassador remarked that an action such as Stefansson was contemplating would meet with 'general and hearty sympathy' among the British public.[32]

This encouraging comment was enough for Stefansson. He broached his plans to young Crawford, giving him the impression that the British government was secretly behind the scheme. Its true purpose, Stefansson explained, was to establish a Canadian presence before any opposition from other nations could be aroused. He said nothing about the possibility that the Soviet Union already regarded the island as part of its territory.[33] As for the earlier deaths on Wrangel Island, Stefansson assured the young man that they had been due entirely to the 'incompetence and foolhardiness' of Captain Bartlett. He himself had 'never lost a man.'[34]

Crawford saw the chance to go north on such a patriotic enterprise as the opportunity of a lifetime. He unhesitatingly answered yes 'in a loud voice.' 'I would have done anything to go north and then to get an offer like this,' he marvelled.[35] As the only Canadian, Crawford would be in nominal command of the party, but Stefansson instructed him to rely on the advice of the older and more experienced Knight and Maurer. The fourth member of the party was another American, an enthusiastic Texan youth named Milton Galle. Galle had been enraptured by one of Stefansson's Chautauqua lectures and had subsequently offered to work as his helper and secretary. Stefansson told Galle that he had been chosen for the Wrangel expedition over thousands of other highly qualified applicants, which Galle apparently believed.[36] The four men expected that Stefansson would join them in 1923, once the true nature of the expedition had been made public.

By the middle of August, Crawford, Knight, Maurer, and Galle were all in Seattle. There they were photographed for the purpose of future publicity. Crawford had grown a small moustache, perhaps in an attempt to look older than his twenty years, and he slicked his hair back into a more sophisticated style than he had previously worn. On 18 August the group sailed for Nome on a passenger steamer, taking with them only a year's supply of food. Stefansson had originally instructed them to buy two years' worth, but he changed his mind on 7 August. Their families were not informed of the change.[37] Polar veteran Roald Amundsen was also in Seattle at the time, and he observed the party's preparations with a cynical eye. Amundsen was planning a new Arctic expedition himself, hoping to drift across the polar basin in his ship *Maud*. As he pointedly told reporters, he intended to take enough supplies to last for seven years.[38] Before they left Seattle, Knight and Crawford were persuaded to accept shares in the Stefansson Arctic Exploration and Development Company in lieu of their salaries. The eager Galle had agreed to work without wages; instead, he would receive a percentage of the profits on any furs he might obtain on Wrangel Island.[39]

When the group arrived in Nome, it was widely rumoured that they were following up some oil or gold discovery made during the 1913-18 expedition. They attempted to hire Inuit families to accompany them, but in the end only one woman, Ada Blackjack, was willing to go to so remote and desolate a place as Wrangel Island with only a year's supplies. Blackjack – a petite, soft-spoken, pretty woman in her early twenties – was separated from her drunken husband and had a young son to support. It was later said that she had occasionally resorted to prostitution in order to survive.[40] Whether the allegation was true or not, she desperately needed the money Stefansson offered.

Stefansson, meanwhile, was making a last-ditch bid for official Canadian support. An opening was provided by a dispute over the names given to islands and other geographical features during the 1913-18 expedition. The Geographic Board of Canada had rejected several of the names submitted by Stefansson on the grounds that the places in question had already been seen and named by nineteenth-century British explorers. Stefansson protested vehemently, maintaining that these were his own discoveries and that the Geographic Board's decision was a 'deplorable attempt to pervert history.'[41] He was invited to attend a meeting of the board on 6 September. He agreed, but only after ascertaining that Sir James Lougheed would be in Ottawa on that date. '[I] would like to submit [an] important matter to [the] Minister of Interior at the same time,' Stefansson explained to George Desbarats.[42]

Stefansson reached a satisfactory compromise with the board, though at the cost of adding geographer James White to the list of his enemies in Ottawa. White had long been interested in the Arctic: while working as the Department of the Interior's geographer, he oversaw the production of the first Canadian maps showing sector lines extending as far as the North Pole (see Chapter 3). In 1909 White left this post to serve as deputy chair of the Commission of Conservation; in 1916 he also became the first chair of the Advisory Board on Wildlife Protection. White has been described as 'seemingly ready to complain at any slight or inconvenience, sometimes a bit narrow, fussy, and stiff, at best a difficult and prickly personality,' but he was an exceptionally efficient administrator.[43] He was also a stickler for accurate detail and a man of exceptional tenacity.

In the spring of 1921, the Commission of Conservation was abolished, and White – who had offended both Borden and Meighen – was not offered another civil service position. Former Liberal cabinet minister Clifford Sifton believed that White was deliberately being treated in a 'venomous and contemptible' manner by Meighen.[44] In October 1921 he was told that he would be superannuated in six months. In the interim, White still served on the Geographic Board and the Advisory Board on Wildlife Protection, and he was still respected and consulted by his colleagues. The point of contention between White and Stefansson was the naming of islands which White believed to have been discovered by Leopold McClintock and Sherard Osborn during the Franklin search expeditions.[45] Stefansson would have been wise to conciliate White, but he did not bother to do so, perhaps because he believed that White's career was at an end. But in the spring of 1922, White would deftly avoid retirement, instead negotiating a transfer to the Department of Justice at the same rank and salary as before. (In an expert display of bureaucratic skill, he even managed to retain several members of his support staff.)[46]

Stefansson's second and more important mission was a complete failure. In the event, he met not with Lougheed but with Meighen's secretary, C.H.A. Armstrong. There seems to be no record of exactly what was said, but on 7 September Stefansson wrote to Armstrong asking for $2,000 to be sent to the Stefansson Arctic Exploration and Development Company in Vancouver, 'preferably today.' A delay of even one day could 'mean [the] difference between success and failure.'[47] Armstrong replied, 'I placed the matter you mentioned to me yesterday before the Prime Minister ... Mr. Meighen says that, with the present pressure of his duties he could not possibly go into the matter personally, and can, consequently, take no action.'[48]

On 10 September Crawford and the others left Nome on board a chartered ship, the *Silver Wave*. According to press reports, their purpose was to 'establish a supply base at Wrangel Island for future exploration work to be carried on by Stefansson, who, with others, will join them at the base next Summer.'[49] One story stated that Stefansson would then set out over the ice in an attempt to reach the North Pole, taking with him only ammunition.[50] In another interview, Stefansson spoke of a two- or three-year scientific expedition, which would remain in the general vicinity of Wrangel Island.[51]

Crawford and the others landed on Wrangel on 15 September. Early the next morning, they raised both the Union Jack and the Canadian Red Ensign. The excited young commander then read a grandiloquent and obviously carefully prepared proclamation:

> I, Allan Rudyard Crawford, a native of Canada and a British subject and those men whose names appear below, members of the Wrangel Island Detachment of the Stefansson Arctic Expedition ... on the advice and council [sic] of Vilhjalmur Stefansson, a British subject, have this day, in consideration of lapses of foreign claims and the occupancy from March 12th 1914 to September 7th 1914 of this island by the survivors of the brigantine Karluk, Captain R.A. Bartlett commanding, the property of the government of Canada chartered to operate in the Canadian Arctic Expedition of 1913-1918 ... raised the Canadian flag, raised the British flag and declared this land known as Wrangel Island to be the just possession of His Majesty George, King of Great Britain and Ireland and the Dominions beyond the Seas, Emperor of India, etc., and a part of the British Empire.[52]

It was the sort of gesture that Stefansson was in the habit of ridiculing when he preached the gospel of effective occupation to Ottawa bureaucrats.

When the news of the party's safe arrival on the island reached New York, Stefansson told reporters he would announce definite plans for the next year's work in the spring of 1922, 'probably not before March.' He claimed that he had been receiving applications to join the expedition at the rate of twenty-five a day. Many of the applicants, he said, had offered to donate thousands of dollars to the expedition's funds. None of them would be accepted, because 'when [men] once feel that they are financial backers of the undertaking they assume that they have got to help boss it.' The best volunteers were those who were motivated 'simply by an adventuresome desire to see the Far North.' As if in passing, Stefansson mentioned that the eight months' stay of the *Karluk* survivors had 'effected a renewal of the British right of occupation of the island.' Any party on Wrangel would have ample food, since the island was 'the greatest centre for polar bears in the world.' The bears were drawn to Wrangel because of 'the great number of walruses and seals, on which they prey.' In the future, the island would likely become the site of a 'great fur industry.' It would also 'be developed in years to come as a great pasture land for the raising of reindeer. It abounds in meadows, where the grass is green at intervals during a large part of the year.'[53]

STEFANSSON MAY HAVE HOPED that the publication of his narrative of the 1913-18 Canadian Arctic Expedition, with its laudatory preface by Sir Robert Borden, would raise his stock in official Ottawa, whether Meighen's government remained in power or was replaced by a new Liberal administration. An enthusiastic public response to the book could certainly make new Arctic enterprises more attractive to the government. The narrative, titled *The Friendly Arctic,* appeared in late November 1921. On 1 December *MacLean's Magazine* published the first article in a series by Stefansson advocating the 'colonization and development of Northern Canada, along the lines of our growth in Western Canada.'[54] Whether intentionally or not, these publications coincided closely with the federal election campaign.

The weeks after the election of a new government are usually a time when civil servants are even more cautious than usual about arousing any public controversy. So great, however, was the outrage he felt over Stefansson's version of the Collinson Point episode in *The Friendly Arctic* that Rudoph Anderson unhesitatingly made his displeasure public. His statements were strongly backed up by the expedition's geologist, John J. O'Neill, who was then teaching at McGill University.[55] (Three years earlier, Anderson had commented that if asked to record his opinion of Stefansson, 'O'Neill would have to write on asbestos.')[56] For several days, beginning on 14 January 1922, Canadian newspapers were dominated by headlines such as 'Karluk Journey Was Big Scheme to Exploit News: Dr. Anderson Charges Stefansson Using Arctic Expedition for Own Purpose,' 'Declares Stefansson Pursued Policy of Wild Extravagance,' and 'Arctic Explorer Far from Being Popular Hero He Represents.'[57] With regard to

Stefansson's theory of foraging off the resources of the country, Anderson detailed the great expenditure on supplies and then remarked, 'The foraging was done on the people who pay the taxes.'[58] On 21 January an editorial in the *Montreal Standard* suggested that there should be an investigation of Stefansson's conduct. '[O]ur Canadian scientists,' the writer declared, 'are entitled to a square deal from the imported adventurer who makes the Arctic his stock in trade. The more so when Mr. Stefansson goes out of his way to write a book and insult them to our faces.' He made the accusation that Stefansson had 'been behaving more like an exploiter than an explorer,' seeking to profit from his reindeer lease. Finally, the writer revealed that Stefansson had been blackballed when he was proposed for membership in the Rideau Club, and he claimed that this had occurred because official Ottawa now realized that Stefansson did not 'represent the pure scientific spirit.'[59]

Stefansson sent a copy of his book to Charles Camsell, the recently appointed deputy minister of mines. As deputy minister, Camsell was in charge of the Geological Survey of Canada and the Victoria Memorial Museum, where both Anderson and the Canadian Arctic Expedition's ethnographer, Diamond Jenness, now worked. Perhaps with the possibility of future disputes in mind, Stefansson had taken care to flatter Camsell during and after their previous meetings. For example, he ingratiatingly informed Camsell that he was the only man in civilization to fully understand Stefansson's theories.[60] However, Camsell immediately made it clear that Stefansson would not have his support in any campaign against Anderson and the others. 'I may as well say frankly that I feel that it would have been better to have left some of the things unsaid,' he told Stefansson after reading *The Friendly Arctic*. Stefansson's accusations of mutiny had 'resulted in stirring up feelings which were gradually dying down.' It would be 'very regrettable if things [went] so far' as an official investigation. An investigation could do no good to anyone involved. Since newspaper controversies were 'undignified,' Camsell thought that the scientists should be allowed to place their views on record in one of the geographical journals, 'where they would be in as permanent form as your book.'[61] It was hardly the response Stefansson had hoped for.

By the time Stefansson received Camsell's letter, he had already written to the new prime minister endorsing the idea of an investigation. The results, he insisted, should be released to the public. King referred the matter to Stewart (who was minister of mines as well as minister of the interior), and Stewart concurred with Camsell's opinion. It was quickly decided that there would be no such invitation to further controversy.[62] The scientists were forbidden to make any additional statements to the press, and even O'Neill, who was not a government employee and therefore was under no obligation to do so, obeyed the decree. However, O'Neill was no happier than Stefansson with Camsell's suggestion that the scientists should put their case on record in an academic journal. O'Neill was strongly in favour of an investigation, feeling that without one the public would be 'bound to credit Mr. Stefansson, because he has been placed in a strong position by Sir Robert Borden's introduction to his book.' Camsell was reluctant to risk any public dispute involving the former prime minister, but O'Neill argued that Borden could not have intended to damage the reputations of government employees.[63]

After the minister's press embargo, Stefansson did nothing to calm or conciliate his opponents, and indeed he seemed to go out of his way to inflame their anger. Jenness (whose conduct was not criticized in the book) returned the complimentary copy sent to him by Stefansson because he disagreed with the statements made about Anderson and the others. Jenness's letter explaining this action was polite and low-key. Stefansson's answer, while evidently intended to give the impression of imperturbable good humour, nevertheless had a spiteful, jeering tone. He assured Jenness that the scientists' public statements would not injure him in the long run; instead, the controversy would only add to the sales of his book, thus increasing Stefansson's royalties. 'This pleases me well, not only from the point of view of income, but more especially from the point of view of getting a wide audience for the message I am trying to convey to the people of Canada,' Stefansson concluded triumphantly.[64]

Jenness had already issued a relatively restrained statement in support of Anderson and O'Neill. He avoided their flamboyant rhetoric, but clearly stated his belief that Anderson's 'firm attitude' in the face of Stefansson's demands for the scientific party's resources had been fully justified.[65] The only effect of Stefansson's letter was to drive Jenness further into the enemy camp. (Decades later, an acquaintance remarked that Jenness – a man of exceptionally magnanimous character – 'came as close to bitterness towards Stefansson' as he was capable of.)[66] In late February Jenness joined Anderson, O'Neill, and topographer Kenneth Chipman in signing a letter to Stewart. The four men complained that Stefansson had slandered them ever since his return in 1918, and had now put the slanders into print in his book, 'which, to the uninitiated, seems to be an official publication.' The scientists were 'determined' to clear themselves, and they too asked for an investigation.[67]

From the other side, Stefansson continued to place pressure on Camsell. He insisted that he had made accusations against the scientists in his book only because Borden had encouraged him to do so. According to Stefansson, the scientists were continually circulating rumours about him in Ottawa, and Borden felt that, 'seeing my enemies would not keep quiet, it was better to force them into the open.' Stefansson then made the entirely false claim that the American geographical societies had transferred the 1913-18 expedition to the Canadian government only on the condition that Stefansson should be in full, unquestioned command. He hinted that if the Department of Mines stood by the scientists, it might find itself the object of unfavourable publicity. 'It seems to me that I ought probably to ask the Department of the Naval Service to issue to the press a statement of my full command of the expedition and a denial of the assertion that the Government has ever officially endorsed any of Dr. Anderson's disobediences of my orders,' Stefansson wrote.[68] According to Anderson's later account, George Desbarats had decided to back Stefansson strongly because he was afraid that a prolonged controversy might eventually make the public aware of how lax his department's control over the explorer's actions and expenditures had been.[69] But Camsell remained firm, and a few weeks later Stefansson denied that he wanted to 'push' the matter. He now asked Camsell to publicly announce that he had been in command of the entire expedition; in return, Stefansson would 'not press for vindication' on the charges made against him.[70] This arrangement would, of course, give Stefansson all the benefits of a finding in his favour without the disadvantages of an actual investigation.

Stefansson had arranged for his supporter Dr. Raymond Pearl to review *The Friendly Arctic* in the respected American journal *Science*. The review was an enthusiastic endorsement of Stefansson's theories about living off the resources of the country. Pearl explained that the 'mutiny' had arisen because the scientists lacked the vision and imagination of their leader. But even though the scientists had 'actively hinder[ed]' Stefansson's plans, the leader had achieved 'brilliantly successful results.' Jenness then provided a solution to Camsell's problems by taking up the idea of a statement in an academic publication. With Chipman's help, he wrote a highly effective rebuttal to Pearl's review for publication in *Science*. His letter cited the expedition's orders and other documents from government files to show that Anderson had simply been following his official instructions. Jenness suggested that both 'a sense of justice' and 'a little critical acumen' should prevent discerning readers from 'accepting Mr. Stefansson's charges at their face value.'[71] In return for their public vindication, the scientists all undertook to make no more attacks on Stefansson through the press. Anderson was reasonably satisfied with this outcome; as he observed, *Science* was read by approximately ten thousand highly educated people. He and Jenness also sent 'quite a number' of offprints to non-subscribers whose opinions they wished to influence.[72]

Nor was *Science* the only journal to publish a critical account of Stefansson's book. James White was still angry over Stefansson's victory at the Geographic Board's September 1921 meeting. White sent a detailed list of criticisms of *The Friendly Arctic* to Hugh Langton of the University of Toronto, who wrote the notice that appeared in one of the most influential Canadian academic publications, the *Canadian Historical Review*. Langton made effective use of the material provided by White, pinpointing Stefansson's errors and misrepresentations. He observed in conclusion that if, as Stefansson claimed, some of his subordinates had been 'mutinous or disobedient,' this fact could only produce the impression 'that Mr. Stefansson's conduct of the expedition and management of his men left much to be desired.'[73]

STEFANSSON WAS THUS AT a considerable disadvantage when he approached the new government on the subject of Wrangel Island. By March 1922 it had become imperative that he do so. He had no money with which to finance a relief expedition, but because Crawford's party had taken only a year's supplies, a ship would have to be dispatched in the summer. If it was sent under government auspices, the public would receive the impression that Stefansson's venture was an official one.

As he had done in the autumn of 1920, Stefansson made a direct approach to the new prime minister. He employed what seems to have been a characteristic ruse in order to prevent Mackenzie King from realizing that an attempt was being made to force his hand. As Stefansson recounted, the American crew of the *Silver Wave* had witnessed the raising of the British flag, an unfortunate 'tactical blunder' by Crawford. On their return to Nome, the crew talked freely of what they had seen. An article was published in the *Nome Nugget*, and many Alaskans waxed indignant at the idea of a British claim to the island.[74] Stefansson informed King that the matter was currently under investigation by the State Department.

The *New York Times* had heard about this development and was supposedly determined to publish the news. However, thanks to Stefansson's close friendship with the managing editor, Carr Van Anda, the *Times* was willing to present the story from Stefansson's, and Canada's, point of view.[75] In fact, the possibility of a British claim to Wrangel Island had been brought to the State Department's attention by John Burnham, president of the American Game Protective and Propagation Association. The American government had therefore made a few inquiries on the subject. However, the interest in Washington was as yet very slight.[76] It was hardly the type of story that could not wait.

Stefansson then gave King his own version of the events of the past few years, claiming that international law set a definite limit of five years before an inchoate title gained through discovery or mere proclamation of ownership would expire.[77] The claim created by the raising of the Canadian flag on Wrangel by the *Karluk* survivors had therefore lapsed in 1919. Stefansson had accordingly urged Borden, Meighen, and Lougheed to renew Canada's title by sending another expedition. (Here Stefansson was careful to blur the distinction between the temporary presence of an exploring party and actual occupation.) However, for 'highly complicated reasons, which it would take too much space to explain,' Meighen had 'at length decided that there was no hurry about following up this claim and that the Government, for reasons of economy, could not send an expedition there.' Although Stefansson was convinced that neither Russia nor the United States had a better claim than Canada, he had come to fear that the Japanese, 'a very enterprising and highly original people,' would soon realize the strategic importance of Wrangel Island. 'This made it to my mind certain that within a year or so they would quietly move into this no-man's land of which they could get possession without entering into any conflict as to sovereignty with any nation,' Stefansson recounted. Therefore, he had taken matters into his own hands and spent $15,000 of his own money, plus $5,000 in borrowed funds, to send Crawford and his party to the island.

'Through this small expenditure I have secured to the British Empire its possession of this valuable outpost,' Stefansson told King. The *Times* would soon make the facts public. Stefansson assured the prime minister that he was not asking to be reimbursed for the money he had spent. His sole purpose was 'to urge most strongly that the Government, through its utterances and actions (for I believe that all we now need is a statement of policy and no further immediate action or expense), should back up what I have done and thus not relinquish our clear rights.' He added that the new government should give priority not only to the matter of Wrangel Island, but also to its policy on the Arctic in general. Stefansson's own discoveries in the Arctic archipelago might be at risk, for he had left them in June 1917. Therefore, by his interpretation of international law, Canada's inchoate title to these islands would expire in June 1922.[78]

Other historians have taken Stefansson's statement about the pressure put on him by the *Times* at face value.[79] However, given his close relationship with the newspaper over the years, it does not seem likely that its editor would have insisted on publishing against his wishes. Stefansson himself had previously named March 1922 as the date for the publication of his plans. It is hardly plausible that the story of the flag-raising brought by the *Silver Wave*'s crew had taken six months to travel from Alaska to Washington and New York. Instead, it is far more likely that Stefansson wanted the truth to be made public in the spring of 1922, and that he had been the one to contact the *Times,* rather than vice versa. His account was certainly disingenuous in other ways. In contrast with his claim to the prime minister that he had no intention of putting forward any monetary demands, Stefansson told another correspondent: 'I am hoping that in a year or two when people begin to see the enterprise in its true light, the Government may feel like returning the money.'[80]

Stefansson's approach to King seems to have been carefully timed, coming as it did only a few days after the opening of the new parliamentary session on 8 March. A letter from Stefansson to Harkin on 7 January reinforces the impression of advance planning: Stefansson wrote that he was 'hoping to be able to come up to Ottawa in five or six weeks and learn in detail about the things which mutually interest us and how the prospect is for them under the new Government.' Harkin's reply was brief and noncommittal, and Stefansson made no further attempt to work through him.[81]

King, inexperienced as yet in both the art of governing and the pitfalls of dealing with Stefansson, met briefly with the explorer on 17 March. He was apparently much impressed by Stefansson's arguments.[82] King did not consult with Pope, Christie, Cory, or Stewart, either before or after the meeting. The story of the expedition appeared on the front page of the *Times* on 20 March under the headline 'Stefansson Claims Wrangell Island for Great Britain: The Expedition He Sent Out Last Fall Has Established Possession, Says Explorer.' Wrangel was described as 'one of the most important islands in the Arctic region, because strategically it dominates Northeastern Siberia.' Stefansson, the *Times* reported, had put his entire personal savings into the expedition, but the explorer remained unsure 'whether his country will accept from his hands the gift of Wrangell Island.'[83]

The announcement came as a complete and highly unwelcome surprise to the civil servants who had long been involved in Arctic policy. Pope lost no time in making his views known to the prime minister. The alarmed and irate old man expressed himself in the strongest possible terms. 'In my judgment,' the under-secretary wrote, 'no more far-fetched claim could well be imagined, and any attempt to associate Canada with such fantastic pretensions could scarcely fail to prejudice us in the eyes of the world, besides weakening our legitimate claim to certain Arctic islands adjacent to our own territory, in respect of which we have a strong case.'[84] However, this protest made little impression on King. A fundamentally insecure man who demanded constant reassurance that his decisions were right, the new prime minister distrusted many of the civil servants whose careers had flourished under the Conservative governments of the past ten years.[85] Christie, who had been so close to both Borden and Meighen, and Pope, who had once been Sir John A. Macdonald's private secretary, were particularly suspect as Tory supporters. Christie joined Pope in advising strongly against any action on Wrangel, and he later heard that King, when recounting the story to others, had spoken 'very loftily about the insolence of civil servants setting up their opinion against Ministers.' From March 1922 onwards, Christie was kept in the dark about important policy decisions, and he was denied access to secret Foreign Office telegrams and letters. '[The Liberals] evidently want to freeze Christie out, and are succeeding,' wrote journalist John Stevenson, a close friend of both Meighen and Christie.[86]

Stefansson had assured King that only a declaration in support of the Crawford party would be required. To do this may have appeared to the prime minister as an excellent (and inexpensive) way to assert both his independence from the civil service and his determination to follow different policies from those of the previous administration. King was later known for his exceptionally slow and cautious approach to policy matters, and indeed he was sardonically remembered after his death as a man who never did things by halves when he could do them by quarters. According to poet F.R. Scott, King 'never let his on the one hand / Know what his on the other hand was doing,' and his motto was 'Postpone, postpone, abstain.'[87] Wrangel Island provides one of the few instances in King's political career of a quick and poorly judged decision.

The *New York Times* did all it could to promote the impression that Stefansson's secret acts were favourably regarded in the Canadian capital. Under the headline 'Canada Expects to Hold Wrangell,' it reported on 21 March that the announcement made in New York 'gave rise to the assumption [in Ottawa] that the island was formally claimed in the name of Great Britain to avoid complications with a view to its being eventually transferred to Canada as part of the Dominion.' Unnamed government officials were said to have agreed that the 'occupation' by the *Karluk* survivors provided a valid 'basis for ownership.' Although Russia might 'possibly have some claims,' Canadian authorities reportedly believed that the island was 'sufficiently far to the north of Siberia to be outside of Russia's sphere of influence.'[88]

Canadian newspapers told a different story. Accurately reflecting the views of Pope, Christie, and other civil servants, the Toronto *Globe* pointed out that 'Capt. Bernier hoisted the British flag over some Arctic islands, but they were directly north of Canadian territory, and his claims were recognized. But Wrangel Island is remote from Canada, and adjacent to Siberia.' Wrangel would be 'expensive and difficult for us to administer,' and a Canadian claim to it 'might lead to international complications.'[89] As even the *New York Times* admitted, the British response was still more unfavourable from Stefansson's point of view. 'Inquiries in official quarters [in London] showed much indifference as to the fate of Wrangell Island,' the paper reported.[90] (Stefansson attempted to remedy this situation by sending a letter to Arthur Hinks of the Royal Geographical Society, in which he wrote enticingly that Wrangel Island would be an ideal addition to the British Empire, since there were no 'temperamental Irish' or 'refractory aborigines' inhabiting it. He asked Hinks to make representations on his behalf to 'the proper officers of the Government.')[91]

Officials in the Department of the Interior were initially no better informed about Stefansson's plans than were Pope and Christie. Like Pope, Cory learned about the occupation from the *New York Times* article. He could do no more than instruct Finnie to keep track of events by following press reports.[92] Craig, for his part, noted uneasily that Stefansson's actions had 'drawn the attention of the world to the Arctic regions' and that it was 'quite possible that the interest awakened may result in some definite action this year on the part of the Government of the United States or of some other country.' It was therefore 'imperative' that a Canadian expedition should soon go north.[93]

The Department of the Interior was brought into the discussions when Stefansson came to Ottawa in early May, but External Affairs was still excluded. Stefansson met first with the prime minister and Stewart and then with Finnie and Craig. As he had done with Charles Camsell, Stefansson had already made attempts to cultivate Finnie's good will. For example, he had presented Finnie too with an autographed copy of *The Friendly Arctic*. The inscription read: 'I envy you your job of developing the country this book tells about.'[94] Finnie reported to Cory that Stefansson was now demanding compensation for the money he had spent on Wrangel. When told by Stewart that there would be no cash payment, Stefansson suggested that a lease of the island might be granted to his company.[95] In a letter to Finnie, Stefansson outlined terms far more favourable than those in the Baffin Island lease and noted that he considered this arrangement fair because of the money he had spent for the patriotic purpose of claiming Wrangel. He asked that a decision be made quickly, since he intended to use the lease as a means of borrowing enough money to send a relief ship that summer.[96] To an American friend, Stefansson wrote more frankly that he hoped to 'sublet the island to some fur company for enough to get a handsome annual return on the money so far "invested."'[97]

Finnie asked for Craig's opinion. The latter replied that Stefansson's terms were reasonable enough under the circumstances, but the 'important point' was to consider the implications of granting a lease. 'Is Canada prepared to stand behind the implied assumption of sovereignty? Does Canada realize that this assumption may result in acts of retaliation by some other nation regarding some of the vastly more important islands of the Northern Archipelago?' Craig asked. 'If Canada can go a thousand miles outside her boundary and claim Wrangell Island, the United States can apparently, with equally strong justification, go and claim one or more of our Northern islands. It would appear to be a rather risky proceeding, and might conceivably lead to serious international complications.' Craig recommended that nothing should be done 'without a full investigation as to the possible consequences, and if we are not prepared to stand behind [Stefansson's] act in every way, we should ignore it.'[98] As a first step towards such an investigation, Craig asked the RCMP to make confidential inquiries about possible American plans, while Pope made a similar request to Merchant Mahoney, Canada's agent at the British Embassy in Washington.

King and his ministers were in a less cautious frame of mind. Two days after Craig made his recommendation, the subject of Wrangel Island was raised in the House of Commons. Meighen asked for a statement of the government's policy. The minister of the Naval Service, George Graham, at first answered evasively that it was a 'delicate matter,' but some expert needling from Meighen soon brought the rash reply, 'What we have we hold,' from William Fielding, the minister of finance. This remark was backed up by Graham. King then revealed that he had 'had interviews with Mr. Stefansson.' It would not, he insisted, be in the public interest to give a full account of them, but he stated that the Canadian flag was flying on Wrangel Island and that, in his government's view, the island was 'part of the property of this country.' Meighen then pointed out the possibility that a claim would be made by Russia. He also remarked that in his opinion, Ellesmere Island was 'of greater consequence' than Wrangel.[99]

Needless to say, Stefansson was delighted by this exchange.[100] But as it turned out, even though the government finally took action on Ellesmere soon afterwards, its ringing declarations about Wrangel proved hollow. Perhaps because Stefansson had roused fears about the archipelago with his talk of a five-year limit on titles gained by discovery, the cabinet gave its approval to a 1922 voyage by the *Arctic* less than two weeks after the 12 May Commons debate. However, no decision was made about Stefansson's lease. Not only did Craig and Finnie advise caution, but both Merchant Mahoney and Sir Auckland Geddes reported from Washington that the State Department was seriously considering an American claim, and Belle Anderson provided a formal statement from William McKinlay, who firmly denied that the 1914 flag-raising on Wrangel had been intended as an affirmation of Canadian sovereignty.[101]

Nor was this all. The American papers soon published the news that, after a thorough investigation, the State Department believed Russia had the best claim to the island. State Department records show that American officials had been prodded into action by the *New York Times* story. As the press reports stated, their conclusion was that they themselves could not put forward a strong case for ownership of Wrangel. Nevertheless, the Americans objected to the idea of any other nation gaining 'undue economic advantage in any part of Russia.'[102] On 29 May the Conservatives made a point of mentioning the press stories in Parliament.

Then on 2 June came the news from London that the government of the Soviet Union had asserted its ownership of Wrangel Island and was demanding to be informed whether Stefansson's action 'took place with the cognisance and sanction of the British Government and whether or no the British Government considers Wrangel Island to be a British possession.'[103] According to Stefansson, there could be no basis for Russian ownership other than proximity. 'If you allow proximity as the basis of a claim for sovereignty, Denmark will get thereby a strong claim to Ellesmere Island because Ellesmere Island is only ten miles away from Danish territory,' he wrote to King, with a touch of desperation in his tone.[104] To another Canadian correspondent (Arthur Ford of the London, Ontario, *Free Press,* who had published a favourable review of *The Friendly Arctic*) Stefansson vented some of his anger and frustration at the government's slowness. 'The cause for alarm is in the fact that the same message which is sinking into the minds of other nations is being shed by Canadians like the proverbial water off a duck's back. I gave my views and observations to the Canadian Government two years before I published them in book form. It is not my fault, therefore, if other nations act on my information and if Canadians fail to do so,' he declared. 'After calling me a fool for thinking that the remote North is valuable, they may next blame me for causing them the loss of that territory through arousing the interest of other nations ... we cannot expect to hold our Northern possessions by merely coloring them red in Atlases published at Ottawa.'[105]

On 20 May 1922, responsibility for the enforcement of the game laws in the Northwest Territories was transferred from Harkin and the Dominion Parks Branch to Finnie's new Northwest Territories Branch. This change brought Harkin's formal involvement in Arctic sovereignty matters to an end. Nevertheless, he must have been filled with relief and pride when the decision to send the *Arctic* north was at last made a few days later, on 23 May. The files contain only scanty evidence about the events surrounding the decision, but it seems that delays were caused by Craig and Finnie, who wasted considerable time and energy in fruitless efforts to secure a legal opinion from the Department of Justice. An opinion had been requested from that department early in 1921, but the deputy minister, W. Stuart Edwards, had demurred, arguing that he did not have sufficient evidence to work with (see Chapter 4). In March 1922 Craig sent Edwards extensive documentation on the 1880 transfer of Arctic sovereignty from Britain to Canada, which had been gathered by H.P. Biggar in London.[106] Edwards, overwhelmed by the material now placed at his disposal, apparently did nothing. Even after the expedition had been approved by the cabinet, Cory, Finnie, and Craig still felt that a legal opinion was imperative, so that the actions taken would be in full conformity with international law. On 14 June Craig asked Edwards for his report. Edwards coolly replied that it was not 'profitable at the present time' to give a 'categorical answer' on sovereignty questions. Canada was free to occupy Ellesmere, and the decision that an expedition would indeed be sent had already been made. 'It is therefore unnecessary that I should deal further with this aspect of the matter,' Edwards concluded.[107]

Owing to the late date at which the decision was finally reached, preparations for the *Arctic*'s voyage were extremely rushed. Equipment and supplies for three police posts were sent to Quebec, but because the amount of cargo was larger than anticipated and the loading of the ship had to be carried out as quickly as possible, only enough for two posts found its way on board. The plan to establish a post on North Devon Island was therefore abandoned. One post would be placed at Pond Inlet, in order to supervise both the Inuit population and the white traders. Harkin and Craig had originally wanted to put this post on Bylot Island, but the killing of Robert Janes had raised serious concerns about interactions between the traders and the Inuit. Janes was well known to have been a domineering, bad-tempered man who to a large extent had brought his fate on himself, and it was feared that without a police presence other traders might behave in a similar manner. From Pond Inlet, a small station could be established on Bylot Island and visited by the police 'from time to time.' The other post would be on Ellesmere, as far north as possible – perhaps at Cape Sabine. The aim was to establish Canadian occupation somewhere not far from Thule, in order to intercept any Inughuit who might attempt to hunt in Canadian territory.[108]

The hastily assembled group of Mounted Police included Inspector Charles Wilcox, two corporals, and seven constables. 'None of us knew exactly where we were going, or could visualize what lay before us on the new frontier of the North ... Ellesmere Land! The name was almost unknown ... I shall never forget the feeling when I gazed at the map and ran my eye north from Labrador to Baffin Land, thence three hundred miles farther north to Ellesmere, running from the 76th degree of latitude to within four hundred miles of the Pole itself,' wrote Constable Patrick Lee.[109] Craig swore a formal oath of allegiance to King George V and was appointed a justice of the peace, coroner, police constable, game officer, and fisheries officer. His commission under the Great Seal of Canada could not be prepared in time, and instead he travelled north carrying only a certified copy of the order-in-council appointing him to command the expedition.[110]

Work on the *Arctic* began on 9 June and was concluded, after a fashion, just over five weeks later. On 13 July a Canadian Press story revealed that the expedition's 'primary purpose' was 'the maintenance of the Canadian title to islands north of Labrador and facing Greenland across Davis straits. The islands are already Canadian, but the expedition, by actual occupation, will set at rest any doubt as to Canadian title.'[111] The *Arctic* left Quebec on 18 July. 'About eleven o'clock on the evening of July 17th,' Craig recorded, 'the engines turned over under their own steam for the first time in several years and at 5.45 o'clock the following morning, without any trial trip or any further delay, the ship ... started on her voyage.'[112] Lee wrote that although the *Arctic* looked 'insignificant enough' as it glided past the huge Canadian Pacific and White Star ocean liners in Quebec harbour, the men on board felt their little ship was 'ten times more important.' Captain Bernier was 'happy as a boy to feel the old vessel beneath his feet again.'[113]

However, Craig's pleasure in being on his way at last was somewhat dampened by the many troubles resulting from over-hasty preparations. As the *Arctic* proceeded down the St. Lawrence, the engines developed one problem after another, prompting Craig to observe after the first few weeks that 'no boat but a Government boat would have been allowed to sail in the condition the "Arctic" was in. A great many little weaknesses have developed since we sailed and these could only have been found out before sailing by a reasonable trial trip.'[114] In fact, the engines were so prone to trouble that for much of the voyage they could not be depended on. This problem was all the more disturbing because, for economy's sake, the *Arctic*'s original barquentine rig had been replaced by inadequate 'fore and aft' sails. As a result, the ship was unable to beat to windward. 'Without steam the ship is helpless in any kind of a breeze, and the constant menace of being caught on a lee shore, with the engines out of commission as they are apt to be when most needed, gives Capt. Bernier many an anxious hour,' wrote Craig.[115]

The ship passed through the Strait of Belle Isle and then turned towards the coast of Greenland, where navigation was easier than on the Canadian side of Davis Strait and Baffin Bay. On 30 July Bernier gave an inspirational address to the crew. 'The time has come when Canada is to occupy her own territory ... It is a happy trip for everybody in the expedition for they are called upon to write a page of history,' he declared. 'It is a great mission we are on,' concurred Craig.[116] Bernier followed the Greenland coast northwards into what the enthusiastic Constable Lee called a 'mystic land of romance and adventure.' In Melville Bay the ship was surrounded by 'dozens of fantastic bergs,' coloured not white but 'every shade of blue and green' and 'towering up ... like the spires of magnificent cathedrals.'[117] It soon became clear that the pack was too thick for the old *Arctic* to force its way through, and Bernier therefore advised that there should be no attempt that year to establish a post at Cape Sabine. Craig could only agree.

On 15 August the *Arctic* arrived at Pond Inlet, where the newcomers were greeted by Bernier's nephew, Wilfred Caron, an employee of the Arctic Gold Exploration Syndicate. Later Sergeant Joy, looking 'brown and fit' after his year in the North, came aboard to report that he had successfully apprehended the killer of Robert Janes.[118] Here, too, the ice was an obstacle, and it was impossible for the ship to reach the spot chosen for the police post. There was nothing to be done but to proceed to the southeastern coast of Ellesmere, where Bernier and Craig hoped that a landing could be made at Fram Fjord. The party took with them an Inuk man, Kakto (Qattuuq), who had been persuaded to spend the next year at the Ellesmere post along with his family.

The newcomers' first impressions of the Aboriginal population, gained during hurried visits ashore, had not been very favourable. To their eyes, the settlement at Pond Inlet was repulsively dirty. They were also bemused by the sight of Inuit wearing battered Scotch tam-o'-shanters. Bernier explained that the caps had been 'obtained ... from the whalers years before and handed down from father to son.'[119] These odd heirlooms seem to have caught Craig's attention, and to have turned his mind to the effects of what he called 'promiscuous intercourse between the whalers and traders and the Eskimos' over many decades in the absence of any government control.[120]

Ellesmere Island was sighted on 20 August, and the Mounties hurried on deck to see their new home. As Lee recorded, it looked like 'one gigantic glacier which dropped its broad mouth into the sea near Cape Tennyson ... Of living creature there was absolutely no sign.' It was the most desolate and inhospitable-looking place any of them had yet seen in the North. Quipped Constable William MacGregor, 'It ain't no peach orchard exactly.'[121] Craig's report praised the beauty of nearby North Devon and Coburg Islands, but was discreetly silent on his first impressions of Ellesmere. However, he noted with pleasure a small, previously uncharted island off the Ellesmere coast, which was promptly named in honour of Minister Stewart.

Like Pond Inlet, Fram Fjord was inaccessible because of heavy ice. Bernier therefore sailed southward along the coast, and on 21 August a small bay was sighted. It was far from an ideal spot, but it was free from ice, and it would have to do. The supplies were hastily unloaded and work began on the buildings. Significantly, there was no formal declaration of sovereignty at the new post. Instead, the performance of scientific and administrative work affirmed the fact of Canada's ownership. Craig carried out a quick survey and, as part of this task, 'a bronze tablet was set in solid rock, the markings of the tablet being "Canada, N.W.T.1," signifying the first tablet set in Franklin district under the direction of the North West Territories Branch.'[122] Nearby glaciers were given the names Cory and Wilcox. The new settlement itself was named Craig Harbour on the suggestion of Inspector Wilcox. It combined the functions of a police station, a customs house, and the most northerly post office in the world. Filled with pride by these solid, tangible manifestations of Canadian sovereignty, Craig wrote a letter to Harkin from the post office, which he described as the 'direct result' of Harkin's 'strenuous efforts for Canada's good in 1920-21.'[123]

The *Arctic* left Craig Harbour at midnight on 28 August, a mere seven days after it had arrived. The bay was beginning to fill with heavy ice, and snow was falling so thickly that the land could not be seen. Winter was clearly at hand, and the *Arctic* was in no condition to risk severe weather. Inspector Wilcox, Corporal Jakeman, and Constables Fairman, Fielder, Anstead, Must, and Lee remained behind, 'the sole inhabitants excepting a single Eskimo family of a million square miles of Arctic wilderness.'[124] On their way back to Baffin Island, Craig and Bernier inspected Dundas Harbour on the island of North Devon, which they felt was an ideal spot for a future post. Situated near the entrance to Lancaster Sound, the gateway to the Northwest Passage, Dundas Harbour was strategically located where foreign expeditions to the archipelago would have to pass. Pond Inlet was finally reached on 6 September, and the police were landed with their supplies. Leaving matters in the 'capable hands' of Sergeant Joy (now supported by Corporal McInnes and Constables Friel and MacGregor), the *Arctic* then paid a quick visit to the Greenland settlement of Godhavn (Qeqertarsuaq).[125]

Although the Canadians stayed for only two hours, the visit made a lasting impression on Craig's mind. His diary is filled with favourable comments on the cleanliness and civilized behaviour of the Greenlanders, which he compared to the dirt and disorder found at Pond Inlet. 'The contrast between conditions here ... and conditions among the Eskimos we left only a few days ago at Ponds Inlet is not only marked but painful,' Craig wrote. 'Instead of the filth, squalor and horrible smells of the settlement at the Hudson Bay Post and Captain Munn's Post with the squabbling dogs, dirty youngsters and piles of refuse, we have a neat clean settlement and everything spick and span.' The wooden houses were attractive, painted white with red trim. The inhabitants, too, showed European characteristics, and Craig unhesitatingly attributed this fact to a generous admixture of Scandinavian blood. Both men and women were 'mostly very good looking.' The women made a particularly favourable impression, being 'just as clean as the Baffin Land women are unclean and they are perfectly ... straight limbed, whereas most of the Baffin Land Eskimos ... appear to be bow-legged.' The Greenlanders had been trained by the Danes in useful arts, for example the production of dyed leather goods for export. All in all, Craig felt that the Native people of Godhavn showed what results could be produced through 'loving care by a paternal government.'[126]

Craig blamed most of the problems at Pond Inlet on the whalers and traders. He left Godhavn convinced that the time had come 'when the Canadian Government must step in and protect not only the caribou and the musk-ox but the Eskimos themselves.' Already, Craig was envisioning an ambitious program for the future: 'Schools must be established and the police must take a sincere interest in these people and try to better them. It does not seem so important that they should have religion pounded into them as they should have habits of cleanliness taught to them ... They must be taught to do things of value to the outside world so their product will be salable; they must be made industrious and by sending among them only the best type of white man as teacher and preceptor they must be made to realise that the white man is not always a scoundrel.' Craig hoped that the photographs taken on Baffin Island and in Greenland would make the need for changes on the Canadian side clear to those in Ottawa.[127]

The contrast between Canada and Greenland was painful to Craig in another way. He had already felt embarrassed and indeed humiliated when the HBC's fine new ship *Bayeskimo* put in an appearance at Pond Inlet during the *Arctic*'s second visit. Now the Danes were given the opportunity to observe the *Arctic*'s sad appearance. In a particularly galling touch, contact between the ship's sides and the ice had rubbed away much of the new white paint applied at Quebec, revealing the bright red used during the

Arctic's days as a navigational beacon on the St. Lawrence. The parsimonious government had provided only cheap, inferior soft coal for the engines, and the ship's sails were almost black from the sooty smoke this coal produced.[128] 'The Government having committed itself to a policy of renewed interest in the north should have a ship that would be a credit to all concerned, if only to impress strangers, company officials and natives alike. They should all realize that the Government is the biggest thing in the north, and that all others are subservient thereto,' Craig noted sharply.[129]

From Godhavn the *Arctic* returned south, safely reaching Quebec on 2 October. Craig went back to Ottawa filled with plans for an ongoing program of northern work that would embrace both sovereignty concerns and the welfare of the Inuit. He believed that more police posts should be established in future years, at spots chosen for their value in asserting and maintaining Canadian sovereignty. The posts could also be used as bases at which 'trained scientific observers' would remain in the North for at least a year, gathering 'that detailed and accurate knowledge concerning the country which is so necessary to its future development.' The resulting body of knowledge would stimulate change, foster the growth of new industries, and generally bring about 'a great and much needed improvement in the general living conditions of the present population.'[130] Craig's vision of the Arctic was beginning to resemble Stefansson's. However, unlike the flamboyant explorer, he clearly anticipated that government workers would be in firm control throughout the early stages of northern development.

STEFANSSON, MEANWHILE, HAD SPENT most of the summer pleading with an unresponsive government. At the end of May, he informed Finnie that the relief expedition would cost from $8,000 to $10,000, and he reiterated that he had expended all his personal funds and could borrow no more without a lease.[131] In June, Stefansson desperately tried to enlist Christie's help. He was promptly and very curtly rebuffed.[132] At Finnie's request, the Department of the Interior's recently appointed legal advisor, T.L. Cory (the deputy minister's son), was instructed to make a study of the matter. The younger Cory concurred with Craig that, while Canada could theoretically make a claim, to do so might 'prejudice her claim to the vast territory to the North and also her standing and good will with regard to the other powers interested.' In Cory's opinion, it was not worth taking such a risk 'for the sake of one man or group of men.'[133]

A few days after Cory's memo was written, Finnie left for the Yukon. With Craig soon to sail in the *Arctic* and Harkin no longer officially involved in sovereignty issues, his departure meant that there were no employees of the Department of the Interior still in Ottawa who had any reason to press for a decision about Stefansson's request for a lease. Despite a promise to the explorer from Stewart on 24 June that the matter would be settled 'in [a] day or two,' nothing was done.[134] On 9 July Stefansson telegraphed: 'Lives of men on Wrangle [sic] Island may be in danger and our enterprise will fail unless Canada promptly gives lease ... my money and credit all used up in service of Canada and I can do no more unaided.'[135] A month later there had been no change. Stefansson's agent in Nome, Carl Lomen, had by this time made a tentative arrangement with Captain Joseph Bernard to carry out the relief expedition in his schooner, the *Teddy Bear,* for $1,000 in advance and $2,000 more if he reached the island. (Bernard, a naturalized American born in Quebec, was a friend of the Andersons and hence no admirer of Stefansson, but he 'was sorry for the boys' on Wrangel.)[136] Lomen was also offering 'a complete equipment for trapping and half of the furs they should trap' to anyone who would spend the winter of 1922-23 on the island. Both Stefansson and Bernard were of the opinion that the ship should leave Nome on or around 20 August.[137]

A personal visit to Ottawa was clearly necessary. Stefansson arrived on Sunday, 6 August, and spent the Monday and Tuesday conferring with Cory senior and J.H. King, the minister of public works. In this episode, the portly deputy minister for once appears as a man capable of decisive action. Moved by Stefansson's pleas on behalf of the four men who had gone to Wrangel 'to hold it for the Empire and Canada,' Cory arranged for the matter to be placed before the cabinet on Friday, 10 August.[138] Stefansson had now reduced his previous estimate of $8,000-$10,000 to $5,000. However, a telegram from Alfred Taylor in Vancouver (sent in ignorance of Stefansson's inflated claims) disclosed the true cost of the *Teddy Bear*'s voyage – $3,000. Cory therefore recommended that the government provide a $3,000 loan, which Stefansson undertook to repay. The details of repayment were to be settled at a later date.[139]

With few advisers knowledgeable on the subject available to him, the prime minister turned, for once, to Christie. Christie produced an exceptionally blunt memo, written in a stiff, frosty tone that speaks volumes about his poor relations with King. He began by pointing out that his opinion had not changed since March. Stefansson had sent his men to

Wrangel Island without any authorization or encouragement from Meighen's government, and everything the explorer did was 'accompanied by a great cloud of embarrassing newspaper publicity.' In short, it was clear that Stefansson was making a deliberate, carefully planned attempt to force the government's hand. Christie characterized this behaviour as 'impudent' and recommended that Ottawa should 'decline to give either support or recognition' to Stefansson's venture.[140] However, a telegram from Allan Crawford's worried parents in Toronto helped to wrest a favourable decision from the cabinet, and the $3,000 was accordingly advanced to Stefansson. The *Teddy Bear* left Nome as planned on 20 August – the same day that Craig, on board the *Arctic,* first sighted the coast of Ellesmere Island.

From Ottawa, Stefansson had gone on to visit Toronto, where he announced that the era of Arctic exploration was over and a new era of commercial exploitation was about to begin. He himself intended to devote the rest of his life to 'the building up of a greater Canada.'[141] Stefansson later explained to Finnie that after 'many conversations with authorities on air navigation and submarine navigation,' it had become clear to him that 'all the polar regions will be opened to our knowledge as a mere incident of commercial development within the next decade or two.' This change, he explained, 'puts exploration more than ever upon a basis of sport and lessens the force of the motives which had been back of my work up to that time. I accordingly decided not to go to Wrangel Island myself and not to make any attempt to raise large funds for exploration, but merely to continue holding the island by maintaining a permanent colony there.'[142] In fact, Stefansson's change of plan likely stemmed from the realization that large amounts of government money would not easily be obtained. But to the men on Wrangel Island (who, of course, believed that Ottawa and London had been behind the scheme from the beginning), it might well have seemed like a betrayal of the promises made to them.

In both Ottawa and Toronto, Stefansson gave press interviews in which he spoke of Wrangel as a part of the British Empire and implied that his expedition had the full support of the Canadian government.[143] At least in public, he managed to preserve his confident demeanour, even when the *Teddy Bear* returned to Nome on 22 September with the news that severe ice conditions had made it impossible to reach the island. To the Crawford family he insisted that Allan was in no more danger than he would be in any other part of the world. In a letter to the *Toronto Star,*

Stefansson denied that the *Teddy Bear*'s voyage had been in any way a rescue expedition. The ship, he said, had carried only additional supplies for men who were already abundantly provisioned, and who were 'as safe as any city dweller.'[144] Nevertheless, he must have felt serious qualms when he received the news. Having taken only a year's food supply with them, the five people on the island would now be forced to put Stefansson's 'friendly Arctic' theories to the test. If they survived, he would be vindicated; but if tragedy ensued, it would be certain to attract the attention, and most likely the censure, of the world. Public perceptions of Stefansson and of his plan to secure a vast Arctic empire for Canada thus to a large extent hung on inexperienced Allan Crawford and his party.

FIGURE 20 At the Franklin memorial on Beechey Island, 1923 (Captain Joseph Bernier on right) | LAC, e010752416

FIGURE 21 Launch from the Danish ship *Islands Falk* alongside *Arctic*, Godhavn, 1923 | Photo by Lachlan Burwash | LAC, PA-990404

FIGURE 22 Canadian delegation en route to the Imperial Conference, September 1923. William Lyon Mackenzie King third from left, O.D. Skelton fourth from left | LAC, C-002132

FIGURE 23 Knud Rasmussen with the Danish and British flags at the grave of some of Sir John Franklin's men, Adelaide Peninsula, 1923 | Originally published in *Across Arctic America: Narrative of the Fifth Thule Expedition;* courtesy Canadian Museum of Civilization

FIGURE 24 Filming northern scenery from the bow of *Arctic*, 1923 | LAC, PA-118127

FIGURE 25 Rasmussen's visit to Ottawa, spring 1925. Rasmussen at back on left, Oswald Finnie at back on right | LAC, e010752399

▸ *Facing page, top to bottom:*

FIGURE 26 Frank Henderson surveying at Rice Strait, Ellesmere Island, 1924 | Photo by Roy Tash | LAC, PA-102285

FIGURE 27 On board *Arctic* just before departure from Quebec, July 1925. Richard Finnie stands to the right of the two uniformed RCMP constables | Photo by Oswald Finnie | LAC, PA-100443

FIGURE 28 One of the MacMillan-Byrd Expedition's planes at Etah, August 1925 | Photo by Richard Finnie | LAC, PA-100618

FIGURE 29 Charles Stewart | LAC, C-052281

FIGURE 30 *Beothic* in the ice, 1926 | LAC, e010752414

FIGURE 31 Bache Peninsula post under construction, 1926 | LAC, e010752413

FIGURE 32 A.Y. Jackson sketching at Beechey Island, 1927 | LAC, PA-212886

FIGURE 33 Muskoxen on Devon Island, photographed during the 1929 Eastern Arctic Patrol | LAC, PA-048029

FIGURE 34
Vilhjalmur Stefansson, 1947 | Photo by Yousuf Karsh | LAC, PA-165878

FIGURE 35
J.B. Harkin in 1937, the year following his retirement | Photo by Yousuf Karsh | LAC, e010767606

7
Stefansson in London

There are imperial considerations of which acc[oun]t sh[ou]ld be taken.

– *William Lyon Mackenzie King's diary, 7 April 1923*

'Canada's Northern Empire within 850 Miles of North Pole, Making Our Sovereignty Certain,' proclaimed a headline in the *Ottawa Journal* on 17 October 1922. The Toronto *Globe* informed its readers that Canada could 'now boast of the farthest north post office in the world, as a result of the work of the expedition which explored the northern archipelago this summer under the direction of J.D. Craig and on the good ship Arctic.' Portions of Craig's report had just been made public, two weeks after the *Arctic*'s return to Quebec. The report explained that after years of unfortunate official neglect, the Department of the Interior had 'taken the first steps in an extensive program calculated to efficiently maintain Canada's sovereignty in that vast region.' Elsewhere in the report, Craig described the Arctic islands as Canada's 'north Empire.'[1]

As befitted a civil servant, Craig did not embark on a nationwide lecture tour or produce any articles for popular consumption. Instead, he turned to the work of planning the 1923 patrol. During the winter of 1922-23, the *Arctic* was overhauled far more carefully than before. Although its engines were too old to perform really well, the *Arctic*'s average speed during the 1923 voyage would be 25 percent faster than in the previous year. Craig also took care to have all the old red paint scraped away before a fresh coat of white and black was applied. He and Captain Bernier agreed that a second ship would be needed in 1923. Having heard that Shackleton's *Quest* was for sale at a bargain price, Bernier travelled to London to inspect it. He quickly decided that the *Quest* was in too dilapidated a condition. The government was not prepared to pay the cost of a new ship as large as the *Arctic*. However, Bernier found an alternate possibility in the sturdy sea-going tug *St. Finbarr*. Although too small to fully replace the *Arctic*, the *St. Finbarr* could act as a tender, carrying supplies to the police posts even in ice conditions too severe for the *Arctic* to face. The *St. Finbarr* was accordingly purchased and renamed the CGS *Franklin* in honour of British explorer Sir John Franklin.

In December, with the plans for the 1923 voyage proceeding well, Craig turned his attention to Stefansson's expedition. He found that in July 1922 the British government had forwarded notes by the hydrographer of the navy which contained detailed comments about the value of Wrangel Island. Neither the Admiralty nor the Air Council considered Wrangel to be of any practical use in the present, although they thought it might possibly have some strategic significance in the future. In general, Stefansson's views on both the strategic and the commercial possibilities of the island now seemed 'over-sanguine' to the naval authorities. British officials did not believe they had a strong claim to the island, but they also doubted the validity of the Russian claim. The Foreign Office was awaiting Canadian views before it replied to the Soviet note.[2]

No statement had yet been conveyed to London. In November the British had passed on the news that the American Embassy had sent the Foreign Office a copy of a note received from the Imperial Russian government in 1916. At that time, the Russians had notified the United States that Nicholas II Land (Severnaya Zemlya), Tsarevich Alexei Island (Malyy Taymyr), and Starokadomskiy Island, all discovered by Boris Vilkitsky in 1913-15, were now part of the Russian Empire.[3] The Russian authorities had taken the opportunity to assert that they considered Wrangel and other northern islands as forming 'une extension vers le nord de la plate forme continentale de la Sibérie.' Even though some of the islands had been discovered by foreigners, they were Russian territory.[4] The Americans themselves considered that their failure to protest the contents of this note prevented them from now making any claim to Wrangel Island.[5] They assumed that the British government had received a similar notice, as indeed it had.[6]

The new British letter had prompted an exchange between Charles Stewart and the Department of External Affairs. Although Stewart did not express himself as emphatically on the subject as had Sir Joseph Pope and Loring Christie, he too felt that 'under existing conditions it would be inadvisable to press Canada's claim to Wrangel Island.' Stewart sent Pope a revised copy of T.L. Cory's memo, asking him to bring it to the prime minister's attention at the earliest opportunity. This Pope immediately undertook to do, and Stewart apparently believed it would settle the matter.[7]

Craig, too, seems to have taken it for granted that the government would give no further support to Stefansson. His main concern was with the $3,000 advanced for the relief expedition. To his dismay, he found that his department's file on Wrangel contained only 'a rather surprising lack of information of any kind' on this subject.[8] Stefansson had made no report on the *Teddy Bear*'s voyage, nor had he provided any accounting of how the government's money had been spent. The only information the department had came from the newspapers. Oswald Finnie and W.W. Cory agreed that the time had come to make inquiries.

In reply to Cory's letter, Stefansson expressed surprise that his company had sent no account. He explained that he himself had not yet heard from Captain Bernard. Bernard was soon to visit New York, and further news could be expected then. Meanwhile, he insisted it was 'clear that the failure of the ship to reach the island was due to its sailing too late in the season.' The delay 'was due to my inability to borrow money,' Stefansson wrote, smoothly placing the responsibility for the situation on the dilatory government.[9] Nothing more was heard from Stefansson for a month, although there were a few vague press reports about Bernard's visit.

A letter from Finnie produced the information that Alfred Taylor, not Stefansson, had spoken to Bernard in New York, and that Stefansson (then in Minnesota on yet another lecture tour) could make no report until he returned to the east in March.[10] At the end of February, a note arrived from John Anderson, the secretary of the Stefansson Arctic Exploration and Development Company. Anderson explained that the company had just moved its offices from Vancouver to Toronto; the move had caused an unavoidable delay in sending the financial statement. According to Anderson, $1,500 of the government's money had been spent on chartering the *Teddy Bear*, $200 on telegrams, and $900 on wages due to Fred Maurer and Ada Blackjack, for a total of $2,600.[11]

With considerable restraint, Finnie politely requested a more complete report and a justification of the use of government money for wages.[12] It seems that he also arranged for Bernard to visit the Canadian capital, since the Wrangel file contains a statement from the captain, dated 'Ottawa, 6th March 1923.' Bernard emphasized that both he and Stefansson had considered 20 August the best date for the *Teddy Bear* to sail because September was well known to be the most favourable month for ice conditions around Wrangel Island. When Bernard heard about Stefansson's claim that the *Teddy Bear* should have set out earlier, he was outraged by this attempt to deceive government officials, and he immediately decided to have no more to do with Stefansson.

Stefansson informed Finnie on 8 March that some bills for the relief expedition had only just been received, and that a final statement of the expenses should soon be sent by Anderson. 'Of course,' he concluded, 'you understand in a general way that the three thousand dollars advanced by the Government is but a small part of the total amount of money that has been spent on the Wrangel Island enterprise.'[13] Six days later, after his return to New York, Stefansson sent a much longer letter, enclosing a copy of Captain Bernard's report to Taylor. The letter, written in Stefansson's usual confident, plausible tone, must have enraged everyone in Ottawa who read it. Stefansson wrote as if there had never been any question of his repaying the money, and he commented that he had 'understood the allowance of $3,000 by the Government last year was to help us continue the Wrangel Island enterprise.' Nor can the enclosed record of Bernard's meeting with Taylor have raised Stefansson's credit in the Department of the Interior. Bernard had told Taylor about a conversation with Captain Hammer of the *Silver Wave.* According to Hammer, Allan Crawford's party had taken insufficient supplies for even one winter. Stefansson, however, insisted in his letter to Finnie that he had given the men 'a free hand' in choosing their own supplies and that they had taken 'an outfit which in their opinion was ample for two years.'[14]

Despite these confident assertions, Stefansson knew his position in Ottawa was weak. He therefore tried to persuade Washington officials that an American claim to Wrangel might succeed. There was some interest from the navy, but the State Department rejected his arguments.[15] He turned back to Ottawa, reiterating in another letter to Finnie that the $3,000 was 'a contribution to the general expense' of the expedition. If there was any balance left over, he thought it should be applied to the 1923 relief effort. Finnie doggedly replied that the balance must be refunded. Stefansson intended to visit Ottawa at the end of March, and Finnie suggested that they should meet to discuss the matter.[16] Stefansson eagerly seized on this offer, but he took care not to show his desperation. Ignoring the fact that he had secured the $3,000 by claiming the lives of his men were in danger, he described the party as 'a colony, no more stranded than were the early settlers of Nova Scotia or Plymouth Rock.' He wrote loftily, 'I cannot control the policy or actions of the Government but I can at least do my best to secure private support for the colony if public support is unavailable. Should I fail, my record will in any case be clear and will be remembered ...when every nation in the world covets such air bases as ... Wrangel Island.'[17]

These statements were more than enough to goad Finnie and others in his department into renewed efforts to block the Wrangel plan. In this endeavour they had the enthusiastic cooperation of External Affairs. On 22 March Pope wrote a memo to Prime Minister King, stating that he and Stewart were both strongly against a claim to the island.[18] To reinforce their arguments, King also received a letter from Joseph Bernard. 'In case Mr. Stefansson approaches the government for an appropriation for aid in his Wrangel Island adventure, I wish you to understand clearly that Stefansson has *not* chartered my vessel, nor will I have anything to do whatsoever with him. I have many good and sufficient reasons for this decision,' Bernard wrote.[19] When Stefansson arrived in Ottawa on 27 March, Finnie confidently informed him that if he wished to discuss sovereignty matters, he must meet with Stewart and King.[20]

As King's diary records, at the meeting Stefansson was firmly told there would be no claim because Wrangel Island was not even in the western hemisphere; moreover, 'the claim by which he w[oul]d assert our right to it w[oul]d defeat [the] claim we have to valuable islands off [the] Greenland coast.'[21] But Stefansson was not prepared to accept defeat. He saw King again the following day. Somehow, he managed to arrange for another discussion of Wrangel within the next few weeks. Press items disclosing that there had been talks between Stefansson and the government then began to appear. Pope addressed another strongly worded memo to the prime minister, but to no avail. Stefansson returned to Ottawa on Saturday, 7 April. After a lunch at King's home, he addressed several members of the cabinet in the prime minister's office, arguing with great eloquence that the Wrangel plan should not be rejected without first consulting the British government. There were broad strategic issues at stake, about which only the imperial authorities were fully qualified to judge. Stefansson was then asked to leave the office. He was called back after about an hour and asked whether he himself would be willing to go to England. He immediately agreed, provided that he could be sure of a prompt hearing in London. This King undertook to arrange. Stefansson's fare to England was to be paid by the government, along with a living allowance of fifteen dollars a day.[22] He was given an advance of $700 – more than enough for his travelling expenses and a month in

London.[23]

King's diary account seems to be the only record of the meeting from the government side, and unfortunately it sheds very little light on his reasons for reversing the earlier decision. 'There are imperial considerations of which acc[ount] sh[ou]ld be taken,' King noted, observing also that Stefansson was 'was much pleased with the outcome. We had nearly or rather actually [decided] previously not to press Canada's claim.'[24] King's reservations about too close a relationship between Canada and Great Britain in matters of foreign policy are well known, and historians have generally assumed that the Chanak crisis of the previous September had greatly increased his wariness.[25] In the autumn of 1922, King had unhesitatingly rejected Britain's call for active Canadian support in the Middle East. Nevertheless, while under the spell of Stefansson's eloquence King was all too ready to defer to the opinion of British experts, even at the risk of involving Canada in difficulties with the increasingly strong and self-confident government of the Soviet Union.

On Monday morning (9 April), the *Ottawa Citizen* reported that Canada would back Stefansson's plan to make Wrangel Island a base for aerial operations by Great Britain. According to this article, a strong point in favour of the plan was that Wrangel lay on the 180th meridian, which also passed through Greenwich on the other side of the globe. The island was thus on a direct line from the mother country, showing its potential as a stop on the future air route from London to the Far East. This neat reversal of Pope's warning that Wrangel was 'an Asiatic island' may well have been one of the arguments Stefansson used in his meeting with the cabinet. The *Citizen* announced that Stefansson would travel to London with 'the endorsation of the government here' and that he was anxious for Canada to establish a police post on Wrangel, similar to the one on Ellesmere Island, 'so that active possession be taken.'[26]

Craig was the first official to see this item. Aghast, he sent it to Finnie, pointing out that if the government's policy could not be changed, 'we should certainly be provided with funds sufficient to complete immediately our program for maintaining our sovereignty in the north instead of spreading it over a number of years as is the present intention.'[27] Finnie, in turn, passed the clipping on to the deputy minister's office with the resigned comment that there seemed to be little use in making 'any further representations to the Minister' about Wrangel.[28] He and Craig may have derived some comfort from a sardonic editorial by Grattan O'Leary, which appeared in the *Ottawa Journal* the next day. With 'millions of leagues of land' to exploit in Canada's own northwest, the Liberal government had chosen instead to 'give Mr. Stefansson a free trip to England and to put him up in a sumptuous London hotel to find out whether we shouldn't take over an iceberg off the Russian coast.' O'Leary suggested that 'not one Canadian in a thousand cares a tinker's curse about Wrangel, or would be willing to spend a nickel on it.'[29] For Loring Christie, the news may have been the last straw. After over a year of frustration at King's lack of confidence in him, Christie submitted his resignation in May.

A TRIUMPHANT STEFANSSON DECIDED to make a public appearance in Ottawa before leaving on his journey to England. A lecture was scheduled for 1 May, and Stefansson's new press agent, Ottawa native Donat Le-Bourdais, made strenuous efforts to obtain favourable publicity. The flyer advertising the event proclaimed that this particular talk held a 'special significance' because Stefansson had 'occupied Wrangel Island in the name of Canada and at his own expense financed an expedition to hold the island ... He has spent or raised by the aid of a few friends nearly $30,000 in the process, and has been misrepresented and ridiculed in certain quarters of Canada for his pains. In his lecture here, he will speak of Wrangel Island and reply to criticisms that have been levelled at him.' Stefansson was 'a big man with a big theme,' whose subject was 'of first importance.'[30]

LeBourdais approached an old acquaintance, fellow journalist Madge Macbeth, and asked for her help. Macbeth was 'incensed' by what she saw as a blatant attempt to manipulate her, and she promptly described the entire episode to Belle Anderson. LeBourdais had spent an afternoon recounting the story of the Canadian Arctic Expedition and the *Friendly Arctic* controversy from Stefansson's point of view. 'He told her,' Belle wrote, 'that the Southern Party mutinied up in the North and would not give Stefansson any food at all, and Stefansson had to go out and live on the ice on what he could kill. She said, "Do you mean to say that just because they did not agree with him, that Stefansson vilified these officials of the government in his book?" He said, Yes, it was necessary to do it to protect himself. He did it on purpose so that they would be discredited and that they would not be believed if they opposed him in any of his projects.' Stefansson 'had great ends in view at present in regard to Wrangell Island,' and he was determined that Anderson and the others should not be allowed to undermine them. LeBourdais went on to speak calmly of the possibility that the party on Wrangel might be dead if they had not lived off the country as Stefansson had instructed them to do. LeBourdais admitted that Stefansson had not as yet received any support from the Canadian government. Macbeth retorted that if so, it was 'disgraceful' for Stefansson to deliberately give the impression that he did have official backing. She decided to lend her own support to the other side.[31]

Stefansson's enemies were therefore ready for him. Both Rudolph Anderson and James White had extensive collections of material detrimental to Stefansson. White was by this time again employed in the civil service: from the spring of 1922 until his death in 1928, he served as the minister of justice's adviser on boundary issues. Anderson scrupulously observed the letter if not the spirit of the promise given to Charles Camsell in 1922 (see Chapter 6); however, White was under no such prohibition. Stefansson was certainly aware of their information gathering. Indeed, he had recently written to White on the subject. Late in 1922 Stefansson had given a lecture in Lansing, Michigan, and a copy of the talk was printed in the Michigan State Teachers' Association bulletin for February 1923. White heard that it contained a number of far-fetched statements, such as a claim that the climate of Iceland was little different from that of northern Italy. He requested a copy from the association. Stefansson was notified of the request, and the explorer accordingly addressed a letter to White on 18 April – just over a week after his successful meeting with King and the cabinet. Stefansson claimed that the stenographer who transcribed the talk (a marked copy of which he enclosed) had made numerous errors.

In the first flush of victory, Stefansson went out of his way to be offensive to White. 'I hope that some of the more ridiculous ... errors will suit you exactly and I hope you will make the widest use of them,' he wrote. 'It has occurred to me that your grasp of the subject dealt with may be insufficient to enable you to detect some of the errors. I have, accordingly, indicated some of them by a red check mark. In cases where I have thought the check mark would not be enough, I have added an annotation pointing out exactly wherein the error consists. Should you at any time become interested in the views which I really hold, they are to be found in my books.'[32] Evidently believing that Harkin was still on his side, Stefansson sent him a copy of this reply. In a two-sentence acknowledgment, Harkin merely commented that it was 'some letter.'[33]

White responded that he 'would be lacking in courtesy' if he did not 'acknowledge the urbane tone' of Stefansson's letter, which he mockingly claimed was similar to that in the letters of Lord Chesterfield – long considered the very model of gentlemanly politeness.[34] When White heard about Stefansson's 1 May lecture, he simply drew on his hoard of documentation. He had a letter from Pope stating that Stefansson had never been naturalized as a British subject, and another from an official in the North Dakota state government confirming that in 1902 Stefansson had run for public office in the United States – something he could not have done without being an American citizen.[35] On the day of the lecture, the *Journal* published a second anti-Stefansson editorial by Grattan O'Leary, which revealed these facts and ridiculed the explorer's claim to be a patriotic Canadian with the best interests of the British Empire as his only goal.[36]

This evidence of intense and abiding hostility in Ottawa enraged Stefansson, even though he had brought it on himself. He urged everyone who attended his lecture to read the editorial, citing it as a perfect example of the slanders he had to endure.[37] To both Harkin and Camsell, he wrote that he intended to give the *Journal*'s criticisms of him the widest possible publicity; perhaps they would be printed as an appendix to the next edition of *The Friendly Arctic*. In order to present himself as an innocent victim, Stefansson now told Harkin that he had not sent the offensive letter to White after all. Harkin replied soothingly that the *Journal* was only a local Ottawa paper, and Stefansson had nothing to gain by giving international publicity to its views.[38]

Clearly, by this time Stefansson had acquired so many enemies that, even with the prime minister's support, it was highly unlikely he could get his way on a plan involving as many difficulties as the occupation of Wrangel Island. Even as he set off for England full of optimism, the slow, determined campaign to undermine his reputation continued. On 30 March, as part of Stefansson's own campaign to maintain his credibility, *Science* had published two responses to Diamond Jenness's comments on *The Friendly Arctic* (see Chapter 6). Stefansson had already written tauntingly to Camsell that the Jenness piece 'played remarkably well into my hands' because 'others' would 'reply to it in a way that will leave my cause ... in a better position than it was before. If their enemies (Dr. A[nderson]'s) and not mine had written the "Science" article, they could scarcely have made a better job of it.'[39] One of the responses to Jenness was by Stefansson himself; the other was written by two members of the 1913-18 expedition, Burt McConnell and Harold Noice. McConnell had found work with the *Literary Digest;* Noice, after a period of drifting, had once again joined up with Stefansson and would soon go north to assist Carl Lomen with preparations for the 1923 Wrangel relief effort.

Stefansson stated that he had always preferred not to reply to any of the charges made against him, resting his case instead on 'the preface to my book written by the prime minister of Canada.' Sir Robert Borden's approval, he claimed, was the same thing as the approval of the Canadian government. But now Stefansson had been goaded into the revelation that the government had refused to investigate the charges made by Anderson and the other scientists. According to Stefansson, this decision was made because the prime minister and the cabinet felt the complaints were unworthy of serious consideration.[40] That Stefansson himself had also asked for an investigation, and that the government had refused on the grounds that such controversies did no one any good, was not mentioned.

Like Stefansson, McConnell and Noice referred repeatedly to Borden's 'endorsement' of *The Friendly Arctic.* Their statement included a letter from George Desbarats, which was printed in full. Desbarats insisted that Stefansson had been in full command of the expedition. He explained rather lamely that, though the quotations from the expedition's orders given by Jenness were accurate, when taken together the extracts gave 'a different impression to that which they were intended to convey.' According to McConnell and Noice, this statement proved the scientists were officially considered mutineers. 'Disobedience of orders on a polar expedition is as serious a thing as disobedience in time of war,' they commented.[41]

'I thought I had this affair all quieted down ... Stefansson has now opened the case again,' Camsell wrote in exasperation.[42] He foresaw a strong response from the scientists, and he was not wrong. Rudolph Anderson decided that the time had come to confront Borden directly. He later admitted that he was actually pleased when the McConnell and Noice letter appeared, because it gave him 'a good opening' for a direct approach to Borden.[43] Anderson wrote to the former prime minister on 14 April, inquiring whether the statements made in *Science* were a fair representation of Borden's views. The reply was short and somewhat equivocal. Borden told Anderson that he had had 'correspondence and also interviews with Mr. Stefansson' while Stefansson was writing *The Friendly Arctic,* but he stopped well short of saying that he endorsed Stefansson's version of the events at Collinson Point. Borden carefully noted that he was no longer prime minister at the time the preface was written. The preface, he told Anderson enigmatically, 'must speak for itself.'[44]

Anderson immediately returned to the charge. He asked whether the preface should 'be interpreted as answering any alleged charges or rumors' circulated in Ottawa by members of the Southern Party. Borden replied, 'As already explained the preface to "The Friendly Arctic" must speak for itself. It had relation to the achievements of the expedition therein detailed. I cannot recall [any] conversation respecting the difficulties that arose between Mr. Stefansson and other members of the expedition, except on one occasion.' On that occasion, Stefansson had 'explained that certain charges had been made against him, and he felt some hesitation in putting forward his side of the case in the forthcoming volume. He asked my opinion as to the course [he] should pursue, and I advised him that he should set forth the circumstances as he understood them.' Borden himself 'had not had the opportunity nor would I think it desirable to enter into examination [or] discussion of the difficulties alluded to.'[45]

This tight-lipped, laconic refutation of Stefansson's claim that Borden fully accepted his version of the so-called mutiny was more than enough for Anderson and the others.[46] Clearly, Borden's private letters could not be made public, but it was equally clear that Stefansson could no longer hide behind his former patron's prestige. A letter from Camsell accordingly appeared in the 8 June 1923 issue of *Science*. It was blunt and to the point. Camsell stated that both Stefansson and the scientists had been eager for an inquiry; Minister Stewart had refused their requests because '[t]oo much publicity, unfortunately, had already been given to the differences that had arisen between the members of the expedition and any more was highly undesirable.' As for the 'so-called mutiny at Collinson Point,' to call it a mutiny was 'using too strong a term.' It was 'a decision on the part of the so-called mutineers to adhere to instructions originally given them ... One has only to reflect on what a mutiny means and he will realize that if there were sufficient grounds for such a charge steps would have been taken long before this by the government to punish the offenders.' This much having been said in justice to the scientists, Camsell hoped that no more would be heard of the matter.[47]

Camsell's letter was a major blow to Stefansson's cause, at least among those who read journals like *Science*. Stefansson's determination to present himself publicly as the injured party had resulted only in a damaging statement from an official who would have been happy to see the controversies of 1913-18 forgotten. At the same time, one of Stefansson's misleading public comments about Wrangel Island also came back to haunt him. In mid-May Captain Bernard arrived in Montreal from the United States. There he was shown the press articles stating that the RCMP might establish a post on Wrangel. Bernard immediately wrote to Cortlandt Starnes at police headquarters to warn him of the likely dangers. He told Starnes that he had spent considerable time in northeastern Siberia and had often come in contact with the Soviet authorities there, who 'were quite free in expressing their opinions concerning foreigners occupying Russian territory.' Since few Canadians were familiar with Russia, Bernard felt it was his duty to inform the government 'that very great danger will be involved in any occupation of Wrangell Island ... I feel that Russia would not hesitate at wiping out any post established on that island with a political motive behind it. I know the sort of men around East Cape and in that neighbourhood, and they would stop at nothing.'[48] Starnes thanked Bernard warmly for this information. He immediately forwarded it to Pope, who sent it on to the prime minister. Pope told Starnes that Bernard's statements agreed with 'the general information I have on the subject. I think we should make a great mistake to interfere in any way with Wrangel Island, which does not belong to us, and with which we have no legitimate concern.'[49]

Stefansson arrived in the imperial capital on 23 May with an exhilarating sense of having reached a new world where all things were possible. The Duke of Devonshire, now back in England as colonial secretary (he was replaced as governor general by Lord Byng of Vimy in August 1921), had been requested by Ottawa to provide Stefansson with 'an opportunity of expressing his views to the appropriate officials promptly after his arrival in England.'[50] Stefansson was contacted by the Colonial Office less than an hour after he checked in to his London hotel, the Savoy. Meetings with representatives of the Admiralty, the Foreign Office, and the Air Ministry were quickly arranged. Stefansson was overwhelmed by the heady sensation of being at the centre of things and by the general atmosphere of power and privilege that official London still possessed. 'On the whole I have never had a pleasanter experience,' he later wrote. 'Although I came there ostensibly to bring information and recommend action, I learned more than I was able to teach. They were technical experts in the true sense, men of scholarship and wide outlook. To few of them were the subjects I presented new, and in many cases their range of information was wider than mine.'[51] Stefansson was particularly impressed by the wealth of information about the Arctic in Admiralty files. When he came to write his own account of the Wrangel Island affair, Stefansson could not resist the temptation to contrast the wide knowledge and cosmopolitan outlook of London with the narrower views found in Ottawa, declaring that the British knew far more about the Canadian north than did anyone in Canada.[52]

Because Stefansson came to London apparently as the representative of the Canadian government, his views were at first listened to with respect, and his personal charm quickly won over the first lord of the Admiralty, Leo Amery, and the air minister, Sir Samuel Hoare. This feat was all the easier because Amery had long been an advocate of development in the Canadian north. In 1910, while a journalist with *The Times* of London, Amery had visited the shores of Hudson Bay. From this trip he gained the impression that Canada's middle north had great potential for the future.[53] It was not much of a leap from this conviction to a belief in the potential of the Far North. Amery had already read Stefansson's *The Northward Course of Empire,* and he invited the explorer to lunch a few days after his arrival in London. Over the summer, Amery went on to read all of Stefansson's other books, recording his favourable impressions in his diary.[54]

To create public interest in Wrangel Island, Stefansson wrote articles for *The Times,* the *Spectator,* the *Manchester Guardian,* and the *Observer.* He was particularly pleased by the opportunity to publish in the *Spectator,* observing that it was read by 'practically all the influential people in Great Britain ... the men who controlled the Empire.'[55] Stefansson now claimed that the orders given to him in 1913 had authorized him not only to take possession of new lands, but also to reaffirm British sovereignty over any that had previously been discovered by officers of the Royal Navy. (In fact, the instructions referred only to lands lying directly north of Canada, and there was no clause about reaffirming earlier rights.)[56] 'In accordance with this our men raised the Union Jack on Wrangell on July 1st, 1914, as part of a formal ceremony of reasserting British rights,' Stefansson wrote. Their act had made the island British until 1919. Therefore, the Russian claim made in 1916 could not be valid. The island had considerable commercial value: it was 'probably the greatest polar bear country in the world,' and there were 'no locations better for the trapping of white and blue foxes.' In a new touch, Stefansson added that his men had seen 'considerable quantities of the remains of the prehistoric elephants –

mammoths – so that fossil ivory may be of some importance.'

But, according to Stefansson, Wrangel's greatest value was as the site for an air station. He had therefore made 'representations at Ottawa' in 1919. By 1921 'a series of expeditions was being secretly planned ... with the double aim of discovery and confirmation of ownership.' Stefansson left it to be inferred that these expeditions were to go to Wrangel Island and the surrounding area. However, he recounted, there were unfortunate complications over the choice of a commander. A 'deadlock between my supporters and those of the other candidate' resulted in 'postponement of action for that year.' Stefansson claimed to have received a message from Meighen, 'saying that an expedition could not be sent in 1921, but would go in 1922.' Fearful of Japanese expansionism, Stefansson decided to act immediately, using his own resources rather than the government's. Government action need not involve any expense, since there were plenty of commercial companies ready to step in and develop the island. Stefansson did not hope to share in the profits that were sure to ensue. Instead, he preferred to be known as someone who had acted 'entirely on grounds of public spirit.'[57] The notes of British patriotism and imperial pride were sounded throughout the articles. To a reader who suggested the internationalization of Wrangel Island as the solution, Stefansson replied that his great desire was to see 'a really united Empire powerful enough to dictate the peace of the world.' Those who had faith in the Empire as 'a force for peace and stability' should 'try to make good our control of the Polar Ocean.'[58]

Within a month of Stefansson's arrival, the authorities in London had decided that the question of Wrangel Island must be handled by the Foreign Office. A memorandum written in that department summed up the arguments in favour of a British claim put forward by Stefansson, Amery, and Hoare. Stefansson had repeated his usual claims about the island's commercial value, adding that the Hudson's Bay Company was eager for a lease. He insisted that there were problems in reaching the island only 'about once in twenty years.' Wrangel was 'the only territory in a vast area to which Great Britain has any claim.' Therefore, it would be a very short-sighted policy to surrender the British title. The climate would be no handicap to air transport, since the atmosphere was 'generally clear and still,' with continuous daylight in summer. In the future, the island might become a useful link on the air route from London to Hong Kong and Singapore. As to ownership, the American government had never made a formal claim. The island had not been discovered by a Russian, so the Soviet claim could only be based on proximity. There was a strong feeling in Alaska against British or Canadian ownership, but Stefansson was sure that 'responsible naval and military opinion in the United States is prepared to see this happen.' All in all, it seemed feasible for a claim to be based on British discovery and Canadian occupation.[59]

Whether such a claim would be politically expedient and worth the inevitable conflict with the Soviet Union was quite another matter. Lord Curzon declined to make a firm recommendation. Instead, he insisted that the matter was one for the entire cabinet to weigh. On 7 July Hoare notified Stefansson that the final decision should be made soon. Stefansson accordingly wrote to King: 'Since the Imperial Government seems to have concluded that the matter is one of considerable importance, and since I have always been of that opinion myself, I consider it necessary for me to stay until everything is concluded.'[60]

While Stefansson was in London, the responsibility for a 1923 relief effort fell heavily on Alfred Taylor. The Stefansson Arctic Exploration and Development Company had no funds available for the purpose. Stefansson did not respond to Taylor's telegrams, leaving Taylor to do what he could on his own. King declined to meet with him to discuss a grant, and Cory was away from Ottawa. At a meeting with Roy Gibson, the acting deputy minister of the interior, Taylor urged humanitarian considerations and the fact that Allan Crawford was a Canadian citizen. Gibson coldly responded that until 'the decision of the Imperial authorities' was received, it would be 'impossible to consider any application for a grant to aid in an expedition to Wrangel Island for the very good and sufficient reason that any steps in this direction would be interpreted as giving an official significance to the expedition and might complicate the issue at present under consideration in the Old Country.' Taylor could only reply that Stefansson would have to either procure a decision in London or raise the money by a public appeal.[61]

Stefansson was able to obtain some money through an appeal published in *The Times,* and more was provided by aviation enthusiast Griffith Brewer, the managing director of the British Wright Company. (Stefansson was a close personal friend of Orville Wright.) Publicly, Stefansson presented the venture as a purely humanitarian one, designed to ensure the safety of the party on Wrangel Island. Privately, however, he was determined to continue his 'colonization' – as he now called it – if any way could possibly be found to do so. It was necessary to purchase extra supplies in case the relief ship had to winter in the ice. Stefansson reasoned that if the ship did not have to winter, the supplies could be used to continue the occupation of Wrangel. Despite reports from Nome that the Soviet authorities were threatening to arrest anyone who entered Russian territory without permission, Stefansson gave orders to Carl Lomen and Harold Noice that the rescue plans 'were to be converted into plans of continued occupation if and when the rescue proved successful.'[62] Lomen and Noice hired an American, Charles Wells, and several Inuit families to spend the winter of 1923-24 on Wrangel. Wells and his group would share in the profits from any furs they collected. The relief ship, the *Donaldson,* sailed on 3 August with Noice and the new 'colonists' on board.

On 25 July the British cabinet discussed Wrangel. Amery recorded in his diary that Curzon's reluctance to proceed with a claim was very clear. However, no final decision was made. The ministers agreed that 'cautious steps' should be taken to ascertain the American government's position. They also thought there should be consultation with the Canadians.[63] In response to an inquiry from London, on 15 August the embassy in Washington reported that the United States would certainly issue a strong protest against any Canadian or British claim. This information gave the lie to Stefansson's earlier statement that high American officials were prepared to accept British ownership of the island. Two days later the chargé d'affaires in Washington, H.G. Chilton, sent a further report on recent articles by Robert Bartlett in the New York papers. In the articles, Bartlett firmly asserted that when he landed on Wrangel Island in 1913 he considered it American territory, and that he had never received any instructions either from Stefansson or from the Canadian government to claim it for Canada.[64]

Then, on 25 August, the Foreign Office received yet another Soviet note. The government of the USSR, 'being wholly unable to understand the absence of the required explanations, and having in the meantime learnt that new expeditions are being planned by British subjects to the Isle of Wrangel,' considered it 'necessary again to state, that it regards the Isle of Wrangel as an integral part of the Union of Soviet Socialist Republics ... Therefore ... it regards the raising of the British flag on the Isle of Wrangel as a violation of Russian sovereign rights.'[65] Unaware that Stefansson was, in fact, sending new 'colonists' to Wrangel, the Foreign Office instructed the British agent in Moscow to reply that the expedition in the *Donaldson* was 'a private one organized by Mr. Stefansson for the purpose of rescuing Mr. Crawford and his party, that the question of ownership of the island is not thereby raised, and that any attempt to interfere with the relief expedition would be viewed most seriously by His Majesty's Government.'[66]

Despite these strong words from Britain, Curzon was now convinced that his initial assessment of the situation had been the correct one. There were no realistic grounds for hope on Stefansson's part; however, the explorer clung to the belief that ultimate success was still possible, as long as he had the support of Mackenzie King. King would arrive in late September for the 1923 Imperial Conference. Stefansson had every reason to believe that he possessed an unusual degree of influence over the Canadian prime minister. A personal interview would be imperative, and it was in Stefansson's interest to see King alone, without his ministers or other advisers. This might be easier to achieve in London than in Ottawa. Stefansson decided to gamble on remaining in England for another month, even though he was completely without funds to pay his own expenses. The allowance provided by the Canadian government had been given on the assumption that his visit would last only one month in total, and two had already passed since his arrival.

Stefansson wrote to King at the end of August, putting the best possible face on the situation. He declared that in view of the great issues at stake, he 'could not afford to consider' the question of expenses. He could only hope the government would support him in his decision to stay. Stefansson claimed that 'most or all of those members of the Cabinet who expressed themselves' at the 25 July meeting were 'favourably disposed' to his plans. However, the consensus among the ministers was that it might be 'better to carry on such enterprises as that of Wrangell Island privately for the time being and that in any case final decision had better be held over for the Imperial Conference when you yourself would be here, and able to make clear the Canadian attitude.' Stefansson thus deftly transformed his own hope that King's presence might bring about a change in policy into a flattering hint that Curzon was ready to defer to Canadian opinion. In a second letter written on the same day, Stefansson relayed more 'semi-confidential news' which he felt sure would interest King: the reindeer herd established on Baffin Island by the Hudson's Bay Reindeer Company was reportedly doing well. In a gratified, complacent tone he would soon have reason to regret, Stefansson concluded: 'Irrespective of what happens in the particular case of Wrangell Island I feel rather well satisfied with my summer in England, for a number of the more prominent men are taking a keener interest in northern development and viewing it from an angle new to them.'[67]

Before the Soviets could respond to the British note, and before Stefansson's letters to King were received in Ottawa, front page headlines around the world reported the news brought by the *Donaldson* on its return to Nome: Allan Crawford, Lorne Knight, Fred Maurer, and Milton Galle were dead. Of the five people who had landed on Wrangel Island in 1921, only Ada Blackjack remained alive.

The expedition's first year had passed harmoniously and successfully enough. The four men got along well, and they took scientific observations to keep themselves occupied. Ada Blackjack, as the only woman and the only non-white person in the group, was terrified by the appallingly isolated situation in which she found herself. At first the men considered her moody and lazy, and they were made extremely uncomfortable by her apparent determination to marry one of them, preferably Crawford. Eventually, despite the many barriers to cross-cultural understanding, accommodations were made. Blackjack applied herself to tasks like sewing and cooking. During the first summer, game was reasonably abundant, but in the second year the seals and bears deserted the area. Knight fell sick, probably with scurvy. In January 1923 Crawford, Maurer, and Galle set off across the ice towards Siberia to bring help and supplies. They were never seen again. Showing great resourcefulness, Blackjack brought in just enough game to keep herself, Knight, and the expedition's pet cat alive. Knight died of his illness in June, leaving Blackjack entirely alone for two months. 'I don't think I could have pulled through if it hadn't been for thoughts of my little boy at home,' she later explained. 'I had to live for him.'[68] The story of Blackjack's ordeal quickly made her a newspaper heroine.

From the very beginning, Stefansson flatly refused to admit that shortage of food was the cause of the tragedy. The news broke on 1 September; the 2 September edition of the *New York Times* contained a statement by Stefansson that there had to be 'some special explanation' for the deaths. Knight and Maurer were experienced in Arctic survival techniques, and there were 'seals, bears and walrus in plenty at Wrangell Island.'[69] Stefansson argued that the trip to Siberia must have been undertaken simply because the men were too impatient to wait for relief. Noice more or less backed up this story, although his hints that Knight and Maurer had possessed too little Arctic experience infuriated Stefansson.

When he came to Toronto at the end of September to visit Crawford's parents, Noice told a reporter that the 'boys' had underestimated the difficulties of a journey across the ice. On Wrangel they allegedly had two years' supplies and 'a good game country full of walrus, geese and ducks.' Therefore, the disaster was not due to 'any neglect on the part of Stefansson.'[70] The next day's papers, however, printed a letter from Fred Maurer to his wife, Delphine, written just before the party set out across the ice. 'The chief reason for our leaving is the shortage of food. There is not adequate food for all, there being only ten twenty-pound cases of hard bread and three cakes of seal oil to last until next summer. The prospects for getting game between now and next summer or sealing season are very poor,' Maurer recounted.[71]

As far as Stefansson's reputation in Canada was concerned, the death of Allan Crawford was a fatal blow. With the exception of geologist George Malloch, the men who died during the 1913-18 Canadian Arctic Expedition had not actually been Canadians, and in the autumn of 1914 there were so many other deaths to mourn. Now, however, young Crawford was quickly elevated by a sensation-hungry press to the status of a Canadian hero. He was idealized as the very embodiment of the nation's young manhood. 'Canadians will always remember that [Crawford] went forward with faith ... he deserves a place alongside the youth that fell in the Great War,' declared University of Toronto president Sir Robert Falconer.[72] Inexperienced he might have been, but Crawford had cheerfully followed the advice given by Knight and Maurer, and on the whole his leadership was entirely satisfactory. Photographs taken during the expedition's first year showed him as handsome in his Arctic garb, competently performing various practical and scientific tasks. Crawford appeared relaxed and confident, and he seemed to have taken well to life in the North. He had been sent to Wrangel despite his youth and inexperience, and he had acquitted himself well. His death was all the more poignant because Crawford had believed he was serving his country.

'The call to the Arctic mission came to him in the flush of his young manhood, midway in a brilliant scholastic career, with all the rich promise of life before him. He responded without thought of the perils and hardships, except as an opportunity of service, a broadening of his mental horizon, a challenge to his courage and endurance,' declared the *Globe*. But Wrangel Island was 'not worth the brave lives it has cost ... The value of the island, present or potential, is not to be weighed against the precious lives of Allan Crawford and his companions.' Other papers were even more severe. According to the *Montreal Standard*, 'the blood of these three young men is on the head of Mr. Stefansson. While they were starving to death on a frozen island to prove the Stefansson theory that an explorer could live off the country, Mr. Stefansson himself was prancing about from one luxurious spot of civilization to another, lecturing and telling what a hero he was.' The *Mail and Empire* lamented the death of Allan Crawford, 'a patriotic Canadian,' and accused Stefansson of having 'humiliated' Canada by his outrageous schemes. *Saturday Night* agreed that Stefansson was to blame: Crawford and the others had 'perished while the author of the wild enterprise was far away talking nonsense about the supposed strategical importance of Wrangel. That adventurous lads of so fine a type as Crawford and his companions should needlessly perish is something difficult to forgive.'[73]

King, Stewart, and Cory all managed to evade the press on the day the Wrangel story broke. It was left to Finnie to state that 'the Crawford expedition ... had nothing to do with the Canadian government' and that Wrangel was 'not a British possession.' Finnie added that the 'entire responsibility for sending Crawford and his associates to Wrangel Island in the first place lay with Mr. Stefansson.'[74] The next day Stewart expressed his sympathy for the Crawford family, but he emphasized that the government had had no connection with the flag-raising on the island. Wrangel, he said, 'was more a matter of concern for the British Parliament than for Canada.'[75] Not long afterwards, King left Ottawa to attend the Imperial Conference, which began on 1 October.

Despite the dramatic turn for the worse in his fortunes, Stefansson was still in London, in accordance with the plan he had announced to the prime minister at the end of August. The *Donaldson* had left Charles Wells and his party on Wrangel Island, and Stefansson was as determined as ever to continue with his colonization scheme. A few weeks earlier, he had sent Alfred Taylor a wire, assuring him that Amery and Hoare were still 'strongly behind' the plan and that they had been 'delighted [to] learn Noice left people [on] Wrangell.' The Foreign Office was 'not yet committed to Wrangell,' but it soon would be 'if Noice does [the] right things.'[76]

On 5 October Stefansson sent the prime minister a plea for money, noting that he owed large sums for the *Donaldson*'s voyage and for the payment of salary arrears to the dead men's families. He complained of 'grave inaccuracies' in Noice's reports and other press accounts, which had given rise to 'the most painful misunderstandings' about the cause of the tragedy.[77] Unmoved by any of this, King refused to see Stefansson. The explorer was later observed sitting forlornly in the lobby of the Ritz Hotel, where King was staying. A member of the Canadian delegation spoke to Stefansson, who seemed 'terribly "down."'[78] Stefansson may simply have turned up without an appointment. He managed to see one of the prime minister's advisers, O.D. Skelton. Skelton recorded in a letter to his wife that he 'liked [Stefansson] very well personally but didn't agree with his Wrangel Island proposals.'[79] There was quite obviously no hope that King would plead Stefansson's case with the British cabinet. Admitting defeat, if only for the moment, Stefansson sailed back to New York on 8 October. In spite of his debts, he took a first-class passage on the liner *Leviathan* in full confidence that the Canadian government would pay the entire cost of his trip.

The day after Stefansson's departure, the London *Times* printed his version of events. Stefansson disclaimed any responsibility for the shortage of stores on Wrangel. He recounted that he had simply given all the money he had to Knight and Maurer, 'telling them to use it as they thought best.' If only the *Teddy Bear* had left earlier in the summer of 1922, he claimed, the men would have received ample additional food supplies. And there could be no doubt of the suitability of Wrangel Island for colonization, since Noice had reported 'beautiful green fields stretching away to the mountains in the interior.'[80]

By the time Stefansson reached New York, the *Arctic* was back in Quebec after another successful voyage. There had been no difficulty in securing the funds for a northern patrol in the summer of 1923; however, the *Franklin* had not yet arrived in Canada because of a dockworkers' strike in England. Disagreements with the RCMP also posed a threat to the plans made by Finnie, Craig, and Bernier. Cortlandt Starnes, now police commissioner after the retirement of A. Bowen Perry, had received a report from Sergeant Joy to the effect that 'unscrupulous white traders' at Cumberland Sound were exerting a bad influence over the Inuit, to whom they gave 'considerable intoxicating liquor.' Starnes, who felt duty-bound to consider the matter 'from a Police point of view,' therefore wanted to place a second post on Baffin Island. The supplies already purchased in July 1922 and left behind in Quebec for lack of space on board the *Arctic* could be used for this purpose. The police would thus get what they wanted without incurring any fresh expenses. Starnes thought that the establishment of a post on uninhabited North Devon Island for sovereignty purposes could safely be postponed until 1924.[81]

Craig took immediate exception to this plan. No doubt a police post at Cumberland Sound was desirable, but Craig did not 'see any reason why this should be allowed to interfere with our scheme of covering the Archipelago with police posts, custom houses and post offices at what might be called strategic points in the sense of international law.' There could be no doubt of Canada's title to Baffin Island. In Craig's opinion, a more northerly post on Ellesmere should take priority over both North Devon and Cumberland Sound.[82] Bernier agreed.[83] After a good deal of negotiation within the Department of the Interior, Finnie secured Minister Stewart's agreement that three posts should be established in the coming season: one at Cumberland Sound, one on North Devon, and a second post on Ellesmere in addition to the existing station at Craig Harbour. Cory informed Starnes of the decision late in May and asked that the RCMP secure the funds for the necessary building materials and provisions. He added the assurance that, with better planning, enough supplies for all three posts could be carried in the *Arctic*.[84]

Starnes promptly replied that his minister, Sir Lomer Gouin, could not 'see his way to ask for any further appropriation for new Mounted Police detachments in the Far North this year. He is, of course, willing to utilise supplies and buildings now in store at Quebec for the establishment of a new Police detachment on Cumberland Sound.'[85] But Finnie and Craig were adamant that if any part of their proposed program were to be dropped, it should be the post at Cumberland Sound, which lay 'far south of any territory that may be disputed.'[86] Starnes and Stewart met to work out a compromise. They agreed that only one post would be established, using the supplies already at Quebec. An attempt would be made to place this post on Ellesmere, somewhere in the vicinity of Cape Sabine. If the attempt was unsuccessful (as Starnes must have known it likely would be), the post would be placed at Cumberland Sound.[87]

Craig was pleased to find that work on the *Arctic* had been done 'carefully and thoroughly during the winter ... and the good results were at once apparent when the ship sailed.' During the first week of July, the supplies were neatly stowed, without the rush and confusion of the previous year. Besides the police and the ship's crew, the party included surveyor Frank Henderson, naturalist Joseph Dewey Soper, medical officer Leslie Livingstone, mining engineer Lachlan Burwash, and a judge and jury for the trial of Robert Janes's killer at Pond Inlet. Soper, a young graduate of the University of Alberta, claimed to have been 'born with [a] longing for boreal latitudes.' His childhood reading had included Franklin, Nansen, Amundsen, and Shackleton.[88] He was recommended for the job by Rudolph Anderson, and he would justify Anderson's confidence by doing excellent work in the Arctic over the next eight years. Livingstone, too, was destined to become a renowned worker in the Far North.[89] Burwash, who already had northern experience in the Yukon, later carried out mapping surveys on Baffin Island and led a government expedition in search of Franklin relics.[90]

Bernier had arranged for his nephew, Wilfred Caron, to be hired as the ship's third officer. He hoped that Caron, with his northern experience, would be able to take over some of the watchkeeping when the ice was heavy or the wind unfavourable – an exhausting task that had fallen entirely on the elderly Bernier in 1922. The plan was thwarted by Caron's untimely death: he was washed overboard and drowned early in the voyage. An unusual feature of the 1923 voyage was the presence of Mrs. Craig. Following the example of Robert Peary's wife, Josephine, Gertrude Craig became one of the few white women who had ever visited the northern archipelago.

The expedition left Quebec in the early evening of 9 July 'amid the farewell shouts of the crowd assembled on the deck, the waving of handkerchiefs and hats, and, as the ship proceeded down stream, the salutes of other craft by steam whistle and siren.'[91] Again, the *Arctic* followed the coast of Greenland northwards. This time the Canadians called at Godhavn on their outward voyage, in order to pick up thirty dogs ordered the previous year. Craig and the others were invited to visit the scientific station run by Dr. Morten Porsild and his son Erling. There they met explorer Lauge Koch, who had recently completed his three-year expedition to northern Greenland and was waiting for a ship to take him back to Denmark. The Canadians were then invited to a 'most enjoyable' tea on the Danish naval ship *Islands Falk*, with 'dainty refreshments, fine linen and attractive table furnishings.' In return, Craig and Bernier invited the Danish scientists and officers on board the *Arctic* for a showing of the films taken in 1922. 'The Danes,' Craig recorded, 'showed great interest in the pictures, and it was evident that they were quite aware of the real objects of our expedition and that they were impressed with the efforts of the Canadian Government to establish and maintain our sovereignty.' Koch gave a short talk on his expedition, described by Craig as 'a very modest account of a trip necessitating very great hardships.' Following refreshments, toasts, and a singsong, the Danes left shortly after midnight. Craig had found time to discuss administrative matters with Dr. Porsild, and he left Godhavn more convinced than ever that 'a system modelled somewhat after that of the Danes in dealing with the Greenland natives' would be 'advantageous' in the Canadian north.[92] Koch, for his part, would later inform the Royal Geographical Society that he and his countrymen were happy to have a 'congenial neighbour' in the Far North, and that they looked forward 'with pleasure to an amicable, peaceful work in the Arctic country with the British Empire.'[93]

On the *Arctic*'s arrival at Craig Harbour, the police were found in good health and spirits after their first northern winter. Several were ready to stay for three years instead of the two they had signed on for. Craig observed with pride that the buildings 'presented a most pleasing appearance in a fresh coat of gray paint with black trimmings, and a neat terrace of sand and whitewashed rocks round the foundation.' Over the door of the main building was a neatly lettered sign reading: 'Royal Canadian Mounted Police. Customs. Post Office.' However, neither the police nor Kakto's family were happy with the prospect of spending the next winter even farther north at Cape Sabine. Indeed, it soon became clear that Kakto 'could not have been induced under any circumstances' to do so. Inspector Wilcox was placed in a dilemma by the orders from Ottawa: he felt that if the more northerly post were established, it would be his duty to remain there, but on the other hand, Craig Harbour was officially the police headquarters in the archipelago.[94]

As it turned out, Wilcox did not have to make the choice. The *Arctic* arrived at Etah without difficulty. There the Canadians encountered Donald MacMillan, who was once again in Greenland. With MacMillan's assistance, they persuaded two Inughuit families to accompany them to Cape Sabine or, if no post could be established there, to spend the winter at Craig Harbour. The *Arctic* then attempted to cross Smith Sound and got within ten miles of the goal. But the threatening ice conditions again made Bernier nervous, and the judicial party had to be landed at Pond Inlet. Reluctantly, Craig agreed to turn south.

On the way to Pond Inlet, the *Arctic* stopped at Craig Harbour, Dundas Harbour, and Beechey Island, where the ill-fated 1845 Franklin expedition had spent its first winter. One of the Franklin search expeditions had erected a monument in memory of the lost explorers on Beechey Island, and it had become something of a tradition for Canadian expeditions to pay their respects at this site. The 1903-4 Low expedition in the *Neptune* was the first to do so, and Bernier (who took a keen interest in the feats of his British predecessors) had visited Beechey Island on several previous occasions. RCMP historian Harwood Steele called the Franklin memorial the 'keystone of Canada's Arctic empire.'[95] The police and the ship's company went ashore and stood at attention while the Union Jack 'was slowly hoisted to the top of the flag pole above the cenotaph, and after it had fluttered in the breeze amid complete silence for a minute, we gave three hearty cheers ... for His Majesty the King.' Craig found the ceremony 'most impressive.' He had brought along another Union Jack from his days on the International Boundary Survey, and Gertrude Craig was photographed holding it beside the Franklin monument. The party then spent the afternoon searching the island for relics of the old expeditions and watching an 'immense' school of white whales. The visit to Beechey Island turned Craig's thoughts to a new possibility: that part of the cost of a ship to replace the *Arctic* could be paid by carrying tourists on the annual patrols. Such a chance for ordinary Canadians to see the natural wonders and inspiring historical sites of the Far North could, he thought, provide very useful publicity for the government's Arctic program.[96]

At Pond Inlet, Craig was gratified to find that the police had made a 'vast improvement' in the visual aspect of the settlement: 'The Eskimos appeared cleaner, better clothed, and much healthier, while their dwellings had been placed in line along the shore, and the surroundings cleaned up giving a much neater and more attractive appearance to the entire post.' Whether this regimentation was in conformity with their cultural heritage was a question the well-meaning Craig was incapable of asking. Craig was also pleased to meet Dr. Therkel Mathiassen of the Fifth Thule Expedition. Mathiassen had spent the spring and summer of 1922 doing ethnographic and archaeological work on the Melville Peninsula and Southampton Island, while Rasmussen, Kaj Birket-Smith, and Helge Bangsted studied the inland Inuit. After a second winter at Danish Island, the members of the expedition took widely different directions in the spring of 1923.

Rasmussen set out on the long journey west to the Mackenzie River, accompanied by an Inughuit couple, Arnarulunguaq and Miteq. Having given up his plan for a sledge journey through the archipelago back to Thule, he now intended to end the expedition in Alaska. He would take a steamer from Nome to Seattle, cross the United States by rail, then return to Denmark. Birket-Smith and Jacob Olsen took a passage south from Churchill to York Factory on the Hudson's Bay Company ship *Nascopie,* then made their way to The Pas by way of the Nelson River and the track of the as yet unfinished Hudson Bay Railway. From there they travelled to Winnipeg, Ottawa, Montreal, and New York. While they were in Ottawa, Birket-Smith took the opportunity to call on Finnie, and made a very favourable impression on him.[97] Birket-Smith and Olsen reached Copenhagen on 25 September. Only Peter Freuchen (who was suffering from the effects of a frostbitten foot), Bangsted, and some Inughuit members of the expedition remained at Danish Island. When Craig met him at Pond Inlet, Mathiassen was awaiting the arrival of the *Nascopie* in order to begin his journey home, carrying with him about two thousand Inuit artifacts.

The Danes had established excellent relations with the Canadian police. In response to Harkin's demand for surveillance of the expedition, both Sergeant Douglas at Chesterfield and Sergeant Joy at Pond Inlet had made highly favourable reports. Douglas described the Danish explorers as 'very fine men to meet,' and he reported that Rasmussen was 'very popular with the Natives.' The expedition appeared to be purely scientific. There had been one moment of slight concern when the Danish flag was spotted on Rasmussen's tent. Douglas gave him a Union Jack to fly above it, which he cheerfully did. Rasmussen later asked whether he could keep the British flag for use on future occasions; naturally, the request was granted.[98] Joy's report confirmed that the expedition's aims were scientific.[99] Nevertheless, during 1923 rumours reached Ottawa from several sources that the Danes were trading for valuable furs. The most damaging allegations came from Henry Toke Munn, who claimed that the Danes had secured all the white fox skins available in the Fury and Hecla Strait area.[100] The various reports were patiently refuted by Joy, and yet another favourable account of the Danish expedition was received from Inspector E.G. Frere at Chesterfield Inlet. In the end, Munn withdrew his statements.[101] Not surprisingly, the Danes considered the police a 'splendid' group; Freuchen later described Joy as 'one of the finest men I ever met in the Arctic.'[102]

Craig recorded that he was 'very impressed' by Mathiassen, 'a very conscientious worker who had spent two years practically alone with the Eskimos, adopting their language and their manner of living, although to a man of his education this latter could not have been particularly attractive.' Whatever the perceived disadvantages of going native in this way, there could be no doubt that it produced a great deal of useful knowledge, and Craig found it 'rather discouraging to think that we have had to wait for the Danes to take the lead in this urgent work which we should have had in hand many years ago.' He continued to make optimistic plans for the future, when 'young men in their junior year at college, physically and temperamentally fit for the work, might be selected and engaged to go north ... The field is new and there is plenty of opportunity for men of the right stamp to do good work for the country and at the same time make names for themselves in the scientific world.'[103] Both Craig and Bernier hoped that after the *Franklin* joined the *Arctic* on the annual patrols, the older ship might be able to winter in the North, serving as a base for scientific work.

The *Arctic* remained at Pond Inlet for two weeks. Soper was delighted by the opportunity to do more extensive work than had been possible during the ship's brief calls at other spots. He also found time to make a fascinating photographic record of the Janes murder trial. His pictures show the Mounties, resplendent in their full dress uniforms, and crowds of Inuit outside the building where the trial was held.[104] Craig proudly noted in his report that the trial was conducted with exactly the same formalities used in southern Canada. The killer, Nukudla (Nuqullaq), was found guilty of manslaughter rather than murder. The *Arctic* carried him south to serve his sentence in the Stony Mountain Penitentiary.

After duly establishing the new police post at Pangnirtung on Cumberland Sound, the expedition returned to Quebec, arriving there on 4 October. Soper asked one of the reporters who had gathered to meet the ship for the latest news of Stefansson. 'He's dead,' was the startling reply.[105] Once the group realized that Stefansson's demise was of a purely figurative nature, they were highly amused, and the story went the rounds as soon as they were back in the capital. Among other souvenirs, Gertrude Craig brought with her a gigantic narwhal tusk, which she later had made into a bridge lamp. This peculiar artifact was long remembered by her acquaintances in Ottawa, including Richard Finnie. Hearing that the post of assistant radio operator for the 1924 patrol was open, young Finnie (still a high school student) spent the next winter assiduously practising Morse code.

MACKENZIE KING RETURNED TO Canada in mid-November, after the end of the five-week Imperial Conference. It was not long before Stefansson attempted to open the question of Wrangel Island once again. He wrote to the prime minister early in January 1924, enclosing a copy of the December 1923 issue of the *Geographical Journal*. The issue contained an anonymous article on Britain's right to Wrangel Island, written by Arthur Hinks. Although the article was unsigned, it was no secret to Stefansson that Hinks was the author. Hinks had disliked Stefansson on their first meeting (see Chapter 2), but after years of correspondence on geographical matters, the two men had become good friends. The article therefore voiced strong support for Stefansson's cause.

Stefansson also asked whether King had seen the newspaper announcements of a proposed transarctic flight by the US dirigible *Shenandoah*. 'The Americans are about to carry out the plans which I began as early as 1919 to urge Canada to undertake in 1921. They are carrying them out by air. My plan was to forestall such inevitable Air exploration through previous exploration by my method of "living off the country" so that we should already be the owners of any strategically located islands before their value became obvious,' Stefansson wrote. 'I hope you will get someone to go into these matters thoroughly again and see if there is not something which Canada can yet do to prevent her being set back too far by the enterprise and foresight of the Americans.' Though the State Department might not want Wrangel Island, the navy did, 'but our colony is there.' Stefansson claimed that an American admiral had agreed with him about Wrangel's great strategic value. 'I had been hoping for the chance to come to Ottawa to talk over the Wrangell Island situation with you but I am so busy writing and otherwise trying to earn money to pay some of the Wrangell Island debts that I have not been able to find the time,' Stefansson hinted in conclusion.[106] King's secretary replied blandly that the prime minister would 'take occasion to see that the members of the Government particularly concerned in the matters of which you have written are informed of the representations you have made.'[107]

Although there is no direct evidence about King's views on Wrangel Island at this time, all indications are that he had firmly decided not to give the plan another hearing. Both the Department of External Affairs and the Department of the Interior were more adamant than ever that no claim to Wrangel should be made, and King was aware that even Sir Robert Borden now had doubts on this score. Not long after his return from London, King had attended a League of Nations luncheon at which the Norwegian explorer Fridtjof Nansen spoke. King sat beside Borden, and the conversation turned to Arctic matters. As King recorded in his diary, Borden '[did] not speak too confidently of Steffanson [sic],' and he was 'against us asserting a claim to Wrangle [sic] Is. on [the] score we may lose other islands we claim on basis of it.'[108] Given Borden's very strong support for Stefansson in earlier years, even these cautious words indicated a major change of heart.

Stefansson came to Ottawa in late March 1924, but he achieved only a highly unsatisfactory one-hour meeting with Stewart. The minister enraged Stefansson by asking that he submit his proposals about Wrangel Island in writing. Stefansson was still 'in a rage' when he called at the Department of Marine and Fisheries for a discussion of the long-delayed official report of the Canadian Arctic Expedition. He vehemently declared that he would write no reports for the government because, despite all his past services to Canada, 'he had not been treated right in the Wrangel Island affair.'[109] Stefansson had been accustomed to meet at least briefly with George Desbarats on most of his visits to Ottawa; the year before, he had been invited to Sunday lunch at Desbarats' house on the day after his successful meeting with the prime minister and the cabinet.[110] Significantly, there is no reference to Stefansson in Desbarats' diary for March 1924.

As a last hope, Stefansson travelled to Toronto to visit his old supporter, Sir Edmund Walker – the only Canadian who had contributed to the Wrangel Island relief expedition. To his dismay, he found on his arrival that Walker had died the previous night. Soon afterwards, while Stefansson was still in Toronto, his activities were the subject of an unexpected discussion in the House of Commons. The $700 given to him for travelling expenses in May 1923 was singled out for comment by the Opposition during the debate on the Department of the Interior's expenditures for 1923-24. Stewart gladly seized the chance to state that, although he did not 'know officially' what the British government's position was, 'so far as Canada is concerned we do not intend to set up any claim to the island.'[111] Harkin, who happened to be in the House when the matter came up, thought that 'the expression on Mr. Stewart's face whenever he mentioned Stefansson spoke volumes.'[112]

Journalists pounced on this interesting bit of Arctic news.[113] According to Stefansson's later account, the press articles he saw did not mention that there had as yet been no definite statement from the British government to Ottawa.[114] In May Stefansson sold his Arctic Development and Exploration Company to Carl Lomen. At a meeting witnessed by Donat LeBourdais, Stefansson told Lomen that the Soviets 'had put up a bluff, and apparently the British Government had been fooled by it.' Patriotically, Lomen asked whether 'by taking over your interest I could claim for the United States the benefit of your three years' occupancy?' Stefansson answered that it was 'worth trying.' Lomen accordingly arranged for a new relief effort, this time in Captain Louis Lane's ship *Herman*. Lane was instructed to 'raise the Stars and Stripes' and to bring back the furs collected by Charles Wells and his party. Fired by Stefansson's tales of Arctic adventure, LeBourdais asked to go along.[115]

Early in June, Stefansson heard about Stewart's qualifying statement that Britain might yet decide to make a claim. The news placed him in a quandary. Evidently hoping that there were some vestiges of support for him in London, Stefansson attempted to explain his action to the Foreign Office. Britain, he wrote, had a clear legal claim to Wrangel Island, while the Russian case was weak. But Stefansson could not possibly afford a third relief expedition. Therefore, he had had no other option but to 'withdraw in favour of the United States.' The Americans had the 'second best claim, and ... through a residence of forty years in their territories I have their interest at heart almost as much as ours.' However, Britain could still 'stand on her rights,' calling on the League of Nations for arbitration if the Americans or the Russians made difficulties. Stefansson hoped to be repaid for the money he had spent, but this was 'no condition laid down by [me] nor any essential part of what ought to be done just now.'[116] On the same day, Stefansson sent a letter to Stewart, pointing out that either Britain or Canada could still claim the island, 'especially if British action affirming ownership is taken this summer.' He proposed to the minister that Canada should pay him for his past services, thus enabling him to clear his Wrangel debts.[117]

The letter to the Foreign Office had a very different effect than the one Stefansson hoped for. Britain's first Labour government had come to power in January 1924. The new prime minister, Ramsay MacDonald, felt no doubt that good relations with the Soviet Union were more important than the possession of Wrangel Island. The Colonial Office therefore informed Ottawa that 'His Majesty's Government would be unwilling to adopt an attitude calculated to create difficulties with the Soviet Government, unless substantial interests were at stake.' The Air Council had now learned that the island was located in a region of frequent fog. Therefore, it would not be suitable as an air base. The Admiralty did not 'consider that the interests at stake are sufficient to justify the reference of the question of ownership to arbitration.' But before making a final decision, British officials wanted to know what Canada's views were.[118]

Again, Stewart seized the opportunity presented to him. He acted with great speed, obtaining concurrence from the heads of the other departments concerned and providing the prime minister with a draft letter to the governor general. The draft asked Byng to inform the Colonial Office 'that the view taken by the Imperial authorities as to the undesirability of laying claim to Wrangel Island, is shared in by the Canadian Government.'[119] King signed without demur, and a telegram was sent to London on 18 July. The Foreign Office promptly informed Stefansson that 'after due consideration and consultation with the Canadian Government, His Majesty's Government do not propose any initiative in advancing a claim to Wrangel Island or to contest any claim preferred by the United States or Soviet Government.'[120] When a copy of this document reached Craig's desk, he noted gleefully that it 'should hold [Stefansson] for a while!'[121] All Stefansson could do to rebuild his career was to work on his new book, *The Adventure of Wrangel Island,* while hoping that the result of Lomen's expedition would justify his earlier actions.

The *Herman* left Nome on 17 June and spent the summer battling with heavy ice. Repeated attempts to reach Wrangel Island failed, though Lane did make a grisly discovery on nearby Herald Island: the bodies of four men from the *Karluk,* who had evidently become disoriented during their attempt to reach Wrangel from the sinking ship. They appeared to have died of cold and exposure on the rocky, barren islet. Lane raised the American flag on Herald Island and then turned back towards Alaska on 4 October.[122] On his arrival there, he learned that the Soviets had not been bluffing. The gunboat *Krasnyy Oktyabr' (Red October)* had fought its way through the ice, reaching Wrangel in mid-August. The Russians had raised the red flag and arrested Wells's party.

The Americans were taken to Vladivostok, where they were interrogated by Mr. Fonshtein of the Commissariat for Foreign Affairs. Fearing that Wells might later give inaccurate reports of his treatment to the foreign press, Fonshtein asked J.C. Hill of the British commercial mission to be present. According to Hill's report, Wells testified that Harold Noice had told him no passports or other official documents were needed, because 'all was settled between the two Governments (i.e. Great Britain and Soviet Russia) ... All that would be required, [Noice] said, would be for the party to have the British flag with them and, in case of need, to hoist it, and they would not be molested. Such flag was supplied them.' Wells denied that this Union Jack was intended as an expression of British sovereignty. Since no supply ship had arrived, he and the others were 'very glad' to be taken off by the Russians. Fonshtein was convinced that Wells had been duped by Noice and Stefansson.[123] He was willing to let the party go, but complications arose from the fact that the United States had not yet recognized the Soviet government. In addition, the State Department felt that Carl Lomen, not the American government, should pay for his employees' transportation home.

In the end, the intervention of the Red Cross was required to free the Americans. Wells died of pneumonia in Vladivostok, and three of the Inuit children fell victim to disease or accident. However, the others reached home safely in the spring of 1925.[124] That summer Stefansson wrote to Lomen, 'I still have the feeling that if you were to put trappers [on Wrangel Island] late in the autumn, you would be doing something not very risky that might lead to results.'[125]

Knud Rasmussen arrived in Nome late in the summer of 1924. Since leaving Danish Island in the spring of 1923, he had done exceptionally valuable ethnographic work. Wherever he went, the Inuit readily gave information to this stranger who 'came to us from far away and yet could speak our tongue.' Rasmussen spent the summer of 1923 in the vicinity of King William Island. On 21 September, to the great amazement of the local inhabitants, a ship appeared. It proved to be a Hudson's Bay Company vessel, sent to establish a post at Gjøa Haven. To Rasmussen's delight, the party was led by Peter Norberg, a Swede, and Henry Bjørn, a Dane. Rasmussen had heard from the Inuit about a spot on the Adelaide Peninsula where the skeletons of several men from the Franklin expedition remained unburied. He and Norberg went to investigate. Finding that the report was true, they gathered together the bones, built a cairn over them, and raised the Danish and British flags, with the latter uppermost.[126]

Rasmussen left King William Island on 1 November and reached the HBC post on the Kent Peninsula two weeks later. There were letters for him at Kent, including one from London, which announced that he had been awarded the gold medal of the Royal Geographical Society. Rasmussen then spent a few months studying the Kitlinermiut or 'Copper Eskimos,' some of whom were so isolated from white society that they had never heard of the 1914-18 war. After a short stay at the RCMP post on Herschel Island (where he was soon on friendly terms with the commanding officer, Inspector Wood), Rasmussen and his Inughuit companions went on to Alaska.[127] On 5 May 1924, they passed the frontier, which was marked by two six-foot-tall posts, with 'Canada' written on one side and 'United States of America' on the other. Rasmussen commented on how strange these monuments appeared 'here in the snowfields, where one sees not a single sign of human life.' The party reached Nome on 31 August. This marked their return to 'civilization': when Rasmussen, Arnarulunguaq, and Miteq went into a restaurant dressed in their caribou-skin travelling clothes, the manager refused to serve them.[128]

But the Fifth Thule Expedition was not quite over. Rasmussen found that the next ship to Seattle did not leave until the end of October. He therefore decided to complete his ethnographic odyssey with a month-long visit to the Chukchi people of eastern Siberia. Joseph Bernard was willing to take him across Bering Strait in the *Teddy Bear*. However, Rasmussen did not have permission from the Soviet government to land on Russian territory, and attempts to contact Moscow by telegraph were unsuccessful. It was not yet known outside of Russia that the *Krasnyy Oktyabr'* had gone to Wrangel Island, so Rasmussen and Bernard were not fully aware of the risks they were taking when they set out on 8 September.

They realized their mistake soon after their arrival in the town of Uelen. The local authorities were polite (Rasmussen found their Old World manners a pleasant change 'after all the informality of Canada and America') and evidently wished to be helpful. However, there was clearly a problem. '"Washington" – that is all I grasp of all their rapid speech – they repeat it incessantly,' Rasmussen wrote in his journal. 'I learn that the reason ... is the strained relations with the surrounding world because no one will recognize ... Russia's right ... to the Wrangel Islands [sic], to which a warship has been despatched.' Rasmussen now realized that, thanks to Stefansson, foreign explorers were 'not ... popular' with the Soviets. That he was Danish rather than Canadian did not help. When Rasmussen mentioned the name of King Christian, the local governor suddenly began to shout in German, 'Caput – Caput, a nephew of Empress Dagmar – a cousin of Czar Nicholas – caput – caput.' Then he changed to 'rapid Russian,' while everyone in the room stared accusingly at Rasmussen. 'For the first time on that long journey westwards from Greenland it was no recommendation to travel under the protection of the King of Denmark,' the explorer noted wryly.[129]

Rasmussen was taken back to the *Teddy Bear* under police escort. Bernard had narrowly escaped arrest 'as an accomplice of V. Stefansson' for his role in the 1922 Wrangel Island relief attempt. Fortunately, he was able to produce various letters critical of Stefansson which he had written to the press.[130] 'A few days later we again lay at Nome, with which the Fifth Thule Expedition had been carried through and brought to an end,' Rasmussen recorded. In total, the members of the expedition had travelled over twenty thousand miles.[131] Rasmussen had accomplished what he set out to do, but it seems ironic that his plans, which had almost been cancelled in 1921 because of his fellow explorer's machinations, were hindered once again at the end of his long journey by Stefansson's unyielding obsession with Wrangel Island.

8
The Sector Claim

Canada's northern territory includes the area bounded on the east by a line passing midway between Greenland and Baffin, Devon and Ellesmere Islands to the 60th meridian of longitude, following this meridian to the Pole; and on the west by the 141st meridian of longitude following this meridian to the Pole.

– *Charles Stewart to the press, 12 June 1925*

'We would ... like to see Dr. Rasmussen,' Oswald Finnie wrote cordially to Otto Nobel, the acting Danish consul, on 30 October 1924. 'The local press reports he reached Seattle a few days ago ... if the Doctor [is] passing through Ottawa on his way to the Atlantic coast, we would appreciate his stopping off and giving us the benefit of his observations and advice.'[1] A great deal had changed in the Canadian capital since 1920: Stefansson had fallen dramatically from favour, while, thanks to the reports from Jack Craig and Sergeants Douglas and Joy, the members of the Fifth Thule Expedition were held in high esteem. Finnie was therefore eager to hear whatever Rasmussen might be able to tell him about the inhabitants of Arctic Canada.

The consul, J.E. Boggild, had gone to greet the triumphant but weary Rasmussen in New York. On his return, Boggild wrote to inform Finnie that, although Rasmussen was eager to visit Ottawa, he could not easily do so at present. Rasmussen would be returning to North America in the following year, and he would then be pleased to stay in Ottawa for a week or two, conferring with Canadian officials.[2] Boggild also wrote to Sir Joseph Pope to convey Rasmussen's 'sincerest and most heartfelt thanks for the hospitality and amiable assistance rendered him and the members of his expedition in all parts of arctic Canada.'[3] After a gratifying but exhausting hero's welcome in the eastern United States (which involved, among other events, a visit to the White House), Rasmussen arrived in Copenhagen on 2 December 1924, and the members of the Fifth Thule Expedition were reunited.[4] Therkel Mathiassen, Kaj Birket-Smith, and Jacob Olsen had returned in 1923 (see Chapter 7); in 1924 Helge Bangsted followed the same overland route as Birket-Smith and Olsen. Peter Freuchen and the remaining Inughuit (Arquioq, Arnanguaq, Nasaitdlorssuarssuk, and Aqatsaq) attempted to sledge back to Thule in the spring of 1924, but they could not cross Lancaster Sound because of open water. Instead, they were picked up by the *Søkongen* at Pond Inlet in August. The Inughuit returned to Thule, and Freuchen reached Copenhagen in September.

In December, Jens Daugaard-Jensen confirmed that Rasmussen was 'very anxious' to meet with Canadian officials. Daugaard-Jensen also proposed a friendly, ongoing exchange between the two governments of 'views and experiences about any question that may arise from time to time of importance to the Eskimo tribes.'[5] A visit by Rasmussen in April-May 1925 was eventually arranged. In the meantime, Rasmussen enjoyed the acclaim of his countrymen and the award of an honorary doctorate from the University of Copenhagen. He also had to consider the future of the Thule station and the effect on the Inughuit of the Canadian post at Craig Harbour. The Danes feared that next the Hudson's Bay Company might establish a station on Ellesmere, which could draw the Inughuit away from their established trading ties. The fact that Inughuit families had been hired to work at the Canadian police post also caused concern. A memo written on the day of Rasmussen's return, probably by Ib Nyeboe, shows uneasiness about the Mounties' presence in the North and the 'effect these six revolver-armed people must have on the Eskimo's fantasies.' Nyeboe thought that the Danish presence at Thule should be enhanced in its turn: 'Whether the Danish government will send six soldiers up is doubtful, but it would be more effective to have a Controller who had a motor boat and crew.'[6] Rasmussen was extremely reluctant to give up the Thule station to the government.[7] A visit to Ottawa could serve the additional purpose of allowing him to assess the likely impact of future Canadian activities on his own fortunes and on the people of Thule.

THE PUBLICATION OF *The Adventure of Wrangel Island* in April 1925 brought the Canadian phase of Stefansson's career to a close. Rasmussen's visit to Ottawa shortly afterwards rounded off the story of the Fifth Thule Expedition. The year also saw momentous new developments, as the government faced the prospect of threats to its northern sovereignty from Norway and the United States. This time the threats were not entirely imaginary, but the well-coordinated and effective Canadian response demonstrated how far the country's bureaucratic apparatus had evolved since the blunders of 1920 and 1921. Thanks to the Eastern Arctic Patrols, Ottawa officials now had a reassuring sense that the acts of occupation required to secure Canada's title were well underway. Even before the new threats became known, the members of the Department of the Interior were determined that this process should continue.

Finding Smith Sound free of ice, the 1924 patrol had finally reached the coast of Ellesmere Island in the vicinity of Cape Sabine, but no spot suitable for a post could be found. Even the best place visited, Fram Havn on Rice Strait, was 'so bleak and exposed and the prospects for hunting were so poor that the Eskimo families who had come north to assist the police could not be induced to stay.'[8] Instead, a small building was erected and stocked with supplies for future use. The *Arctic* then successfully established a post at Dundas Harbour.

Craig did not take a direct part in the 1924 voyage. The sudden illness of Édouard Deville in the spring of 1924 had left the position of surveyor general open. Given the serious nature of Deville's condition, this turn of events presented Craig with the prospect of significant career advancement. He decided not to command the third patrol. In recognition of his past services in the Arctic, Craig was paid a bonus of $1,000. He was replaced as commander by Frank Henderson, a Dominion land surveyor who had acted as his assistant in 1922 and 1923. After Deville's death in September, Craig was appointed director general of surveys, an important senior post that placed him in charge of the Topographical Survey, the Geodetic Survey, and the International Boundary Commission. It also made him the equal of Finnie and Harkin in the civil service hierarchy. As surveyor general, Craig continued to take a keen interest in the annual patrols. Along with Finnie, he remained one of the main Arctic policymakers in the Department of the Interior.

After the *Arctic*'s return in the autumn of 1924, Craig and Finnie were firmly convinced that in 1925 yet another attempt should be made to establish a more northerly station on Ellesmere. Because the area around Cape Sabine had proved so inhospitable, they now considered the nearby Bache Peninsula as the best location for such a post. According to the accounts by Robert Peary and Donald MacMillan, the peninsula was an excellent hunting ground. However, Craig reported that Cortlandt Starnes was 'of the opinion that in addition to being probably a very uncomfortable post, it is not very necessary that his force should be represented so far north.' Inspector Wilcox, in contrast, was 'still in favour of ... our original plan by which Sabine ... was regarded as the most important of all the posts to be established on account of its proximity to Greenland and to the crossing point used by the Eskimos between Greenland and Ellesmere Island.'[9] It was not long before new threats from abroad reinforced Craig's and Finnie's determination to carry out their plan.

In March 1925 the Norwegian consul general made a formal inquiry about the basis for Canada's claim to the Sverdrup Islands, giving rise to fears in Ottawa that Norway intended to assert its own claim. At the same time, American papers printed the news that an expedition led by Donald MacMillan and Lieutenant-Commander Richard Byrd would have the official support of the United States Navy. MacMillan and Byrd intended to make an air search for new Arctic lands from a base at Etah, with a second, intermediate station in Grinnell Land (the part of Ellesmere first discovered by American explorers) and a third station on Axel Heiberg Island – a locality not yet visited by the Canadian patrols. Their ultimate goal was Stefansson's 'region of maximum inaccessibility.' On 11 April the *New York Times* dramatically announced that, according to the experts at the National Geographic Society, the discovery of new land – perhaps even a new continent – was highly likely.[10]

Reports from several sources alleged that MacMillan had flouted the Northwest Game Act during his 1923-24 expedition. These stories were denied by MacMillan, but in view of the substantial evidence to the contrary, the denial only made Finnie, Harkin, and Craig doubtful that the American explorer's word could be trusted.[11] Now, both the sites chosen for the new expedition's advance bases and the planned use of the latest technology seemed to bode ill for Canada. '[T]he best we can do is to try again with the old C.G.S. "Arctic" and let the ... Americans make their attempts with aeroplanes,' lamented Craig.[12] The *Franklin* had finally arrived in Canada in the autumn of 1924, but plans for the two ships to go north together in 1925 were thwarted when the Department of Marine and Fisheries demanded the return of the *Arctic*. The *Franklin*, which had been purchased for use as the *Arctic*'s tender, could not carry enough coal to go north alone. Reluctantly, Finnie and Craig decided to give the *Franklin* to Marine and Fisheries in the *Arctic*'s place.[13] Not only did this decision mean the curtailment of their plans for scientific work, it also made the successful establishment of a second Ellesmere post far less likely.

Nevertheless, in view of the possible American and Norwegian threats, Craig and Finnie regarded it as unthinkable that they should give way to Starnes. As Finnie explained to the acting deputy minister, Roy Gibson, 'it seems urgent a police post be established at Bache Peninsula, Ellesmere, or as far North as the ... "Arctic" can get.' The American expedition was said to be scientific, but 'one never knows. So far we have offered no protest and the American Press is so filled with matters pertaining to the MacMillan expedition that the American people will begin to think that they own the entire North and Canada is an intruder.' Finnie thought a committee should immediately be appointed to deal with sovereignty issues. If important questions of Arctic policy were 'left entirely to one man,' he might 'easily be guilty of a mistake in judgment, or make some error of omission or commission in law which might be far reaching in its effects.' Clearly, Finnie's motive was to outmanoeuvre the obstructive Starnes.

As members of the proposed interdepartmental body, Finnie recommended himself, Craig, Harkin, and W.W. Cory, all from the Department of the Interior; Starnes of the RCMP; James White from the Department of Justice; and the recently appointed under-secretary of state for external affairs, Dr. Oscar Douglas Skelton.[14] Skelton, a professor of political economy at Queen's University, had been chosen by Prime Minister King as his future chief adviser on foreign policy in 1923. Skelton accompanied King to the 1923 Imperial Conference (see Chapter 7), and in 1924 he was appointed counsellor in the Department of External Affairs. Following Pope's retirement in March 1925, Skelton immediately became the new under-secretary. Despite his relative inexperience in government, he already had the prime minister's trust and the ability to significantly influence policy decisions.

Finnie's proposal was promptly submitted to the cabinet by Minister Stewart; the other ministers approved it on 23 April.[15] In addition to the members suggested by Finnie, T.L. Cory from the Department of the Interior, George Desbarats from the Department of the National Defence, and Duncan Campbell Scott, the deputy superintendent of Indian affairs, were also appointed.[16] Charles Camsell and Rudolph Anderson were later added to the list. The first meeting of the Northern Advisory Board, as the new body was called, took place on 24 April.

Finnie had prepared a memorandum for his colleagues, titled 'Statement of Position: Canada's Arctic Islands, April 1925.' He summed up the information indicating that the United States might challenge Canada's authority over Grinnell Land and that Norway intended to claim the islands discovered by Otto Sverdrup. He also noted that, according to newspaper reports, MacMillan and Byrd had applied to Denmark for permission to make their main base at Etah, as required by Danish law.

Finnie then revived a suggestion made by Harkin and Craig in June 1921, which Cory had dismissed at the time: that the Northwest Territories Act should be amended to impose a similar requirement on explorers and scientists wishing to work in the Canadian north.[17] Alarmed by the number of 'aliens' who had 'invaded the North West of recent years,' and convinced that 'before long we should have sufficient police posts scattered throughout the territory to enable us to check [explorers] up very thoroughly,' Finnie and Craig had already consulted the Department of the Interior's legal adviser, K.R. Daly, 'as to ways and means of dealing with this question.'[18] The two officials were eager to ensure that every foreign explorer would be obliged not only to apply for a permit, but also to provide the Canadian government with a report on his work. Daly replied that an amendment to the Northwest Territories Act would indeed be the most appropriate method of achieving this aim. Such legislation would 'be a definite notice to other countries that we claim sovereignty over the northern islands.'[19] The American expedition presented Finnie and Craig with the welcome opportunity to place their idea before a new interdepartmental body dedicated to northern issues.

At their first meeting, the members of the Northern Advisory Board concluded that the question of how to respond to the American plans was 'very urgent.' The possibility that MacMillan might find new islands north of Canada was especially troubling. The Norwegian inquiry was dealt with by the simple expedient of not making any reply. However, immediate action on the American threat was clearly necessary. Cory was then in New York, and the board members agreed that he should be asked to contact MacMillan and Byrd.[20]

It was at this time of exceptionally high public and official interest in the Arctic that *The Adventure of Wrangel Island* appeared. On the morning of 25 April, Belle Anderson received a telephone call from Allan Crawford's mother in Toronto to tell her that the book was finally out. (An advance copy had already been circulating among Ottawa officials for several days; Rudolph Anderson read it on 21 April.) Belle and Helen Crawford had corresponded since November 1923. The Andersons were therefore very well informed about Stefansson's repeated efforts to obtain public support from the families of the men who had died on Wrangel. He had succeeded in persuading the Knights, the Maurers, and the Galles that he was not to blame for the tragedy but, thanks in large part to the information about

Stefansson's character which Belle and Rudolph were only too happy to supply, the Crawfords had firmly refused to cooperate with him or to be associated with the book in any way.

Now Meighen and Harkin were appalled to find that Stefansson had impudently twisted the facts not only about Wrangel but also about the earlier events in which they had both been involved. Stefansson's book gave the impression that the main purpose of the expedition planned in 1920-21 was the occupation of Wrangel Island. He began by stating that the July 1914 flag-raising was intended to reaffirm an earlier British claim based on the sighting of the island by Henry Kellett, in accordance with the instructions given to Stefansson by the Canadian government. But Stefansson had not been told to reaffirm Canada's possession of islands discovered by British navigators, and his orders were intended to apply only to the Canadian archipelago (see Chapter 7).

Worse was to follow. Stefansson gave a vague account of the scare he had created over the Fifth Thule Expedition, stating only that by 'good fortune' an unnamed European nation had 'cast some doubt about the validity of Canadian claims to certain ... islands.' Ottawa had then roused itself to action at last, and 'the plans of an expedition on a great scale took shape.' Harkin was allegedly one of the 'most enthusiastic' proponents of the plan. He and Stefansson had spent the winter of 1920-21 'trying to get everything ready so that the moment we received approval and money from the Government we could push ahead.' Stefansson printed Meighen's 19 February letter informing him that the government would 'assert the right of Canada to Wrangel Island,' and he claimed that it had 'made Mr. Harkin ... happy for a day.' But then he and Harkin had 'received notice that, while the Government had not exactly reversed their decision ... they had placed the matter again under discussion, asking us to do nothing further until we heard from them a second time.' Subsequently, a 'most unfortunate controversy as to who should be the controlling personality' of the expedition led to confusion and then to deadlock. Harkin and Stefansson were informed that the expedition would not be authorized in 1921, but it was likely to go ahead in 1922. Under pressure from Lorne Knight, Fred Maurer, and Allan Crawford, who were all extremely anxious to go north, Stefansson had decided to act in 1921, with the realistic expectation that he would receive government backing in 1922.[21] It therefore appeared to readers that Stefansson had been led on by the government, which then abandoned him and his men when they desperately needed help.

As Harkin angrily observed to Belle Anderson on 26 April, 'V.S. confuses purposely the Wrangel expedition with the Ellesmere Land expedition.'[22] Although there had been ample talk about Wrangel in 1920-21, the idea of annexing the island to Canada had never been taken seriously by Harkin, Cory, Lougheed, Pope, or (after his initial enthusiasm had faded) by Christie. Nevertheless, Harkin may well have been a little uneasy on this point. There is now no way to determine just how far Harkin's deception of Stefansson went, but it is certainly possible that he allowed Stefansson to believe the planned northern expedition might, with enough determination, be turned towards Wrangel. However, Stefansson's claim that he had been given reason to hope for government support in 1922 was a complete fabrication. Meighen was particularly furious about this misrepresentation.[23] Rudolph Anderson saw the book's lies as 'one last gesture of contempt' towards everyone in Ottawa.[24]

It would be extremely difficult for the government to publicly refute Stefansson's account, since he had so successfully woven fact and fiction into a plausible narrative. Meighen had indeed written the 19 February letter printed by Stefansson. To explain the various twists and turns of government policy during 1921 would only cause embarrassment and public discussion of matters that were best kept secret. Meighen and the rest were obliged to remain silent, but Stefansson paid the price of complete exclusion from Ottawa for many years. 'Let him chase his millionaires in [the] U.S.A. they are fair game but here in Canada we are safe from him, I am sure,' Belle Anderson wrote to Helen Crawford after a conversation with Meighen in December 1925. According to Belle, Meighen now regarded Stefansson as a 'wild fanatic,' and she fervently wished that he had been 'a little wiser as early as 1920.'[25] Mrs. Meighen, equally indignant, did not refrain from telling other Ottawa wives 'just what sort of a person' she considered Stefansson to be.[26]

The Meighens' feelings were shared by Mackenzie King and his advisers. A summary of the book prepared for the prime minister emphasized the passages in which Stefansson seemed 'to be attempting to prove that whilst he, at all times, was actuated by high ideals in [the] matter of attempting to save the party on Wrangel Island, others were not.' The implication that the government had delayed the departure of the *Teddy Bear* was carefully noted by the author of the summary, who also pointed out the frequent use of innuendo with regard to the Wrangel Island party and their decisions. According to the summary, Stefansson 'shirk[ed] responsibility,' attempting to lay the blame for all mistakes on others. In short, 'the impression left upon the mind by the reading of the book is not pleasant.' The only bright spot was provided by the character of Allan Crawford, 'an upright trusty Canadian lad with high ideals.' Crawford evidently 'placed great faith in Stefansson,' mistaking the latter's deceitful rhetoric for true British patriotism, and so the unfortunate, heroic young man went to his death.[27]

It may well have been in April or May 1925 that Harkin discreetly removed many of the papers on the planned Stefansson expedition from the official files. He was now only too aware of the use Stefansson, with his talent for distortion and selective quotation, could make of these documents if they ever fell into his hands. Harkin did not want to destroy the evidence of his efforts to maintain Canada's northern sovereignty, but neither did he wish to take the risk of being characterized in future publications as Stefansson's devoted friend and ally. He chose to guard the documents himself, ensuring that they would never be released during Stefansson's lifetime. It seems that Borden also took measures to protect himself. By his own admission, he had corresponded with Stefansson during the writing of *The Friendly Arctic.* He had also initially encouraged the Wrangel plan.[28] However, there are no letters on these subjects among the Borden papers at Library and Archives Canada. Since Borden, unlike Harkin, had no strong reason to preserve the documentary evidence for future generations, it is probable that he burned all the records in his possession.

Rasmussen arrived in Ottawa on 27 April, in the midst of the initial uproar over Stefansson's book. He attended meetings of the Advisory Board on Wildlife Protection, the Geographic Board, and the Northwest Territories Council, 'freely and frankly' answering a long list of questions about the Far North.[29] He politely refused to be drawn into any discussion of his fellow explorers' failings, but he cannot have been unaware of the opportunity presented to him by the disfavour in which both Stefansson and MacMillan were currently held. As Belle Anderson observed, Rasmussen 'took it all in!'[30] To the Advisory Board on Wildlife Protection, Rasmussen stated that he 'did not believe the Arctic was friendly.' In his view, 'it would always be suitable only for Eskimos and ... a few white trappers and ... traders ... life there was too hard for the white man.'[31] Both Rasmussen and Harkin, among others, were invited to dinner at the Andersons' on 3 May. Unfortunately, Belle did not describe their encounter. She recorded only that everyone 'had a wonderful time.'[32] It is unlikely that

Rasmussen ever knew about the role Harkin had played in 1921; the puzzled Danes attributed Canadian sovereignty concerns to the activities of Norwegian explorer Godfred Hansen, who had led an expedition from Thule to northern Ellesmere in 1919-20.[33]

Rasmussen made such a 'fine impression' in Ottawa that even formerly contentious matters – for example, the naming of new islands – were handled with cheerful good will by Canadian officials.[34] Rasmussen's request that two small islands should be named for Crown Prince Frederik and Prince Knud, the sons of King Christian, was readily granted.[35] As Rasmussen himself put it in a letter to Skelton, 'everywhere I have found the most amiable understanding.' Rasmussen took the opportunity to make it clear that he and his government 'acknowledge[d] in every way Canada's right' to occupy Ellesmere. When asked whether supplies for the new post on Ellesmere could be stored at Etah if the *Arctic* was once again stopped by ice, Rasmussen expressed his willingness to assist the RCMP in any way he could, provided that the police would promise not 'to enter [into] any commercial communication with the Greenlandic Eskimos.' He told Skelton, 'Anybody having visited Arctic territories will understand how important it is for two nations in these districts to be on friendly terms, and from this point of view you will agree with me [about] the importance of fixing certain lines for the politic, which will be a basis for a good neighborship.'[36]

To Finnie, Rasmussen expressed hearty 'satisfaction with my stay in Ottawa ... it is a great experience to meet so many people, who are interested in the welfare of the Eskimos. Thus it is both with profit and with my best impressions that I am leaving Canada.'[37] So valuable was Rasmussen's information considered in Ottawa that Finnie and Harkin made a voluntary offer to reimburse him for the cost of his trip ($1,371.70). In a lecture to the Royal Geographical Society later that year, Rasmussen spoke warmly of how 'Englishmen, Canadians, Americans, and Danes all meet' in the field of Arctic exploration. Looking back on the events of 1921 from the perspective of happier times, he expressed 'very hearty thanks to the British Foreign Office and the High Commissioner for Canada, who showed the greatest interest in our undertaking, and helped us in many ways from the very first.'[38]

Good neighbourship with Denmark in the North had certainly taken a major step forward, but good neighbourship with the United States still appeared uncertain. Cory had met with Byrd in Washington (MacMillan was absent on a lecture tour). During the meeting, Byrd suggested that many of the press comments were 'ill-advised' and 'unfounded.' He assured Cory that 'no expedition would go forward without the approval of the Canadian Government' and promised to visit Ottawa in order to explain his aims to officials there.[39] However, on 8 May Byrd reported that this trip would not be possible after all. He claimed that, to his great regret, his naval superiors had instructed him 'not to handle that end of the Expedition, as I had my hands full with details connected with the flying.' MacMillan would therefore 'take it up.'[40] Nothing, however, was heard from MacMillan. On 28 May Byrd promised 'some official communication in the near future,' but Washington remained silent.[41]

At the second meeting of the Northern Advisory Board, held in Cory's office on 13 May, the group decided to send a diplomatic note through the British Embassy in Washington. In the absence of any clear statement from either MacMillan or the American government, this project took on added urgency. The board members also decided in favour of an amendment to the Northwest Territories Act. Thanks to the work already done on this issue by Finnie, Craig, and Daly, a draft of the new legislation was quickly produced.[42] James White was given the task of writing a memo on the legal foundation of Canada's title. Since 1922, when he became adviser to the Department of Justice on boundary issues, White had gained an extensive knowledge of international law (his main concern was the boundary dispute between Quebec and Newfoundland – an international issue, since Newfoundland was not then part of Canada). White was still strongly committed to the sector idea he had developed in 1904, and which he had embodied in the maps printed to accompany W.F. King's report (see Chapter 3). In 1924, when the Northwest Territories and Yukon Branch published its first official map, White had convinced Finnie to place the sector lines on it.[43]

White's views, as expressed in a fourteen-page document sent to Skelton on 25 May 1925, were crucial to the future development of Canadian Arctic policy. His memo is the true source of Canada's official sector claim. Neither the memo nor the public statements later based on it made any reference to Senator Pascal Poirier's 1907 endorsement of the sector theory or to Joseph Bernier's 1 July 1909 proclamation on Melville Island. Instead, White rested his arguments on the 1904 maps (which, in turn, were based on the 1895 and 1897 orders-in-council) and on a summary of the relevant historical and legal facts that was far more thorough than anything else produced in Canada up to that time. White, rather than

Poirier or Bernier, therefore deserves the distinction of being remembered as the author of the sector claim.

The 1904 and 1924 maps, White told Skelton, showed that Canada claimed as its hinterland a huge triangular area. This area was bounded by the 141st meridian on the west. On the east, the boundary was the midway point between the archipelago and Greenland; north of these lands, it followed the 60th meridian to the pole. The 1867 treaty between the United States and Russia specified a boundary line that extended 'due north, without limitation' into the Arctic Ocean. If the Americans accepted such a boundary to the west of Alaska, it only stood to reason that a similar line should mark their eastern border.

White admitted that Canadians had not yet visited the Sverdrup Islands. However, since this part of the archipelago was uninhabited, such visits were not necessary for the purpose of control. With regard to Grinnell Land, White argued that because Ellesmere had proved to be one large island, it could have only one discoverer. This true discoverer was William Baffin, an Englishman. Even if the American explorers were considered discoverers, the United States had never made any formal claim to Grinnell Land or Grant Land. Had such a claim been made, the inchoate title thus gained would have lapsed after so many years without further official action.

According to White, the northern islands formed a clear and unified geographical entity. Canada had established three police posts, 'placed so as to dominate the whole of the archipelago, thus furnishing all the control required to maintain its title.' From his extensive reading of international law texts and cases, White had realized that the requirement for effective occupation was 'not applicable with the same strictness' in the polar regions as it was elsewhere. When due weight had been given to this factor, 'Canada's title may be claimed to be, if not unquestionable, at least much superior to that of any other nation.'[44] White had shrewdly assessed the situation and put forward a coherent, convincing case for Canadian sovereignty over all the islands, based on a combination of the sector theory and effective occupation.

Stewart announced the amendment to the Northwest Territories Act in the House of Commons on 1 June. '[W]e are asking for this amendment in order that we may have authority to notify parties going into that country that they must obtain a permit of entry, thereby asserting our ownership over the whole northern archipelago,' he stated. 'We want to make it clear that this is Canadian territory.' Canada's claims, he continued, extended 'right up to the North Pole.'[45] These widely publicized words at last elicited a response in Washington. As the *Montreal Gazette* observed a few days later, 'It is not often that the mild and deliberate Minister of the Interior strays into the limelight and becomes a national, much less an international, figure, but he has, like [Lord] Byron, awakened to find himself famous.'[46]

Washington was not well prepared for Stewart's bold step. Like the Canadians before them, the Americans were suffering from a lack of co-ordination between different government departments when it came to Arctic policy. The MacMillan-Byrd expedition had the enthusiastic support of President Calvin Coolidge and several senior naval officers, most notably Rear Admiral William Moffett, the head of the Bureau of Aeronautics. However, the secretary of the navy, Curtis Wilbur, was only lukewarm about the plan, and the State Department had not been consulted at all.[47]

Moffett and the others wanted to find new land in the Far North and to claim it for the United States. It was on their initiative that MacMillan's expedition had been given government support, with Byrd in charge of aviation. MacMillan, too, was eager to gain new Arctic possessions for his country. The initial outline of his plans, submitted to Wilbur in late February, argued that it was time for the US to end its attitude of passive acquiescence to Canadian claims. (He also reported that, because of the great esteem and affection in which he was allegedly held by the Inughuit, the people of northwestern Greenland strongly wished 'to be citizens of the United States and under our rule rather than that of Denmark.')[48] On 2 June MacMillan stated publicly: 'If we find anything we shall claim it for the United States, and I think that the nations of the world will regard it as just.'[49] A few days later, he told Wilbur that 'all lands discovered in the great unexplored area by the United States with planes' should be 'claimed in spite of Canada's protest.' MacMillan also wanted to annex Grant Land and Axel Heiberg Island for future use as air bases.[50]

MacMillan had considerable support in the American press. Indignant at the sweeping Canadian claim, the *Washington Post* huffed that an American title to the unknown land had already been as good as established 'by right of discovery ... because American explorers and scientific observers have so strongly indicated the continent's existence in advance of actually surveying it.'[51] The *Washington Star*, in contrast, showed more sympathy for the Canadian case, and (unlike many other American papers) it publicized the fact that Canada had already established police posts on Baffin and Ellesmere.[52]

Secretary Wilbur was inclined to regard Ellesmere as Canadian territory, and he felt that it would be best to comply with the new regulations, at least to the extent of securing a permit to fly over the island. He now requested an opinion from the State Department. At this point, the navy's plans for territorial expansion in the Far North were confronted by a serious obstacle. A year earlier, in reply to a Norwegian announcement that any land discovered by Roald Amundsen on his planned transpolar flight would be claimed by Norway, the State Department had made a firm declaration of US policy. The American note insisted that 'rights similar to those which in earlier centuries were based upon the acts of a discoverer ... are not capable of being acquired at the present time.' Rather, new discoveries, and even the formal taking of possession by an explorer, 'would have no significance, save as he might herald the advent of the settler; and where for climatic or other reasons actual settlement would be an impossibility, as in the case of the Polar regions, such conduct on his part would afford frail support for a reasonable claim of sovereignty.'[53] A month later Anson Prescott, a citizen who had inquired about the status of Wilkes Land in the Antarctic, was told that, even though the land was discovered by an American naval officer, 'this Department would be reluctant to declare that the United States possessed a right of sovereignty over that territory' because there had been no act of Congress officially claiming it.[54]

The State Department was far from certain that Canada's claim should be recognized, especially after Stewart had asserted ownership as far north as the pole. But, even if the archipelago was considered a no man's land, it would be impossible to countenance American claims based either on nineteenth-century exploration without official backing or on the mere sighting of new land from an airplane by US citizens. Because there was no time to make a careful study of the matter, the officials involved decided not to apply for a Canadian permit. On the other hand, neither would the MacMillan-Byrd expedition be authorized to make a territorial claim. If a new continent or islands were in fact discovered, the matter could be dealt with later; if not, no action would be required. This outcome was reported in the American papers but, given the discrepancy between Byrd's soothing assurances and Washington's evident reluctance to recognize Canada's title, suspicions in Ottawa were not appreciably allayed. At the same time, a Norwegian official visiting Minnesota took the opportunity to assert that Axel Heiberg Island belonged to his nation.[55]

'I CONSIDER THAT this year is very critical and that many of the benefits and advantages that we have gained by our expeditions of 1922, 1923 and 1924 may be lost if we do not make a particularly strong showing in the North this summer ... we should show no signs of weakness this year particularly,' Craig wrote to Finnie on 1 May. Therefore, 'all possible pressure should be brought to bear' on Commissioner Starnes. 'It is possible,' Craig pointed out coolly, 'that the apparent participation of the American Government in the MacMillan expedition may be of assistance to us.'[56] Finnie agreed, and the next day he applied for permission to take the matter up with the RCMP. He noted that from a post on the Bache Peninsula, 'the Police could make yearly patrols to Axel Heiberg Island. This is the Island about which the Norwegians have been communicating with us.'[57] The new post was discussed by the Northern Advisory Board on 13 May.[58] The members approved of the idea, and by the 19th Starnes had secured his minister's permission to submit a supplementary estimate for the cost of a post. Finnie informed the police that Bernier would 'make every effort ... consistent with the safety of the ship' to reach Bache. Then he added a new suggestion: if Bernier's efforts were again unsuccessful, the supplies intended for Bache could be used to establish another post farther to the west, perhaps on Cornwallis Island or Bathurst Island.[59] To this plan, however, Starnes would not agree.[60]

Given the strong probability of an encounter with the Americans, it was clear that the 1925 patrol required a leader of long experience and forceful character. George Patton Mackenzie, the gold commissioner of the Yukon, was chosen for the task. His appointment was discussed in the House of Commons on 10 June, along with other Arctic matters including Wrangel Island. Stewart brought one of the 1904 maps for his fellow MPs to examine, so that they might 'learn something of the immensity of that territory and the heritage which this country has up there.' When pressed about the reasons behind Mackenzie's appointment, Stewart would say only that the government was 'quietly and unassumingly trying to maintain our right in the territory.'[61]

The discussion then turned to Wrangel. The statements made on this occasion were the only public response to Stefansson's book by Canadian politicians. Thanks to Allan Crawford's parents, the government now held a piece of evidence that was damning to Stefansson, and Stewart unhesitatingly made use of it at a time when press interest in Arctic matters was at its height. Like so many Ottawa officials, the Crawfords were both amazed and deeply offended by *The Adventure of Wrangel Island.* Stefansson insisted that Crawford, Knight, and Maurer, though fully aware of the possible complications with Russia, had urged him to go forward with his plans in 1921. Stefansson also hinted throughout the book that the Wrangel party, not he himself, was responsible for all mistakes, including the fatal decision to take only a year's supplies.[62] The Crawfords' anger was soon intensified by a letter from Harold Noice. In return for a promise from Stefansson that he would not be criticized in the book, Noice had made a grovelling written apology for any inaccurate statements contained in his press reports. After reading the book (in which he was vilified), he concluded that Stefansson had deliberately 'double-crossed' him.[63] Noice had prudently kept correspondence damaging to Stefansson, including letters and telegrams from the summer of 1921. These papers proved beyond any doubt that the decision to take only a year's supplies had been Stefansson's. Noice gave the documents to the Crawfords.[64]

The Crawfords, who had long debated the advisability of making a public statement, wrote to the press on 28 April[65] and to Stewart on 8 May. They informed Stewart that, in justice to their son, they wished it to be publicly known how 'Mr. Stefansson led [Allan] to believe that Great Britain had a legitimate claim to the island, and that the British government would back up his action ... Our son was told nothing of Russia's claims ... Stefansson, who had no other British subject available to lead his expedition, appealed to our son's patriotism in such a way that the boy felt it his duty to go and raise the flag on Wrangel, and he died in the belief that he was dying for the empire.'[66] Stewart now read this letter aloud. After an unsuccessful attempt by Meighen to cast some blame on the government for not having made a larger contribution to the relief efforts, the leaders of the three parties all paid tribute to Crawford's courage and heroism. It was a clear and public rejection of Stefansson by both the government and the Opposition.

Following these revelations, Stefansson had few contacts with official Ottawa. Among the few was correspondence related to his stay in London. Stefansson's letters on this subject provide a fascinating glimpse into his state of mind. The government's adamant refusal to pay for more than one month's living expenses (a policy due almost entirely to Stewart's personal animosity) drove Stefansson to uncharacteristic expressions of rage and to dire, if vague, threats of revenge at some future date. It is clear that he considered himself morally entitled not only to the full cost of his 1923 trip, but also to all the money spent on the Wrangel expedition.

Stefansson, it seems, believed in all sincerity that the tentative arrangements for a 'secret expedition' made by Harkin in the spring of 1921 had committed the government to eventually occupy Wrangel Island. His own lies, having been told for what he saw as Canada's good, were in Stefansson's mind nothing to be ashamed of, and indeed it apparently did not occur to him that officials taken in by these falsehoods might justifiably feel resentment.

In 1924 Stefansson had submitted a claim for $2,058.20 in addition to the $700 advance. The sum of $2,758.20 covered travel costs and living expenses for the period from 20 May to 8 October 1923. (Stefansson was suitably vague about the date of the British cabinet discussion on Wrangel, even placing it in early September instead of late July.) Finnie dryly observed that the amount was roughly four times as much as Bernier had spent in a similar period when he was in England looking for a second ship. However, Finnie was not averse to paying the claim, and Skelton suggested that since the order-in-council authorizing the trip was broadly worded, and the delays Stefansson had encountered in London were all too typical of any bureaucracy, the government ought to err on the side of generosity. But Stewart was not to be moved. In a personal conversation with Stefansson just after the trip was approved, the minister had clearly stated that one month was the maximum time for which payment would be made. Exasperated by years of dealing with Stefansson's adroit dodges, Stewart stuck firmly to the original agreement. In September 1925 a cheque for $333.20 was issued, representing the balance owed for Stefansson's travel fares and one month's living expenses.

Stefansson at first appeared to accept this outcome with good grace, though he could not resist observing: 'If a historian ever studies the entire case, he will probably come to the conclusion that it is about $30,000 less than it should have been.'[67] However, he soon raised the subject again. He now claimed that he had mentioned the possibility of a longer stay during a private conversation with King, and that King had agreed it might be necessary. The prime minister could recall no such understanding. Stefansson then declared he had 'never supposed ... the action of the Government in this matter would depend on technicalities or on exact quotation and rigid interpretation of what was said by one person or another ... this is not a technically legal situation but one involving only general questions of justice, equity and good will.'[68] Stewart doggedly reiterated that it was 'well understood ... the trip was to take not more than thirty days.'[69]

Stefansson responded with an angry letter to Finnie, casting himself as the injured party. The decision was not 'morally or technically right'; the government should have been 'generous in this little thing to one who had spent all his earnings for twenty years in scientific and development work for Canada, and is now heavily in debt as a result. True, the press accounts of late years have alleged various bad motives on my part, but the Government should not be deceived by the press since it has proofs to the contrary in its own files.' Stefansson knew that he would be vindicated in the future, 'when jealousies die down and people take the trouble to study the situation carefully.' Perhaps the Liberal government would soon be defeated; in that event, Stefansson would seize the chance to press his case 'for historical or autobiographical reasons.' He would once again submit his financial claim, 'not so much because I want the dollars but because there would be a sort of poetic justice in getting a refund at last on at least the money I have expended for Canada at the instance of the Government.'[70] Curtly, Finnie replied that he had 'nothing further' to say on the subject.[71]

With this unpleasant exchange, Stefansson departed from the Canadian scene. His absence was regretted by few in Ottawa or even in Canada. During a brief visit to Ottawa a few years later, Stefansson appeared without warning at the parliament buildings and attempted to 'buttonhole' Mackenzie King. King deftly avoided him; other politicians simply grimaced and walked away.[72] Yet it cannot be denied that without Stefansson, Canada would have been unprepared to cope with the American threat. If Ottawa officials had been as poorly informed in 1925 as they were in 1920, the outcome might have been entirely different. But thanks to Stefansson and the crises he had provoked, their level of knowledge was more than sufficient. The members of the Northern Advisory Board, made well aware of the issues involved in northern sovereignty by the events of the past seven years, worked efficiently and harmoniously together. White's memo provided a far more sophisticated assessment of the legal situation than had Harkin's analyses in earlier years, and so confident were the members of the board in the case made by White that they used his memo as the basis for a public statement. On 2 May an article in the *Ottawa Journal* had pointed out the contrast between the blaze of publicity surrounding the MacMillan-Byrd expedition and the comparatively secretive attitude on Arctic matters that prevailed in Ottawa. It was no wonder, the author observed, if the American public believed 'that all these vast islands are awaiting occupation by Mr. McMillan [sic] and his merry men on behalf of the United States Government.' The item was brought to the Advisory Board's attention by Finnie.[73]

As a result, Stewart spoke to the press on 12 June. Again, he exhibited the 1904 map as proof that Canada's claims were of long standing. 'With few exceptions all the known insular areas in the Canadian Arctic were discovered and formally taken possession of by British commissioned navigators,' the minister explained. The islands were transferred to Canada in 1880; subsequently, the Canadian government 'sent many expeditions to the archipelago and formal proclamations have been made reaffirming British sovereignty. Police posts and custom houses have been established at various points, detachments of the Royal Canadian Mounted Police make extensive patrols every year and a general administration of the law and of the game regulations has been maintained.' As well as being covered in the Canadian papers, the minister's statement was printed verbatim in the next day's edition of the *New York Times*.[74]

A copy of the statement was forwarded to the British Embassy in Washington, along with a request 'that the attention of the Secretary of State may be drawn ... to the readiness of the Government of Canada ... to furnish all permits required for exploring and scientific expeditions entering the northern territory of Canada, including air permits for flying over Baffin, Ellesmere and the adjoining islands within the boundaries of Canada, and its readiness also to afford any assistance that can be given by the Royal Canadian Mounted Police and other Canadian officers in the North.'[75] The chargé d'affaires in Washington, H.G. Chilton, suggested a few changes to this message in order to strengthen it.[76] In the version which Chilton sent to Secretary of State Frank Kellogg on 15 June, Axel Heiberg Island was explicitly named as Canadian territory. The note also pointedly referred to the recent legislation requiring explorers to secure permits and to the imminent departure of the *Arctic*.[77] In reply, the State Department asked how many police posts had been established, where they were located, and whether they were permanently occupied.[78] This information was duly supplied.

The American expedition left the United States on 20 June. The *Arctic* sailed from Quebec eleven days later, on 1 July. To the great embarrassment of everyone concerned, the engines broke down at the very start of the voyage, in full sight of the crowd that had assembled to wave farewell.[79] After hasty repairs the *Arctic* at last set off. Both Mackenzie and Inspector Wilcox had been instructed that in the event of a meeting with the Americans, they should rigorously enforce Canadian laws.[80] Finnie had continued to push the matter of publicity, informing the other members of the Northern Advisory Board that several magazines had asked for information about and pictures of the *Arctic*'s voyages.[81] As a result, Mackenzie's secretary, Harwood Steele, was assigned to write lively and colourful press reports, to be sent back by wireless at regular intervals. Finnie hoped to release a new report every few days, 'regarding the general happenings on board, and what they are accomplishing,' as a counter to the stories published about the American expedition.[82]

The 1925 patrol was the last voyage of both the *Arctic* and its commander. A highly readable account of the trip was published decades later by Richard Finnie, who had successfully gained a place on the 1924 voyage as assistant wireless operator. Aged only eighteen, he went north once more in 1925. Finnie described the *Arctic* as 'obsolete and worn-out,' with cramped, poorly ventilated, bug-infested sleeping quarters. Bernier he considered 'vain,' 'dogmatic,' and sometimes rather boorish, though overall the old sailor was a 'colourful, warm-hearted, and likable character.' On the 1925 voyage, the ship 'leaked more and more, her timbers strained and her seams loosened by too many encounters with ice and storms.' The *Arctic*'s strong-willed captain 'at last ... was to meet his match' in George Mackenzie. The former gold commissioner was a man of 'imposing stature and bearing' who demanded 'straight answers to his questions.' Both Craig and Henderson had treated Bernier with respect, but Mackenzie openly challenged his decisions.[83] (In this Mackenzie may have had the backing of Craig and Oswald Finnie, who had become increasingly exasperated with Bernier's habit of giving unauthorized newspaper interviews in which he glorified himself while neglecting to so much as mention the police or the Northwest Territories and Yukon Branch.)[84] Relations between the commander and the captain were badly strained when Mackenzie detected an error in Bernier's navigation. After a voyage prolonged by bad ice conditions and repeated problems with the *Arctic*'s engines, the Canadians reached Etah on 19 August. By this time, Mackenzie's sole authority was unquestioned. It was therefore he alone who dealt with the leaders of the American expedition.

The Americans were not faring well. Bad weather and mechanical problems plagued their flights, and although extensive areas on Ellesmere and Greenland had been surveyed, there was little hope that the planes would reach Axel Heiberg Island, let alone the unknown region beyond it. Byrd recorded in his diary that MacMillan seemed to have 'given up,' but he himself still wanted to 'make a showing.'[85] After preliminary courtesies had been exchanged, Mackenzie offered to issue the necessary Canadian permits on the spot. Byrd asked for time to confer with MacMillan. Within an hour he appeared on board the *Arctic* in full dress naval uniform. He stated that MacMillan had applied for and been granted a permit after the expedition left the United States. Because the *Arctic*'s radio was not working, it was possible that a message from Ottawa on the subject had not come through. However, Mackenzie strongly suspected that Byrd was lying (the Americans had been told the previous day about the problems with the *Arctic*'s radio, so Byrd was well aware that Mackenzie had no means of verifying his statement). To avoid a future controversy in which it would be one man's word against another's, Mackenzie summarized Byrd's statements in front of the *Arctic*'s first officer, Lazare Morin, and asked Byrd to confirm that the summary was correct. 'Yes. That is correct,' Byrd replied.[86]

A message instructing Byrd and MacMillan to ask for a permit had, in fact, been sent from Washington by radio that very day (20 August). The message stated that this action was 'essential' in order to 'avoid [an] embarrassing diplomatic situation.' The request could not be made from Washington, but the explorers could do so on an 'informal and personal' basis. The permit should apply only to 'Baffin Land or other territory south of Ellesmere Island.'[87] Possibly the Americans hoped this would be an acceptable compromise, soothing Canadian fears even without formal recognition of Canadian sovereignty over Ellesmere and the Sverdrup Islands. Whether the message did not arrive or arrived too late is not clear.[88] It is also unclear whether Byrd or MacMillan was responsible for the lie told to Mackenzie. MacMillan later told the State Department that he had not received instructions to apply for a permit, and that Byrd 'knew perfectly well' none had been issued. According to MacMillan, Byrd had not consulted him before speaking to Mackenzie; instead, Byrd alone had made the decision to mislead the Canadians, probably as the best way to get out of an awkward situation.[89] To a Canadian friend, MacMillan also professed innocence, claiming that he and Byrd had been told 'not to accept a permit under any conditions' and that he had known nothing of Byrd's lie at the time.[90] In any case, the American expedition left Etah soon after the departure of the *Arctic,* ostensibly because of the weather. Although a Canadian note describing the events at Etah was merely acknowledged without comment by the embarrassed State Department, the 1925 Eastern Arctic Patrol had served its purpose. Never again would the

American government publicly appear to challenge Canada's sovereignty over the Arctic islands.

From Etah the *Arctic* proceeded on yet another attempt to establish the second Ellesmere post. 'There was very little ice on the Greenland side of Smith sound, but as [the *Arctic*] neared the Ellesmere coast heavy field ice was encountered. The ship took heavy punishment from ice ... and it was with feelings of relief that the anchor was dropped in the little harbour of Fram Haven,' Mackenzie recorded. He was delighted to find that an overland patrol from Craig Harbour had visited the storehouse built in 1924. However, there was clearly no chance of proceeding as far as Bache Peninsula. The coal supply was low and the ship was 'leaking so badly that it was feared if she were caught and compelled to winter in the ice, it might be necessary to abandon her.'[91] The stores were unloaded and left at Fram Havn. After visits to Craig Harbour, Dundas Harbour, Pond Inlet, and Pangnirtung, the *Arctic* returned south for the last time.

Canadian officials retained a highly suspicious attitude towards both the American government and American explorers for many years. MacMillan became the focal point of these concerns. Byrd, handsome and socially polished, had made such a favourable impression on both Cory and Mackenzie that those in Ottawa were reluctant to blame him for the deception at Etah. Moreover, Byrd did not return to the Canadian Arctic, while MacMillan made almost annual voyages to the archipelago and Greenland until the early 1950s. When Finnie and Harkin learned of MacMillan's plans for a 1926 expedition to collect specimens for the Field Museum of Natural History, they were determined that on this occasion he must be forced to take out a Canadian permit.

Their suspicions were first roused in April 1926, when the expedition's zoologist, Walter Koelz, applied to Harkin for a permit to collect bird specimens (as required by the Migratory Birds Convention Act, which was still administered by Harkin's branch). Koelz's application referred only to collecting in the Maritime provinces and northern Quebec. Nothing was said about the Arctic islands. Harkin promptly replied that the permit should include the Northwest Territories. Shortly afterwards, the director of the Field Museum wrote to Finnie, asking whether the scientists might take specimens of animals protected under the Northwest Game Act. Finnie's answer was that before any such permits could be issued, MacMillan would have to apply for the permit required by the previous year's amendment to the Northwest Territories Act. (For the regulations that permit holders were obliged to follow, see the Appendix.) Harkin wrote a similar letter to the museum, stating that once MacMillan's application had been made, he would be glad to consider Koelz's request. On 21 May MacMillan finally complied, and Permit No. 15 was issued to him on 27 May.[92] No doubt the two Canadian officials savoured their victory, gained as it was through purely bureaucratic methods. Cortlandt Starnes had been mulling over the possibility of having the police seize MacMillan's ship if it was 'found illegally in Canadian territory,' but fortunately such a dramatic course of action was now unnecessary.[93]

In the years that followed, MacMillan generally complied with Canadian regulations. The police watched him carefully and reported a few infractions to Ottawa. On such occasions Finnie, Harkin, and their successors took MacMillan to task, sometimes quite sharply. Whatever the American's private feelings about this constant surveillance may have been, he always replied in polite and conciliatory terms.

Officially, the American government remained silent about Canada's claims. The State Department's position, as expressed in an internal memorandum, was that Washington approved of the Canadian government's 'endeavours ... to extend the rule of law and order' in the Far North; however, the United States could recognize Canada's sovereignty only where effective occupation had been established.[94] Washington's standard for effective occupation was high (the letter to Anson Prescott called for 'actual settlement'), and indeed some in the State Department doubted that sovereignty could ever be established over uninhabitable far northern islands.[95] One member of the department's legal division belligerently informed James White that Canada's claims were not 'worth a damn.'[96] But, seeing no benefit to be gained from any open controversy with their northern neighbour, most American officials were content to let the matter remain in abeyance. In 1935 a hopeful American explorer asked for authorization to claim any new land he might discover in the archipelago or in the vicinity of the North Pole. The State Department's reply admitted that the American government was 'not in a position to give you a permit of the nature indicated.'[97] Occasional complaints that US policy was 'inert' and 'complacent' were ignored.[98] As the years passed, the absence of public protest or action from the United States could more and more readily be construed as tacit consent.

Moreover, one American legal expert with considerable influence in government circles was extremely well disposed towards Canada's claim. Skelton, White, Finnie, and others in Ottawa were greatly heartened by an article that appeared in the October 1925 issue of the American journal *Foreign Affairs.* The author, David Hunter Miller, was a well-known New York lawyer who had served as President Woodrow Wilson's adviser during the Paris Peace Conference. With Sir Cecil Hurst of the Foreign Office, Miller had drafted the covenant of the League of Nations. Having been invited by the editor of *Foreign Affairs* to contribute an article on Arctic sovereignty, Miller wrote to Stewart in July 1925, asking whether he could discuss Canada's views with someone in Ottawa. When Skelton heard the news he was delighted. He observed that Miller was 'a man not only of high standing and wide knowledge, but of broad views. If there is to be any discussion in the United States of our claims in the Arctic, I do not imagine they could be stated in a better publication than *Foreign Affairs,* or by a more authoritative writer than Mr. Miller.'[99] Miller was duly invited to Ottawa, where he met with Cory and White on 27 July. He and White evidently found one another congenial, and the two men maintained a friendly correspondence until White's death two and a half years later.[100]

Miller's article outlined the steps taken by Canada to establish occupation, accurately describing both the existing posts and the plans for future bases. Canada, Miller noted, had made 'a precise and definite claim of sovereignty,' while no other nation had formally announced 'any claim whatever.' The geographical facts were on Canada's side, since many of the islands were 'quite inaccessible except from or over some Canadian base.' And, if Canada was not recognized as the owner of the archipelago, what would happen? There were only two possibilities: either 'something in the nature of a *terra nullius,* an unsatisfactory sort of ownership by everybody, or else ownership by the United States.' The first option was unappealing, and there was no reason to think that the American government wished to undertake the second. Canadian sovereignty, therefore, was the best solution in terms of American interests.[101]

From this point in the article on, Miller's arguments bore a striking resemblance to those in White's May 1925 memo. Like White, Miller cited the 1867 treaty between the United States and Russia as a precedent for northern boundary lines extending as far as the pole. In his view, the Canadian claim that the 141st meridian marked the boundary between Canadian and American territory in the archipelago was reasonable in itself, and it might be to his country's benefit as well. Why, he asked, should not the Americans make their own sector claim in the North? The Canadian theory 'would give the United States (if we wanted it) a very large portion of the present unknown area. What this would mean in terms of territory we cannot now say; perhaps nothing; perhaps a frozen empire.'[102]

When the article appeared, Skelton noted with satisfaction that it was 'extremely fair' and went 'far towards acceptance of our claims.'[103] He did not remark on its one major inaccuracy: the statement that in 1921 both Rasmussen and the Danish government were notified that any Danish discoveries within the sector 'would not affect Canadian claims.'[104] Miller must have received this impression from White, who, of course, was not involved in sovereignty matters in 1921 and so had no direct knowledge of the Rasmussen episode. The sector theory had never been raised in Canadian communications with Denmark. Whether White deliberately misled Miller on this point in order to enhance the sector claim, or whether he himself genuinely believed what he said, remains a mystery. Skelton, as a newcomer to the civil service, readily accepted White's and Miller's version of events. The *Foreign Affairs* article thus laid the foundation for the many misleading accounts that were to follow in later decades.

EIGHT MONTHS AFTER MILLER's article was published, Roald Amundsen, Lincoln Ellsworth, and Umberto Nobile flew from Spitsbergen to Alaska in the dirigible *Norge*, crossing the 'region of maximum inaccessibility.' The explorers were naturally intent on sighting new land, but they saw only a 'chaos' of rough, closely packed ice ridges, over which 'surface travel of any sort would be impossible.'[105] There was no 'frozen empire' north of Alaska for the United States to claim, and therefore no hope that the State Department would officially accept Miller's arguments about the sector theory. Nevertheless, from Canada's point of view it was the best possible outcome. A large land mass straddling the 141st meridian might well have precipitated a crisis in Canada-United States relations. 'We are delighted to find that there is no land. It is in the interest of Canada that there should be no land,' remarked White.[106] Whatever challenges there might be to the sector claim on theoretical grounds, there could be none based on new discoveries by other nations. The Arctic policy regime established during 1925 would therefore serve Canada well for nearly half a century. Not until 1972, after the voyage of the American supertanker *Manhattan* had set off a passionate public controversy over the status of the Northwest Passage, was it necessary to reaffirm and strengthen Canada's claims to jurisdiction in the Far North through an innovative new piece of legislation, the Arctic Waters Pollution Prevention Act.

Conclusion

Canada of Itself

> *Can Canada of itself, that is without specific instructions from the Imperial Government, take any effective action regarding the sovereignty of lands which may be regarded by other nations as outside of Canada?*
>
> – *Jack Craig to Arthur Doughty, 21 January 1921*

THE LATE 1920S, THE 1930s, and the 1940s were a period of consolidation, when most of the outstanding Arctic issues were resolved, and when images of state authority in the archipelago slowly but steadily gained a place in Canadian popular culture. After the Amundsen-Ellsworth-Nobile flight, Canadians could focus on the task of establishing effective occupation within the area embraced by the sector claim. To that end, the Royal Canadian Mounted Police made extensive patrols through the Sverdrup Islands from the post on the Bache Peninsula, which was finally established in the summer of 1926.

Captain Bernier had retired in the autumn of 1925.[1] The *Arctic* was clearly unfit for further service, and there were no funds to build or buy a new ship. Therefore, Finnie decided to charter a vessel. The *Beothic,* a modern steel ship owned by Job Brothers of St. John's, Newfoundland, forced its way through the ice that had baffled the *Arctic* so many times. On 6 August 1926, George Mackenzie selected 'as favourable a location ... for a post as any in the north – a dry ... well drained building location with a southern exposure, well protected from all winds except from the south, with a stream of good water flowing by.'[2] Sergeant Joy, now promoted to inspector, was placed in command of the new post. In 1927 Joy made a sledge journey of over 1,300 miles through the Sverdrup Islands. Two years later he sledged 1,700 miles from Dundas Harbour to Bache. Along with the other patrols made by the police from all the Arctic posts, these trips firmly established Canada's official presence in the remote and uninhabited northern reaches of the archipelago.

To make assurance doubly sure, Finnie searched for a reason to outline the entire archipelago on the map as a Canadian possession. 'Mr. Cory has asked me to suggest some method of showing ... definitely and decidedly that [all the islands] are Canadian territory,' he told James White in December 1925.[3] At a meeting of officials from the Department of the Interior, it was agreed that the establishment of a game preserve in the Far North 'would serve to emphasize to anyone seeing the map that Canada claimed sovereignty to the whole of this northern Archipelago.' However, 'fear was expressed that unless we had a very good reason for establishing a new game reserve in the north or had some other good reason for outlining the islands, such action might look undignified and would not serve any good purpose with regard to Canada's claim.' The conservation of the muskox could not serve as a pretext, since muskoxen were already protected throughout the country. The officials agreed that they would try to determine 'the most plausible reason for establishing a game reserve.'[4]

Rasmussen's remarks to the Advisory Board on Wildlife Protection provided the answer to this dilemma. He had pointed out that the existence of trading posts on the islands might disrupt the migratory patterns of the caribou.[5] This observation was not intended as a criticism of the Hudson's Bay Company (of which Rasmussen spoke in generally positive terms), but it reinforced Finnie's belief that the traders' presence in the archipelago was detrimental to the Native population. Finnie had long resented the HBC's dominant position in the North, believing that the company wished to be 'all powerful[,] even superior to the Government itself.'[6] He was therefore far from pleased with the plan to establish a line of HBC trading posts across the archipelago.[7] The Inuit, he believed, would fare much better under the government's paternal supervision.

Finnie consulted with James White, J.B. Harkin, Rudolph Anderson, and Cortlandt Starnes, all of whom approved his plan. The Arctic Islands Game Preserve was accordingly created in July 1926, on the grounds that 'unless further steps are taken to protect the areas reserved as hunting and trapping preserves for the sole use of the aboriginal population of the North West Territories, there is grave danger that these natives will be reduced to want and starvation, through the wild life being driven out of said preserves by the exploitation of the same by white traders and other white persons.' The boundaries of the preserve, as outlined in the order-in-council creating it, extended northwards as far as the pole.[8] The order thus formalized Charles Stewart's verbal assertions of the sector principle in 1925.

No trading posts could be established on the islands without the approval of the commissioner of the Northwest Territories. Two years later Finnie reported to Rasmussen: 'We have been working on your recommendations regarding native preserves ... and now have them pretty well established and white men excluded from the Arctic Islands Preserve except at two locations, namely, Cambridge Bay on Victoria Island, and Peterson Bay on King William Island. Traders may establish posts at these points but at no other[s].' The new regulations had not been implemented without resistance. 'We had considerable difficulty in ejecting white traders who had already become established but they are now all out and I think our trouble is over,' Finnie recorded.[9]

The Soviet Union, meanwhile, was following a course very similar to Canada's. Having received no satisfactory explanation from the United States of the flag-raising on Herald Island by Louis Lane's expedition (see Chapter 7), in October 1924 the Commissariat for Foreign Affairs informed the other powers that Wrangel, Herald, and the other islands north of Siberia were 'an indivisible component part of the Union's territory.' Not only did they form 'the northern continuation of the continental Siberian plateau,' but they were also situated 'westward of the line which by the Washington Convention of the 18-30th March 1867 defines the boundary to the west of which the United States of America undertook to present no territorial claims.'[10] In 1926 the Soviets made their own sector claim. A decree published in *Izvestiya* on 16 April proclaimed that all known lands and islands lying north of the Soviet Union which were not already acknowledged to belong to some other power, as well as any that might be discovered in the future, were part of the USSR.[11] In the summer of 1926, a Russian colony was established on Wrangel Island.[12] Like Canada, then, the Soviet Union combined a sweeping territorial claim with occupation at strategic points. This development was highly satisfactory to Ottawa, since a nation which had itself made a sector claim in the Arctic could not logically refuse to acknowledge Canada's title to the archipelago.

By the late 1920s, only one major problem remained: the possible Norwegian claim to the Sverdrup Islands. It was handled with assurance and even aplomb by the Department of External Affairs, which was developing into a true foreign ministry under O.D. Skelton's guidance. Consultation with London remained the norm and, in the absence of Canadian diplomatic representation in Norway, messages from Ottawa to Oslo were still sent by way of the Foreign Office. However, the slow, awkward communications system of earlier years had finally been simplified. 'I feel sure that it is ... up to each Dominion to create some effective mechanism to enable it to know what is going on in the world and to assert its views in time to be of use. The existing mechanism is simply ludicrous,' Loring Christie observed in 1926.[13] Beginning in 1927, External Affairs communicated directly with the Foreign Office. At the same time, the department's capacity to formulate policy was greatly enhanced when Skelton hired a number of bright young men to serve as first and second secretaries.

Norwegian inquiries to Canada about the status of the islands in 1925, 1926, and 1927 had gone unanswered. Finally, in March 1928, Canadian officials were informed that the Norwegian government wished to 'reserve ... all rights under international law.'[14] This message at last brought the promise of a detailed reply from Ottawa. However, the Norwegians themselves had serious doubts about their claim. As one politician observed to his colleagues during a secret meeting, 'I think we have to admit that if we take a quick look at the map, one will promptly see that these Islands geographically belong to Canada and there is little we can do to change that.'[15] In April 1929 Prime Minister King received a letter from Otto Sverdrup, who intimated that he would be able to persuade his government to relinquish its claim in Canada's favour, provided that Canada would compensate him for the cost of his 1898-1902 expedition. Soon afterwards, the premier of Ontario, G.H. Ferguson (who was then in London), informed External Affairs that the Norwegian government did not actually want the islands. Ferguson believed that Canada could easily secure its title by paying a 'moderate sum' to Sverdrup.[16]

One of Skelton's new recruits, Lester Pearson, was asked to write a memo on the subject. Pearson noted that Sverdrup's expedition was privately sponsored, and that the Norwegian government had never publicly asserted a right to the islands he discovered. No attempt at occupation or administration had ever been made by Norway. Canada's claim rested on the sector theory and the police patrols. Pearson thought that the sector theory was 'a practical solution of the Arctic question.' Although the contention that unoccupied polar lands could be claimed in this way went against earlier precepts of international law, 'conditions in the Arctic are unlike any visualized by international lawyers in the past, and new rules might well be considered necessary to meet those conditions.' The police patrols did not meet the criteria for effective occupation in temperate lands, but Canada had clearly done far more than Norway. There was no reason to fear a successful Norwegian claim. Still, it might be wise to pay Sverdrup 'rather than carry on a long argument in the matter, which would eventually have to go to arbitration.' Therefore, 'a grant might be made to Captain Sverdrup of grace, not of right, in return for which, though not as a quid pro quo, Norway might acknowledge the disputed islands as Canadian territory.'[17] Skelton agreed.

The government maintained a firm public stance while the negotiations proceeded, giving rise to headlines such as 'Canada Unyielding in Claim to Arctic.'[18] Sverdrup's representative initially asked for $200,000; the Canadians were able to reduce this sum to $67,000, which was paid in November 1930. In return, Norway formally acknowledged 'the sovereignty of His Britannic Majesty' over the islands (though Norwegian officials insisted on noting that their recognition was 'in no way based on any sanction whatever of what is named "the sector principle"').[19] Journalists greeted the agreement as evidence that Canada had entered 'a new stage in international relations,' and the perception that the External Affairs bureaucracy had attained a new level of maturity and competence was well warranted.[20]

In the space of a decade, Canada had acquired recognition of its sovereignty over the archipelago from two northern European nations, Denmark and Norway. The Soviet Union had adopted the sector theory. Although the United States did not formally acknowledge Ottawa's claim, neither did it make any public protest. All these Arctic nations had thus either explicitly or implicitly conceded that the northern islands were Canada's. The Canadian position was strengthened yet further by the 1933 decision of the Permanent Court of International Justice (PCIJ) in the Eastern Greenland case.

Placing its confidence in the traditional doctrine of effective occupation, in 1931 Norway challenged Denmark's claim to the eastern coast of Greenland. The Norwegians argued that Danish sovereignty was limited to the colonized areas on the southwestern coast. Where there were no Danish settlements, other nations were free to put forward their own claims. Denmark promptly took the dispute to the PCIJ. The Danish case was based on such seemingly flimsy supports as vague eighteenth-century proclamations of lordship over the entire island. Nevertheless, the court decided in Denmark's favour. Given 'the absence of any claim to sovereignty by another Power, and the Arctic and inaccessible character of the uncolonized parts of the country,' the evidence presented by Denmark was considered sufficient to prove the Danish government's intention to exercise sovereignty over all of Greenland. If no other state could demonstrate a superior claim to sovereignty over a thinly populated or unsettled country, then 'very little in the way of the actual exercise of sovereign rights' was needed to meet the requirements of international law.[21] As legal scholar Gillian Triggs observes, the Eastern Greenland decision established 'a flexible standard which depends upon the circumstances of the territory. The more isolated the territory and the fewer the inhabitants, the less stringent are the requirements of effective occupation.'[22]

The Eastern Greenland case had clear implications for Canadian Arctic policy. Above all, the court's decision reassured officials in Ottawa that, despite the vast size of the area to which Canada laid claim, their efforts were more than adequate. Indeed, the Danes had cited the Canadian example in their arguments, observing that a mere nineteen policemen were charged with the task of maintaining state authority in the enormous District of Franklin. According to one member of the Danish legal team, Gudmund Hatt of the University of Copenhagen, the official Canadian presence in the Far North was entirely sufficient, given the nature of the territory involved.[23]

Only the United States still clung to a rigorous set of criteria for effective occupation, even in the polar regions. Early in 1939 the State Department re-evaluated its policy and urged President Franklin D. Roosevelt to fully accept Canada's title.[24] However, when the Canadian Arctic took on a new strategic significance after the Second World War, American military authorities considered making a claim to parts of the archipelago. An internal American memorandum written in 1946 grudgingly admitted that, because of the Eastern Greenland decision, the Canadian case would be upheld by an international court. Moreover, 'any action on the part of the United States which could be interpreted as an usurpation of Canadian territorial rights would be followed by political consequences so grave that, except in the case of a very serious emergency, they could scarcely be justified even in terms of short-run expediency.'[25] There remained the possibility of finding a few small, previously unknown islands, which the US could claim in defiance of the sector principle, and where meteorological stations could be placed. No such islands existed, so the

Americans were obliged to seek Canadian cooperation in building and operating a series of Arctic weather stations.[26] Care was taken to ensure that the joint program did not infringe on Canadian rights. Similar arrangements were made between the United States and Denmark for US activities in Greenland. The Americans built an air base at Thule, which, at the height of the Cold War, was home to as many as six thousand American servicemen at a time.[27] This far eclipsed the scale of any American activities in northern Canada during the postwar era. Nevertheless, the Danes too retained their sovereignty.

The years 1946 and 1947 thus marked the last potential threat of any real significance to Canadian rule over the land areas of the archipelago. The United States now acknowledges the validity of Canada's title to the islands.[28] Tiny Hans Island, located exactly at the midway point between Ellesmere Island and Greenland, is currently claimed by both Canada and Denmark; with this minor exception, all the northern islands are indisputably Canadian soil. Nevertheless, since 2002 the Hans Island issue has given rise to frenzied warnings about a new 'Danish invasion.' These outcries bear a startling resemblance to Stefansson's rhetoric. 'The Vikings have returned to Canada and are trying to take over Canadian territory!' proclaimed political scientist Rob Huebert. According to Huebert, Denmark's Hans Island claim is only the beginning of a new series of challenges to Canada's authority over all the northern islands. Another writer, Dianne DeMille, agreed that the Danes' 'first foray' will soon lead to 'a larger series of territorial claims,' including part or even all of the archipelago. Historians Ken Coates and William Morrison took the opportunity to lament, 'Since the 1880s, Canada has had its sovereignty in the Arctic challenged repeatedly, and has done very little to defend it. The latest threat is the most serious and the most comprehensive.'[29] In fact, apart from Hans Island, the only current challenges to Canada's northern sovereignty relate to water alone. Sovereignty over the sea is determined by a different set of criteria than is sovereignty over land. The extent of Canadian activities on the islands themselves is therefore quite irrelevant to the ongoing debate about the Northwest Passage, which centres on the question of what authority Canada can exercise over the waters surrounding its far northern possessions.[30]

The civil servants of the early 1920s created a strong foundation for Canadian state authority in the Far North. Beginning in almost complete ignorance of international law and of their own Arctic possessions, within the space of only a few years the Canadian bureaucracy developed the knowledge and procedures required to carry out the necessary acts of occupation in the archipelago. In the process, Canadian administrators adopted the Danish model of paternal rule over northern Aboriginal peoples as the pattern they would follow. The events of 1918-25 therefore held great significance for Canada's internal as well as its external policy. From today's perspective there are, of course, many problems with the paternalist model, some of which have recently been discussed by historian John Sandlos in *Hunters at the Margin: Native People and Wildlife Conservation in the Northwest Territories.* As Sandlos observes, during the 1920s Ottawa bureaucrats like Harkin and Finnie began to exert control over almost every aspect of Native peoples' lives, considering them incompetent to make decisions about the management of northern wildlife resources. Deeply influenced by the views Rasmussen expressed during his 1925 visit to Ottawa, civil servants believed that the future of northern Aboriginal peoples could best be ensured by a change from hunting to herding. The government accordingly hired two Danes, Erling and Robert Porsild, to assist with the introduction of domesticated European reindeer to the Mackenzie Delta during the late 1920s.[31]

In dozens of small ways, the authority of the Canadian state had come to pervade the Far North, affecting the lives of Inuit, Dene, and whites. Indeed, it could be argued that every act of scientific or geographical research carried out in the Northwest Territories was a manifestation of the state's power and therefore of Canadian sovereignty. Excellent scientific work was done under government auspices, most notably by naturalist J. Dewey Soper, who was best known to the public for discovering the northern breeding grounds of the rare blue goose. A foreign explorer was barred from entering Canadian territory unless he first obtained a permit and agreed to inform Ottawa about his activities and results (see the Appendix). Traders, even those who were Canadian-born, could not establish posts inside the Arctic Islands Game Preserve without government authorization. Whether whites in the Far North were Canadians or foreigners, their very presence in the archipelago depended on Ottawa's approval.

Most of these developments did not greatly interest the Canadian public. There were no dramatic tales from the twentieth-century Arctic to compare with those provided by the earlier polar expeditions of Parry, Franklin, Scott, and Shackleton. Moreover, some government officials felt a certain degree of uneasiness about northern publicity, fearing it might contribute to the impression that Canada was a barren land of ice and snow. For example, W.W. Cory turned down a suggestion from Jack Craig that the Eastern Arctic Patrol films should be shown at the 1924 British Empire Exhibition. Craig thought that British audiences would be especially interested in the footage from Beechey Island, but Cory retorted that such scenes were 'hardly designed to assist in either immigration or trade' – the government's main priorities.[32] Despite such widely held attitudes, from 1925 onwards stories about, and images of, government patrols in the archipelago by both land and sea gradually became a part of Canadian popular culture.

The earliest attempts at publicity for the Eastern Arctic Patrols were failures. Because of the radio problems experienced during the 1925 voyage, the press stories written by Harwood Steele could not reach their southern audience while public interest remained high.[33] In December 1925 James White offered an article about the previous summer's work to *National Geographic;* the editor blandly replied that the journal was already too 'crowded with material' – hardly a surprising response, given the National Geographic Society's vocal support for the MacMillan-Byrd Expedition.[34] As government employees, it was considered inappropriate for men such as Joy and Mackenzie to publicize themselves and their exploits directly. Mackenzie frequently gave talks to Canadian Clubs and similar organizations, but there were no lecture tours of the type in which Stefansson so gladly engaged. Mackenzie pointedly observed in a 1926 speech to the Canadian Club of Ottawa that if the police patrols had been made by 'other people who frequent the Arctic or the Sub-Arctic,' then 'the story would have been spread in block letters across every paper in America.' The Mounties 'neither seek nor desire publicity yet they are ranging the Arctic day by day with dauntless assurance and efficiency.'[35] The process of publicizing this work had to be carried out by indirect means.

Thanks to Craig's efforts, the films taken on the early patrols were put into a format suitable for educational use and released as part of the Canadian Government Motion Picture Bureau's *Seeing Canada* series.[36] Craig and Finnie both believed that new films must be made each year as 'a permanent record in connection with the maintenance of sovereignty' and also because if the films were widely shown, 'the public would have a much better idea of what we are doing and ... we would have less trouble in securing a larger appropriation.'[37] Charles Camsell did his best to obtain favourable coverage for the scientific work done in conjunction with the patrols. In September 1926 he sent an account of Soper's adventures on Baffin Island to the Canadian Press, with the comment that the story deserved the 'widest publicity possible.'[38] The result was a laudatory illustrated article in the *Toronto Star.*[39] The practice of taking an 'official historian' on each voyage resulted in substantial national newspaper coverage. One such historian, journalist Douglas Robertson of the *Toronto Telegram,* also produced a book, *To the Arctic with the Mounties* (1934). Another popular account came from Constable Patrick Lee, who retired from the force and published *Policing the Top of the World* in 1928. Harwood Steele described the various patrols in his *Policing the Arctic* (1936).

But the most effective depictions of Canada's presence in the archipelago came from artists A.Y. Jackson and Lawren Harris. In 1927 Jackson and Dr. Frederick Banting (who was an enthusiastic amateur artist) accompanied the patrol. Jackson's interest in the Far North had been sparked when he happened to hear New Year's greetings sent over the radio to Inspector Joy at Bache Peninsula.[40] Several of his best-known paintings resulted from this trip, including a large canvas of the *Beothic* at Bache, which Jackson presented to Minister Stewart. Acting on Finnie's advice, Stewart donated the painting to the National Gallery of Canada. The gallery's director, Eric Brown, assured the minister that, 'by means of coloured and other reproductions,' the picture undoubtedly would take its place as 'a unique artistic record of the development and protection of the Canadian north land.'[41]

The trip received extensive press coverage, and Banting published a lavishly illustrated account of his experiences in the first issue of the newly founded *Canadian Geographical Journal* (the forerunner of today's *Canadian Geographic*).[42] Jackson himself wrote a shorter article for the *Canadian Forum.* He observed that on the mental map of most Canadians, Hudson Bay was clearly marked, but the archipelago was merely 'a lot of sprawling islands which are somewhat difficult to sort out.' In a series of vivid word pictures, he evoked their visual qualities. A sketch of Bylot Island gave a concrete example of the 'bold and simple' artistic motifs to be found in the northern landscape.[43]

In the summer of 1930, the example of Jackson and Banting was followed by Lawren Harris. After this trip, Jackson and Harris held joint exhibitions of their Arctic work at the National Gallery and the Art Gallery of Ontario. As Banting pointed out in a letter to Finnie, 'the pictures of such men as Jackson & Harris are an exposition of not only the spirit of the country, but also, of our possession.'[44] Indeed, in their own way these northern paintings were 'acts of occupation.' They unquestionably did much to make scenes from the archipelago a part of the Canadian national consciousness and to promote the idea that the Far North was an integral part of the nation. Finnie gratefully wrote to Jackson that there was 'now much more interest in the Arctic than formerly and undoubtedly your pictures have assisted substantially in creating that interest.'[45] And, if *National Geographic* snubbed articles about the patrols, from 1930 on the *Canadian Geographical Journal* was ready to embrace them. Over the years, the magazine published many accounts of government work in the Far North, always accompanied by a large number of photographs.[46]

Rather ironically, by the time these compelling images of Canada's official presence in the northern islands were familiar to the public, a part of the government apparatus that had evolved during the 1920s had been swept away. In 1931 the Department of the Interior underwent dramatic changes. The department had been created in 1873 mainly for the purpose of administering the vast new lands added to Confederation in 1870. Even after the formation of the provinces of Manitoba, Alberta, and Saskatchewan, the public lands and natural resources of the West remained under federal control. But in 1929-30 control was transferred to the provinces. The department thus lost its main function. At the same time, the onset of the Depression placed economy at the forefront of government concerns.

In the summer of 1931, many employees lost their jobs following an extensive reorganization within the department. Cory retired and was replaced as deputy minister by H.H. Rowatt. The positions occupied by Finnie (director of the Northwest Territories and Yukon Branch) and Craig (director-general of surveys) were eliminated. Finnie and Craig were given the option of staying on with reduced rank and authority, though at the same salary as before. Both chose early retirement instead. Finnie was fifty-five and Craig was fifty-six.

Finnie laboured under a strong sense of injustice for the remainder of his life. His resentment may well have sprung from frustration. In the early 1920s, he and Craig had planned ambitious projects for the welfare of the Inuit. Their hopes were thwarted when responsibility for northern Aboriginal people was assigned to the Department of Indian Affairs. Then the task was again given to the Northwest Territories and Yukon Branch in September 1927. 'Please accept my most hearty and sincere congratulations ... as I am convinced that it means good luck to the Eskimos,' wrote Rasmussen. In reply, Finnie outlined the philosophy on which his program for the future was based: 'We believe here that [the Inuit] should not be brought together into schools and hospitals, etc., as we know these institutions, but rather that teachers and doctors might be sent among them to administer to their material and other requirements, in their own environment.'[47] However, the cutbacks of 1931 shattered Finnie's dreams once and for all. 'An embittered man, he lived in quiet obscurity for fifteen years and died of cancer on the 3rd of January 1948,' his son recorded.[48]

There was a further reorganization in 1935-36. The old Department of the Interior, Department of Mines, Department of Indian Affairs, and Department of Immigration and Colonization gave way to the new Department of Mines and Resources. Charles Camsell, as the senior civil servant in the departments being amalgamated, became deputy minister of mines and resources. Roy Gibson, the former assistant deputy minister of the interior, was appointed director of the new Lands, Parks, and Forests Branch, which was responsible for northern administration. Gibson, who has been referred to frequently in the course of this book, was a competent but colourless bureaucrat who took little initiative in policy matters. He was not much liked by his subordinates, who considered him autocratic and narrow-minded.[49] Under his direction, and hampered by budgetary restraints, it was all but impossible to move in new directions. The Dominion Parks Branch (known after 1923 as the National Parks Branch) had survived the earlier reorganization relatively unscathed, losing only thirty-two lower-level positions. Now, however, it was Harkin's turn to face the alternatives of reduced authority or early retirement.[50] He too preferred the latter.

In the mid- and late 1930s, the Eastern Arctic Patrols were commanded by Major David Livingstone McKeand, a veteran of the South African War and the First World War. To reduce costs, the practice of chartering the *Beothic* ended after the 1931 voyage. From 1933 onwards the patrols were carried in the Hudson's Bay Company ship *Nascopie*. Both the HBC men and the tourists who sometimes booked passages on the *Nacopie* viewed McKeand with a critical eye. In 1936 a young American Arctic enthusiast named Prentice Downes was aboard the ship. Downes described McKeand as 'a dogmatic, puffed up old fool,' and nicknamed him 'the mad Major.'[51]

There were not many posts for McKeand to patrol. In 1933 both Dundas Harbour and Bache Peninsula were closed – Dundas for reasons of economy and Bache because ice conditions there were often too difficult for even a ship like the *Beothic* to surmount. In 1931 Mackenzie had informed Finnie: 'The hazard of servicing this post is very great and, in my opinion ... continuous occupation ... for the purposes of jurisdiction is no longer necessary as effective occupation could be maintained from the post at Craig Harbour.' Finnie concurred, adding that the post building should not be dismantled, 'so that it may be used from time to time in making patrols north and for re-occupation at a later date if thought desirable.'[52] In the summer of 1932, two police patrols in search of the lost German explorer Hans Krüger set out from Bache. Harry Stallworthy circled Axel Heiberg Island, while R.W. Hamilton searched Amund Ringnes and Cornwall Islands. Stallworthy's patrol covered 1,400 miles; Hamilton's distance was 943 miles.[53] These journeys were a fitting end to the first phase of the Canadian government's presence in the high Arctic.

AMERICAN GEOGRAPHER JOHN ADAMS visited Craig Harbour and Pangnirtung in the summer of 1936. He observed that despite the large number of Inuit at Pangnirtung, both settlements remained 'outposts occupied by invaders from the temperate zone and not spontaneous agglomerations of native people. Their alien nature is reflected alike in cultural landscape forms and in situation.' Craig Harbour in particular appeared '[s]mall and frail in the vast Arctic landscape.'[54] It might have seemed that the southern Canadian presence was too slight to maintain effective rule, especially after the Craig Harbour post was closed for the duration of the Second World War. Yet within a few years there would be an impressive resurgence of government activity. Despite the loss of experienced administrators such as Finnie and Craig, much of the knowledge accumulated and the bureaucratic apparatus created during the 1920s remained. Although the growth of the Department of External Affairs had been curtailed during the Depression years, there were no extensive cuts in personnel. Men like Lester Pearson and another of Skelton's recruits, Hugh Keenleyside, had retained both their jobs and a keen interest in northern matters.

'The war and the aeroplane have driven home to Canadians the importance of their Northland, in strategy, in resources and in communications,' Pearson declared in the July 1946 issue of *Foreign Affairs*. He added firmly, 'The Canadian Government, while ready to coöperate to the fullest extent with the United States and other countries in the development of the whole Arctic, accepts responsibility for its own sector. There is no reason for sharing that responsibility.'[55] In the same year, Charles Camsell proclaimed the importance of new northern developments in the *Canadian Geographical Journal*. He wrote that the North, dormant for so many years, had now 'burst into new life,' and Canadians were 'beginning to realize that here is a part of their national heritage which may have vast potentialities.'[56] In 1947, following Camsell's retirement, Keenleyside left External Affairs to become the new deputy minister of mines and resources.[57] The internal administration of the Far North thus became his responsibility, while Pearson, as under-secretary of state for external affairs, kept a watchful eye on the American military presence.

Fulfilling the cherished dream of Craig and Finnie, in 1948 construction began on a new ship for the Eastern Arctic Patrols, the *C.D. Howe*. In October 1949 Keenleyside summarized 'Recent Developments in the Canadian North' for the *Canadian Geographical Journal*. He urged that public attention should shift away from the military and strategic aspects of northern development to the work being done by 'the scientists, explorers, administrators, educators, missionaries, doctors and social workers who have been carrying the benefits, and occasionally and inadvertently the detriments, of our national way of life into the endless reaches of the North ... It is they who will bring this harsh, stubborn and silent land into effective contact with the rest of Canada.'[58] Keenleyside's piece was followed by a list of all the northern articles that had appeared in the journal since 1930. There were seventy titles on the list, most of them written by civil servants or former civil servants. Although Stefansson's name was spoken only with distaste in Ottawa, many of the bureaucrats who had rejected his plans nevertheless subscribed to a northern vision that owed more than a little to his influence.

It was because of Keenleyside's position in the Department of Mines and Resources (after January 1950, known as the Department of Resources and Development) and his long-standing interest in northern matters that Stefansson approached him about Harkin's papers. By this time, the history of Canadian policy between 1918 and 1925 was a distant and rather murky subject even to northern enthusiasts like Keenleyside and Pearson.[59] Of the major actors in the drama, only Stefansson, Harkin, Meighen, King, and the Andersons remained alive. Sir Joseph Pope died in December 1926, not long after he left the Department of External Affairs. James White had intended to write a book about the Canadian Arctic after his retirement, but he suffered a fatal heart attack in February 1928. Joseph Bernier passed away in December 1934 at the ripe age of eighty-two. Jack Craig died in 1936, Oscar Skelton in 1941, W.W. Cory in 1943, Charles Stewart in 1946, and Oswald Finnie in 1948. Loring Christie, who had rejoined External Affairs in 1935, died in 1941 while serving as Canadian minister in Washington.

Meighen and King had many other concerns, and King was in poor health (he died in July 1950). Although the Andersons were extremely well informed on all aspects of the Canadian Arctic Expedition and the Wrangel Island affair, they knew almost nothing about the supposed Danish threat of 1920-21. Harkin therefore held the key. Despite the cordial relations established between Rasmussen and Ottawa officialdom in 1925, Harkin had apparently never relinquished his belief in the Danish plan to colonize the archipelago. He therefore had no wish to correct the version of events that stemmed from David Hunter Miller's article.[60] Indeed, it seems to have been his intention that when his papers were finally released to the public, they would reinforce the belief in a threat. Harkin himself would appear in the role of Canada's saviour.[61] But if Stefansson obtained the papers and used them for his own purposes, he would determine Harkin's place in the resulting version of Arctic history. Secrecy therefore remained foremost among Harkin's priorities, even though almost three decades had passed since the events of 1920-21.

Harkin's fears were not altogether unreasonable. After many wilderness years in both Canada and the United States, Stefansson's fortunes had changed dramatically for the better with the coming of the Second World War. He was hired by the American government as a consultant on northern matters. This job brought him a welcome semi-official status such as he had not enjoyed since Sir Robert Borden's retirement. After the war, with American military interest in the Canadian north at its height and his gift for publicity unimpaired, Stefansson busily promoted an image of himself as an unjustly neglected prophet whose time had come at last. He still had the support of journalist Donat LeBourdais, who proclaimed in a 1946 *Canadian Forum* article that Stefansson had formerly been the victim of a 'small coterie.' According to LeBourdais, the members of a more enlightened generation were now ready to collaborate with Stefansson in affirming the 'dominant place' held by the Far North. He noted that Stefansson must find it immensely satisfying, 'after being for so long a lone performer, at last to be conducting an orchestral symphony on his favourite theme.'[62]

Another dedicated Canadian supporter was Richard Finnie. The younger Finnie shared his father's passionate resentment of the government's actions in 1931. By Richard Finnie's own account, during his years in the civil service Oswald Finnie had seldom discussed policy matters with his family.[63] The younger man therefore apparently did not realize how low his father's opinion of Stefansson had been. Instead, Richard Finnie chose to see both his father and Stefansson as northern visionaries whose plans were thwarted by the narrow-minded, penny-pinching government.[64] Finnie lauded both men in his book *Canada Moves North,* and he wrote a number of popular articles about Stefansson and about northern development generally.

Under these circumstances, there was no telling what Stefansson might be able to do with the information in the Harkin papers. (A hint of his possible intentions is provided by a speech he gave to the Ottawa Canadian Club in 1952. In an attempt to play on Cold War rivalries and fears, Stefansson dramatically revealed that Canada had 'once owned strategic Wrangell Island off Soviet Siberia only to have the British Government give it away.' Foolishly hoping to 'be friends' with the Russians, Ramsay MacDonald had ceded the island to them, and so Canada 'lost her claim to what is now one of the most strategic islands in the north.')[65] When Harkin died in 1955, he left the documents in the care of his former secretary, Dorothy Barber. Barber, in turn, passed them on to her nephew, J.E. Lyon. Stefansson died in August 1962; four months later, Lyon presented the material to the Public Archives of Canada. The Harkin documents and Stefansson's posthumously published autobiography, which appeared in 1964, provided the basic framework for subsequent tellings of the story.

As he had done in *The Adventure of Wrangel Island,* in the autobiography Stefansson deliberately confused the Ellesmere expedition with his own plans for Wrangel, claiming that his 1920 trip to London had been undertaken on Borden's orders. His mission, allegedly, was to gain British support for an expedition intended to discover new islands, which would then become 'stations in a world-wide network of great circle flying routes.' This ambitious enterprise was thwarted when Shackleton 'double-crossed' him. Stefansson recounted that Harkin was 'the best-informed man in Ottawa ... on both overt and secret matters connected with our plans. I was told that the only files our project ever had were in either Harkin's private apartment in Ottawa or his office.' According to the autobiography, Borden swore Harkin to secrecy about all aspects of the scheme. When

Borden died without having released Harkin from his vow, Harkin decided that the information must forever be kept 'inviolable.'[66] Stefansson's story can only have added to the perception among researchers that the Harkin papers were an authoritative (perhaps even *the* authoritative) source on Arctic policy in the early 1920s.

The history of this period, then, has been profoundly shaped by the self-serving narratives of two strong-willed, talented, and extremely ambitious men. Both Harkin's long silence and Stefansson's numerous publications were directed towards a similar end – that of securing a suitably high place in the historical record. When their distortions are removed, what remains? First, it is clear that despite the many important changes which had taken place by 1925, the evolving relationship between the Canadian state and the North was not a linear or entirely rational process. Stefansson's personal ambitions and the deceptions he practised were a factor throughout the period, and Canada's progress towards full northern sovereignty therefore followed a convoluted and seemingly erratic course, making retrospective assessment far more difficult than it would otherwise have been.

Second, it is necessary to reconsider the common idea that Ottawa has always taken action in the North only in response to foreign threats.[67] The story behind the first Eastern Arctic Patrol demonstrates that such threats could be manufactured in order to advance previously existing plans for northern development. Unlike the Danish threat, the American threat in 1925 was not altogether a figment of the Canadian imagination. Nevertheless, it cannot be said that the government merely reacted to the situation created by the United States. In the spring of 1925, Finnie and Craig already had well-advanced plans for a post on the Bache Peninsula and for new regulations to control the activities of foreign explorers. James White's plans for a sector claim were of even longer standing, extending as far back as 1904. When the American threat came, all three men were ready to make the best possible use of it. Far from taking the Canadian government unaware and unprepared, foreign threats were created or exaggerated by advocates of a more active northern policy.

Third, when Stefansson and Harkin placed themselves at the centre of events, other important contributions were obscured. Christie, Finnie, and Craig have always received at least some degree of recognition, but White's role in the 1925 sector claim remained unknown. (Here Bernier's self-promotion also played a part in distorting the record.) When White died in 1928, he left behind neither a collection of personal papers nor any published works on Arctic sovereignty. The evidence about him lies scattered in tiny fragments through dozens of obscure files. Among historians who believed that Stefansson's writings and the Harkin papers provided an accurate enough outline of the story, there was no motivation to search for these bits of evidence.

Finally, when the events chronicled by Stefansson and Harkin are viewed from another perspective, the 1925 sector claim becomes far more comprehensible. Whatever the merits or demerits of the sector principle in terms of international law, it expressed an important truth about practical politics in the early twentieth century: that no northern nation would remain idle while its sector was invaded by another state intent on establishing a territorial claim. The corollary of this fact was that no power – not even one of the great powers – was willing to take the risk of conflict involved in claiming islands outside its own sector. Both Canada and the Soviet Union reacted strongly to possible threats; neither the United States nor Great Britain pressed forward in the face of resistance. Robert Logan, a member of the 1922 Eastern Arctic Patrol, wrote to Ottawa from New York in the early summer of 1925: 'I am convinced that MacMillan has no respect for Canadian jurisdiction over northern Ellesmere or Axel Heiberg and is trying to get this government mixed up over them just as Stefansson tried to get Canada mixed up with Wrangel Island. If Canada draws first and does something definite to the "unoccupied sections" the U.S. Government may not back MacMillan any more than Canada backed Stefansson.'[68] Despite the huge disparity in power between the two countries, for Canada to 'draw first' and do 'something definite' was indeed sufficient. When Canadian civil servants decided to oppose Stefansson's Wrangel Island plans and to focus instead on securing their title to the archipelago, they were not (as the explorer and his supporters so vociferously claimed) rejecting the proud northern vision offered to them by Stefansson. Instead, they were very effectively bringing Canada's policy in line with the international realities of their time.

THIS HAS BEEN A story about Arctic policy, not about the North itself. It is told mainly from the perspective of Ottawa. In the early 1920s that perspective was a very limited one, and though it gradually broadened in later years, many northerners would argue that important Arctic realities are still ignored. Canadians see and celebrate themselves as a northern people, but this country has never fully come to grips with either the problems or the potential of the Far North. An overly triumphalist or complacent tone would not be warranted for the conclusion of this book. Nevertheless, in the midst of the current debates about Arctic sovereignty it is useful to take a fresh look at the debates of the past and to put them in a more accurate historical perspective, acknowledging both the illusory nature of some earlier threats to Canada's jurisdiction and the solid work done in the archipelago under government auspices. It is essential to see these events in relation to the history of international law. Throughout the 1920s, it was becoming increasingly evident that the doctrine of effective occupation could not be applied to the polar regions without some modification. Here the Eastern Greenland case stands as an important milestone. Although the sector theory remained controversial, the 1933 decision of the PCIJ demonstrated the validity of the overall approach taken by Canada.[69]

On a more personal level, researching and writing *Acts of Occupation* has been an unusually enjoyable experience for us because the people in the story inhabited, or at least visited, the city where we live and work. On a snowy February morning, it is difficult not to picture Shackleton striding from the Chateau Laurier to Parliament Hill (with a few passers-by perhaps turning their heads to stare at the charismatic, world-famous explorer), or Stefansson walking along Wellington Street, deep in conversation with Cory and Harkin. Pope, Christie, and later Skelton had their offices in the East Block of the parliament buildings; across the street in the Langevin Block was Cory's office, where the meetings of the Northern Advisory Board were held and the sector claim was formulated. The Andersons lived at 58 Driveway, beside the Rideau Canal. Their house, a relatively modest red brick and stucco dwelling, still stands in an area dominated by larger homes. Here Rasmussen spent a cheerful evening with Harkin, Finnie, Craig, and other civil servants during his visit to Ottawa in the spring of 1925. Farther south is Harkin's home in the Glebe. Perhaps, after finishing his response to Hugh Keenleyside's letter, the elderly Harkin soothed his irritated nerves with a stroll along Clemow Avenue and Bank Street?

So vivid have been the impressions created by reading the archival documents that, time after time, the city we know has transformed itself before our eyes back into the city Harkin and the others knew. It seemed fitting, then, to conclude our work with a visit to Beechwood Cemetery, where many of the figures in this book are buried. The graves of Sir Robert Borden, Charles Stewart, J.B. Harkin, Oswald Finnie, James White, Belle and Rudolph Anderson, Charles Camsell, Otto Klotz, and George Mackenzie are in this pleasant spot; Sir Joseph Pope and Édouard Deville lie in nearby Notre Dame Cemetery. On a warm August Sunday afternoon we were almost the only visitors wandering the shady paths. Borden's grave boasts a large stone cross, a historical plaque, and a flagpole flying the Canadian flag. Finnie is buried beside his wife and parents in a family plot, marked by a large stone in a prominent position beside one of the paths. Harkin – a little more obscurely commemorated than he would perhaps have liked – shares a granite marker with his wife and several members of her family, the McCuaigs. We found most of the graves with little difficulty, but White's proved elusive. Finally we spotted it – a handsome stone between two cedar trees, the lower branches of which partly obscure the names of White, his wife Rachel, and their daughter Helen, who died in 1956. Perhaps, we speculated, Helen White planted the trees in memory of her parents, placing them a little too close to the headstone.

It was pleasant to end, as we had begun, with a sense of discovery. In the five years since we first encountered the Harkin papers, every find in the archives has reinforced our belief that the events of 1918-25 constitute far more than a colourful, dramatic episode in the Canadian past. Without an accurate history of this period, it is impossible to grasp the basis on which Canada's northern sovereignty rests. White and the others who laboured in relative obscurity deserve greater recognition but, beyond this, neither Stefansson's version of events nor Harkin's does full justice to the highly effective Arctic policies that had been put in place by the end of 1925. The process of uncovering and piecing together the documents that record Canada's acts of occupation was both challenging and uniquely rewarding. The unexpected conclusion to which it led us is that the Arctic has not been a chronic area of weakness in Canadian foreign policy. Instead, it was by grappling with northern sovereignty issues that politicians and civil servants developed the competencies required to cast off British tutelage and attain true independence in the conduct of their nation's external affairs.

Appendix

Scientists and Explorers Ordinance

Ordinance Respecting Scientists and Explorers

Assented to [by the Northwest Territories Council] 23rd June, 1926

1 This Ordinance may be cited as the 'Scientists and Explorers' Ordinance.

2 No person shall enter the Northwest Territories for scientific or exploration purposes without first obtaining a license so to do from the Commissioner of the said Territories.

3 Any person who enters the said Territories for scientific or exploration purposes without first obtaining a license for such purposes may be summarily ejected from the said Territories and the Commissioner may refuse to consider any application for such license from such person until after the expiration of one year from the time of such unlawful entry.

4 Any license issued under authority of this Ordinance shall be issued upon and subject to the following conditions: –

(a) That the objects of such entry are exclusively for scientific or exploratory purposes and not commercial or political in any way.

(b) The licensee shall strictly observe the provisions of the Northwest Game Act and the regulations thereunder.

(c) The licensee shall furnish the Northwest Territories and Yukon Branch of the Department of the Interior with a statement showing the number of persons to accompany him, and nationality of each.

(d) The following information shall be furnished by the licensee to the said Branch within a reasonable time after the licensee's return, viz.:

1 A statement setting forth any scientific information, if any, he shall have secured.

2 A report setting forth the localities visited, time spent at different places and a list of the specimens taken under various permits or licenses.

(e) The licensee shall furnish the officers of the Royal Canadian Mounted Police or officers in charge of Canadian Government patrol vessels, with a log of the voyages and also full information of the route taken on land, if any, and full particulars in connection therewith should the same be requested by such officers.

5 Anyone who commits any violation of this Ordinance or license issued under this Ordinance shall be liable on summary conviction thereof to a penalty of not more than one thousand dollars.

(from a copy in LAC, RG 85, vol. 916, file 11036)

Notes

Many of the government documents we cite appear in many files. Whenever possible, we have provided a reference to the original document rather than to a copy. When using newspaper clippings found in files, we have almost always looked for the articles on microfilm in order to verify the source and date and to provide a page reference. Where we give a file reference for a clipping, either the paper was not available to us on microfilm or we could not locate the article in question. If an article could not be located, there may have been an error when the clipping was labelled, or it may have been taken from a different edition than the one that was microfilmed.

Introduction: A Policy of Secrecy

1 From 1911 until his retirement in 1936, Harkin served as parks commissioner. He has long been celebrated by such writers as Mabel Williams and Janet Foster as a founder of the conservation movement in Canada. However, his work and legacy have recently come under critical scrutiny from historians, including Alan MacEachern, Tina Loo, and John Sandlos. See M.B. Williams, *Guardians of the Wild* (London: Nelson, 1936); Janet Foster, *Working for Wildlife: The Beginnings of Preservation in Canada,* 2nd ed. (Toronto: University of Toronto Press, 1998); Alan MacEachern, *Natural Selections: National Parks in Atlantic Canada, 1935-1970* (Montreal/Kingston: McGill-Queen's University Press, 2001); Tina Loo, *States of Nature: Conserving Canada's Wildlife in the Twentieth Century* (Vancouver: UBC Press, 2006); and John Sandlos, *Hunters at the Margin: Native People and Wildlife Conservation in the Northwest Territories* (Vancouver: UBC Press, 2007). In contrast, Harkin's biographer, E.J. Hart, defends him as a 'great Canadian.' E.J. Hart, *J.B. Harkin: Father of Canada's National Parks* (Edmonton: University of Alberta Press, 2010), xxii.

2 Stefansson to Keenleyside, 3 November 1949 (quotes portions of Stefansson to Keenleyside, 16 May 1944), Library and Archives Canada (LAC), MG 30 E169, James Bernard Harkin papers (JBH), vol. 2, 'Canadian Sovereignty of Arctic Islands re: V.O. Stefansson.'

3 Keenleyside to Harkin, 27 February 1950, ibid.

4 Harkin, draft reply, ibid.

5 Stefansson to Meighen, 22 March 1939, LAC, MG 31 C6, Richard Sterling Finnie papers (RSF), vol. 7, file 11.

6 Meighen to Stefansson, 24 March 1939, ibid.

7 Stefansson to Meighen, 30 March 1939, ibid.

8 Stefansson to Finnie, 30 March 1939, ibid.

9 Stefansson to Finnie, 24 April 1939, ibid.

10 Richard Finnie to Robert Logan, 26 May 1972, LAC, RSF, vol. 10, file 13.

11 Aviation was certainly discussed in 1920-21, and in 1922 a member of the first Eastern Arctic Patrol, Robert Logan, was on the lookout for possible airfield sites. However, all this was more a matter of future possibilities than of concrete plans for the use of airplanes in the Arctic.

12 The letter is printed in William Barr, *Back from the Brink: The Road to Muskox Conservation in the Northwest Territories* (Calgary: Arctic Institute of North America, 1991), 87-89.

13 'That McMillan Expedition,' *Ottawa Journal,* 2 May 1925, 6. The article is likely by Grattan O'Leary.

14 David Hunter Miller, 'Political Rights in the Arctic,' *Foreign Affairs* 4, 1 (October 1925): 49-51. Miller visited Ottawa in July 1925. He discussed Arctic sovereignty issues with W.W. Cory of the Department of the Interior and James White of the Department of Justice. See Chapter 8.

15 V. Kenneth Johnston, 'Canada's Title to the Arctic Islands,' *Canadian Historical Review* 14, 1 (March 1933): 37-38. The documents used by Johnston are now contained in LAC, Department of the Interior, RG 15, vol. 2. They include a memorandum written by Harkin in November 1920. Harkin's memo was forwarded to A.G. Doughty, the Dominion archivist, by J.D. Craig of the Department of the Interior in January 1921. Along with a few other documents from the fall of 1920, the memo was bound into a volume titled 'Arctic Islands 1920' and placed in the manuscript room in October 1921. Until the early 1960s, these seem to have been the only records on the subject available to researchers.

16 Loring Christie, memo, 28 October 1920, doc. 534 in Lovell C. Clark, ed., *Documents on Canadian External Relations,* vol. 3, *1919-1925* (Ottawa: Department of External Affairs, 1970).

17 Gordon W. Smith, 'Sovereignty in the North: The Canadian Aspect of an International Problem,' in R.St.J. Macdonald, ed., *The Arctic Frontier* (Toronto: University of Toronto Press, 1966), 207-8. See also J.L. Granatstein, 'A Fit of Absence of Mind: Canada's National Interest in the North to 1968,' in E.J. Dosman, ed., *The Arctic in Question* (Toronto: Oxford University Press, 1976), 16-18; Morris Zaslow, 'Administering the Arctic Islands 1880-1940: Policemen, Missionaries, Fur Traders,' in Zaslow, ed., *A Century of Canada's Arctic Islands* (Ottawa: Royal Society of Canada, 1981), 64-66; William R. Morrison, *Showing the Flag: The Mounted Police and Canadian Sovereignty in the North, 1894-1925* (Vancouver: UBC Press, 1985), 163; William R. Morrison, 'Canadian Sovereignty and the Inuit of the Central and Eastern Arctic,' *Études/Inuit/Studies* 10, 1-2 (1986): 253; William R. Hunt, *Stef: A Biography of Vilhjalmur Stefansson, Canadian Arctic Explorer* (Vancouver: UBC Press, 1986), 163; Morris Zaslow, *The Northward Expansion of Canada, 1914-1967* (Toronto: McClelland and Stewart, 1988), 13-14; William Barr, *Back from the Brink: The Road to Muskox Conservation in the Northwest Territories* (Calgary: Arctic Institute of North America, 1991), 89-90; C.S. Mackinnon, 'Canada's Eastern Arctic Patrol 1922-68,' *Polar Record* 27, 161 (April 1991): 93; Nancy Fogelson, *Arctic Exploration and International Relations, 1900-1932* (Fairbanks: University of Alaska Press, 1992), 112-14; Lyle Dick, *Muskox Land: Ellesmere Island in the Age of Contact* (Calgary: University of Calgary Press, 2001), 273-75; Shelagh Grant, *Arctic Justice: On Trial for Murder, Pond Inlet, 1923* (Montreal/Kingston: McGill-Queen's University Press, 2002), 95-98; William Barr, *Red Serge and Polar Bear Pants: The Biography of Harry Stallworthy, RCMP* (Edmonton: University of Alberta Press, 2004), 113-14; Ken Coates, P. Whitney Lackenbauer, William R. Morrison, and Greg Poelzer, *Arctic Front: Defending Canada in the Far North* (Toronto: Thomas Allen, 2008), 43-45. The only historian to contest this view was Trevor Lloyd. Lloyd's planned study of Canada-Greenland relations was, unfortunately,

never written, and he published only a few short articles on his interpretation of Rasmussen's 'no man's land' letter. See 'Knud Rasmussen and the Arctic Islands Preserve,' *Musk-Ox* 25 (1979): 85-90, and 'The True Story Behind the Ellesmere Island Conspiracy,' *McGill Reporter,* 11 December 1974, 3. We became aware of Dr. Lloyd's publications and his extensive files (now in the Trent University Archives) during the course of our research, and we have made use of the Danish documents he collected. We are indebted to Dr. Lloyd's pioneering, but little known, work in this area.

18 The name was often spelled 'Wrangell Island' in the early twentieth century. The island was named for Ferdinand von Wrangell, a Russian explorer of Baltic German extraction. Since double 'l's are not used in Russian, the Russian name for the island is Ostrov Vrangelya. 'Wrangel Island' is therefore now the commonly accepted English spelling. See E. Tammiksaar, 'Wrangell or Wrangel – Which Is It?' *Polar Record* 34, 188 (January 1998): 55-56.

19 In the later part of the twentieth century, even some Canadian politicians and civil servants dismissed the sector theory as unworthy of serious consideration. Pierre Trudeau and his foreign policy adviser, Ivan Head, considered it too flimsy to rely on during the crisis provoked by the 1969 voyage of the American tanker *Manhattan.* Despite considerable research by Head, the two men knew nothing about the background to the 1925 claim, and indeed they were apparently unaware that an official sector claim had been made in that year – a striking indication of the obscurity that has long surrounded the period covered by this book, even among government officials with access to secret files. See Ivan Head and Pierre Trudeau, *The Canadian Way: Shaping Canada's Foreign Policy, 1968-1984* (Toronto: McClelland and Stewart, 1995), 31, 50-52, and Ivan Head, 'Canadian Claims to Territorial Sovereignty in the Arctic Regions,' *McGill Law Journal* 9, 3 (1963): 207.

20 Miller, 'Political Rights,' 53.

21 Richard Finnie to William Hunt, 26 October 1973, LAC, RSF, vol. 8, file 6. See also Constance L. Skinner, 'Stefansson, Sentinel of Civilization,' *Canadian Forum* 7, 74 (October 1926): 398-401; D.M. LeBourdais, *Stefansson: Ambassador of the North* (Montreal: Harvest House, 1963), 10-12, 15, 160-61, 173-80; and Hunt, *Stef,* 180-89, 204-5.

22 Richard Diubaldo, *Stefansson and the Canadian Arctic* (Montreal: McGill-Queen's University Press, 1978), 209, 214.

23 State formation can be defined as the development of the structures and practices of the state, and the processes by which government comes to function in more ambitious and effective ways. For a useful summary, see Allan Greer and Ian Radforth, *Colonial Leviathan: State Formation in Mid-Nineteenth-Century Canada* (Toronto: University of Toronto Press, 1992), 9-13.

Chapter 1: Taking Hold of the North

1 J.M. Tupper to Commanding Officer, Peace River, 30 November 1917, Library and Archives Canada (LAC), Royal Canadian Mounted Police, RG 18, vol. 2159, file 8.

2 'Another Year of Exploring,' *Globe* (Toronto), 7 March 1918, 11.

3 In 1920 the full cost was calculated at $536,552.10. LAC, Department of Marine and Fisheries, RG 42, vol. 490, file 84-3-8. To give some idea of the magnitude of this sum, in 1918 deputy ministers – Ottawa's most senior bureaucrats – made $5,000 a year. Many clerks and other workers in the lower ranks of the civil service made less than $1,000 a year.

4 'Iowa Scientist Attacks Stefansson,' Des Moines *Sunday Register,* magazine section, 29 January 1922. Clipping in LAC, MG 30 B40, Rudolph Martin Anderson and Mae Belle

Allstrand Anderson papers (RMA/MBAA), vol. 30, file 5-1-93. See also Rudolph Anderson diary, 8 and 10 June 1913, LAC, RMA/MBAA, vol. 17. Further details are provided in Stuart Jenness's introduction to the published edition of his father's diary. See *Arctic Odyssey: The Diary of Diamond Jenness, 1913-1916* (Ottawa: Canadian Museum of Civilization, 1991), xxxiii-xxxiv.

5 G.J. Desbarats to Stefansson, 29 May 1913, LAC, Geological Survey of Canada, RG 45, vol. 67, file 4078C.

6 Stefansson had written to Ottawa for permission to take men and resources from the scientific party for his exploration work. His attempt to push forward with his plans was made on the assumption that it would be granted. However, government officials decided that 'the work of the southern party should be carried out as originally proposed and ... it should not be weakened for the purpose of organizing another northern party.' Desbarats to Stefansson, 30 April 1914, LAC, RG 42, vol. 490, file 84-2-55.

7 William Laird McKinlay, *Karluk: The Untold Story of Arctic Exploration* (London: Weidenfeld and Nicholson, 1976), 117-18.

8 Ibid., 161. For an excellent secondary account which makes use of an unpublished manuscript by McKinlay, see Jennifer Niven, *The Ice Master: The Doomed 1913 Voyage of the* Karluk (New York: Hyperion, 2000).

9 'Karluk Journey Was Big Scheme to Exploit News,' *Ottawa Journal,* 14 January 1922, 12. Such comments were frequent during the course of the expedition; for example, see John Cox to W.H. Boyd, 28 June 1915, LAC, MG 30 B66, Kenneth Gordon Chipman papers (KGC), vol. 1, file January-December 1915. Cox recorded that Stefansson made 'no secret of the fact that his object is personal publicity and what it brings to him.'

10 Richard S. Finnie, 'Stefansson as I Knew Him,' *North/Nord* 25, 3 (May-June 1978): 38.

11 Evelyn Stefansson Nef to Richard Finnie, 14 September 1976, LAC, MG 31 C6, Richard Sterling Finnie papers (RSF), vol. 8, file 6.

12 Belle Anderson to Helen Crawford, 29 December 1923, LAC, RMA/MBAA, vol. 9, file 17, and untitled document by Belle Anderson, June 1922, ibid., vol. 12, file 5.

13 'Says Stefansson Expedition Squandering Public Money,' *Ottawa Citizen,* 25 November 1916, 2. The source cited by the journalist was an anonymous Ottawa official who had gained his information from correspondents in the North. On Anderson's denial of responsibility, see Anderson to Borden, 14 April 1923, LAC, MG 26 H, Robert Laird Borden papers (RLB), vol. 259, file 7. Anderson spoke only favourably of Stefansson in a 1917 press interview. See 'Dr. Anderson Was with Stefansson in the Far North,' *Ottawa Citizen,* 28 December 1917, 3.

14 Rudolph Anderson to Isaiah Bowman, 6 November 1918, LAC, MG 30 B8, Charles Camsell papers (CC), vol. 3, file 19.

15 Vilhjalmur Stefansson, *The Adventure of Wrangel Island* (New York: Macmillan, 1925), 73.

16 'Coal on Nearly Every Island: Stefansson Tells of Some Discoveries in the Far North,' *Globe* (Toronto), 17 September 1918, 3.

17 Alex was conceived during Stefansson's 1908-12 expedition. On Pannigabluk and Alex, see Richard Finnie, 'Stefansson's Unsolved Mystery,' *North/Nord* 25, 6 (November-December 1978): 2-7, and Gísli Pálsson, *Travelling Passions: The Hidden Life of Vilhjalmur Stefansson* (Winnipeg: University of Manitoba Press, 2003). On Betty Brainerd and Fannie Hurst, see William R. Hunt, *Stef: A Biography of Vilhjalmur Stefansson, Canadian Arctic Explorer* (Vancouver: UBC Press, 1986), 151-53, 160-61.

18 Stefansson to Frits Johansen, 18 September 1917, extracts published in 'Stefansson May Come Home Again from the North,' *Ottawa Citizen,* 26 March 1918, 10.

19 See Vilhjalmur Stefansson, *Discovery: The Autobiography of Vilhjalmur Stefansson* (New York: McGraw-Hill, 1964), 212.
20 'Stefansson Reaches Ottawa with Report,' *Globe* (Toronto), 30 October 1918, 6.
21 Stefansson to Desbarats, 1 November 1918, LAC, RG 42, vol. 477, file 84-2-29. The lectures were not, however, for the benefit of the Red Cross. Rudolph Anderson took the trouble to write to the Red Cross on the subject, and he was promptly informed that 'Mr. Stefansson's statement that he is going to lecture for the Red Cross is entirely unauthorized. It is against our policy to permit any such lectures or to permit the name of the Red Cross to be used in connection with lectures, commercial enterprises, etc., of any kind.' Ralph Wolf to Anderson, 10 September 1918, LAC, RMA/MBAA, vol. 2, file 12.
22 See Janice Cavell and Jeff Noakes, 'Explorer without a Country: The Question of Vilhjalmur Stefansson's Citizenship,' *Polar Record* 45, 234 (July 2009): 237-41.
23 Stefansson, *Discovery,* 213.
24 Constance L. Skinner, 'Stefansson, Sentinel of Civilization,' *Canadian Forum* 7, 74 (October 1926): 399.
25 See Jean Barman, '"I walk my own track in life & no mere male can bump me off it": Constance Lindsay Skinner and the Work of History,' in Alison Prentice and Beverly Boutilier, eds., *Creating Historical Memory: English-Canadian Women and the Work of History* (Vancouver: UBC Press, 1997), 129-63, and *Constance Lindsay Skinner: Writing on the Frontier* (Toronto: University of Toronto Press, 2002), 124-28.
26 'Drift in Icy Sea to Top of World,' *New York Times,* 31 October 1918, 5; 'Pole Not Coldest Spot,' *New York Times,* 1 November 1918, 13. See also 'Five Years in the Arctic,' *New York Times,* 3 November 1918, 33.
27 Contract between Vilhjalmur Stefansson and Lee Keedick, 5 November 1918, Stef MS-196 (2) K, Rauner Special Collections Library, Dartmouth College, Hanover, New Hampshire (RSCL), Stefansson Collection. The contract gave Keedick 50 percent of the net receipts. There were to be not less than fifty lectures in each of the two years; if Stefansson went north again, the contract could be terminated after the first year. However, in this case Keedick would have the right to manage Stefansson's next tour.
28 Desbarats diary, 9 November 1918, LAC, MG 30 E89, George Desbarats papers (GD), vol. 5.
29 Stefansson to Desbarats, 13 November 1918, LAC, RG 42, vol. 477, file 84-2-29.
30 Rudolph Anderson to Borden, 21 April 1923, LAC, RLB, vol. 259, file 7.
31 Stefansson, *Discovery,* 214.
32 Ibid.
33 'Big Victory Loan Parade Feature of Celebrations,' *Globe* (Toronto), 12 November 1918, 9.
34 Vilhjalmur Stefansson, 'The Canadian Arctic Region,' in *Empire Club of Canada: Addresses Delivered to the Members during the Sessions, 1917-18, and May to December, 1918* (Toronto: Warwick Bros. and Rutter, 1919), 369; see also Stefansson, *Discovery,* 214-15.
35 Stefansson, 'Canadian Arctic Region,' 372-77.
36 See Edmund Walker, 'The National Peril,' *MacLean's Magazine,* March 1912, 515-17.
37 'Canadian Arctic Region,' 378-79.
38 Earl Parker Hanson, *Stefansson, Prophet of the North* (New York: Harper, 1941), xi-xii.
39 Borden to George W. Fowler, 5 July 1920, LAC, RLB, vol. 185, file 529.
40 Stefansson, *The Northward Course of Empire* (New York: Harcourt, Brace, 1922), 238.
41 On the origin of the belief in unknown land in this region, and on Stefansson's determination to discover it, see Trevor H. Levere, 'Vilhjalmur Stefansson, the Continental

Shelf, and a New Arctic Continent,' *British Journal for the History of Science* 12, 2 (June 1988): 233-47, especially 236-38.

42 Claude-Sussex to Stefansson, 21 January 1919, LAC, RG 42, vol. 468, file 84-2-5.

43 Interestingly, Robert Bartlett wrote that Stefansson's main purpose in 1913-18 was 'to discover new land along the 141st Meridian.' Bartlett, *The Last Voyage of the* Karluk (Boston: Small, Maynard, 1916), 306. There is no hint of any such intention in the plans submitted by Stefansson to Borden. See Stefansson to Borden, 4 February 1913, LAC, RLB, vol. 234, file 2117. However, Bartlett's account is confirmed by a statement Stefansson made to the press in May 1913. See 'To Map Out Last Land,' *New York Times,* 27 May 1913, 3. It therefore seems Stefansson had long planned to persuade the government that the traditional boundary should be disregarded.

44 See Berthold Seemann, *Narrative of the Voyage of H.M.S. Herald during the Years 1845-51, under the Command of Captain Henry Kellett, R.N., C.B., being a Circumnavigation of the Globe, and Three Cruizes to the Arctic Regions in Search of Sir John Franklin* (London: Reeve, 1853), 2:116. Kellett had not approached the island closely enough to be certain of its exact configuration. In the mid-nineteenth century, it appeared on British Admiralty charts as two separate islands: Plover Land and Kellett Land. The name Wrangell or Wrangel Island was given by an American whaler, Thomas Long, in 1867. On the spelling of the name, see Introduction, note 18. Although he himself did not come anywhere near the island, the Russian explorer Ferdinand von Wrangell was the first to hear Native reports of its existence and to place it on a map. See von Wrangell, *Narrative of an Expedition to the Polar Sea in the Years 1820, 1821, 1822 and 1823,* ed. Edward Sabine, 2nd ed. (London: James Madden, 1844), 325-26.

45 See William Barr, Reinhard Krause, and Peter-Michael Pawlik, 'Chukchi Sea, Southern Ocean, Kara Sea: The Polar Voyages of Captain Eduard Dallmann, Whaler, Trader, Explorer 1830-96,' *Polar Record* 40, 212 (January 2004): 1-18.

46 C.L. Hooper, *Report of the Cruise of the U.S. Revenue Steamer Thomas Corwin in the Arctic Ocean, 1881* (Washington, DC: Government Printing Office, 1884), 66.

47 See William Barr, 'The Voyages of *Taymyr* and *Vaygach* to Ostrov Vrangelya, 1910-15,' *Polar Record* 16, 101 (May 1972): 213-34.

48 Alex Stefansson repeated this story to his daughter Rosie, who in turn told it to Gísli Pálsson. See Pálsson, Introduction to *Writing on Ice: The Ethnographic Notebooks of Vilhjalmur Stefansson,* by Vilhjalmur Stefansson and edited by Gísli Pálsson (Hanover: University Press of New England, 2001), 66.

49 McKinlay to Belle Anderson, 15 June 1922, LAC, RMA/MBAA, vol. 8, file 7.

50 See Richard Diubaldo, *Stefansson and the Canadian Arctic* (Montreal: McGill-Queen's University Press, 1978), 162-63.

51 The statement was allegedly made to James White. White repeated it to the Andersons, and the incident is referred to many times in their correspondence. For example, see Belle Anderson to Helen Crawford, 31 November 1924, LAC, RMA/MBAA, vol. 9, file 18; Belle Anderson to John Maurer, 30 May 1925, LAC, RMA/MBAA, vol. 8, file 13; and Belle Anderson to Helen Crawford, 3 August 1951, LAC, RMA/MBAA, vol. 9, file 15.

52 For an example of this kind of thinking, see Robert Borden, Introduction to *The Friendly Arctic,* by Vilhjalmur Stefansson, 2nd ed. (New York: Macmillan, 1943). See also Janice Cavell, 'The Second Frontier: The North in English-Canadian Historical Writing,' *Canadian Historical Review* 83, 3 (September 2002): 364-89.

53 On the decline in the muskox population on the mainland due to over-hunting instigated by the HBC, see William Barr, *Back from the Brink: The Road to Muskox Conservation*

in the Northwest Territories (Calgary: Arctic Institute of North America, 1991), 14-26, 96-98. The skins were used mainly for sleigh robes.

54 See John Sandlos, *Hunters at the Margin: Native People and Wildlife Conservation in the Northwest Territories* (Vancouver: UBC Press, 2007), 115-18.

55 On Hewitt's activities as a conservationist, see C. Gordon Hewitt, *The Conservation of the Wildlife of Canada* (New York: Scribner's, 1921). Hewitt's death from influenza in February 1920 at the age of only thirty-five deprived Stefansson of an important ally.

56 Maxwell Graham to O.S. Finnie, 2 December 1922, LAC, Northern Affairs Program, RG 85, vol. 1203, file 401-3. The Barrymore Cloth Co. of Toronto offered to produce some sample fabric if enough wool could be provided to them, but this proved impossible. In the end the small amount of wool available in Ottawa was carded and spun by hand by a woman living in East Templeton, Quebec.

57 See LAC, RG 42, vol. 481, file 84-2-38.

58 Meighen to J.S. McLean, 9 May 1919, LAC, MG 26I, Arthur Meighen papers (AM), I1, vol. 7, file 43.

59 Memo from Stefansson to Meighen, copy in LAC, Department of Justice, RG 13, vol. 239, file 1919-1781. Stefansson's statement is undated; Meighen forwarded it to several of his cabinet colleagues on 22 March 1919. The minister of trade and commerce, Sir George Foster, replied that he had 'read the statement with a great deal of interest and it certainly opens up a field for beneficial conservation if not for development and cultivation.' Foster to Meighen, 17 April 1919, LAC, MG 27 II D7, George Foster papers (GF), vol. 25, file 5054.

60 John Stevenson, quoted in Roger Graham, *Arthur Meighen,* vol. 2, *And Fortune Fled* (Toronto: Clarke, Irwin, 1963), 81.

61 Meighen to McLean, 9 May 1919.

62 Diubaldo, *Stefansson and the Canadian Arctic,* 145.

63 See Moreau S. Maxwell, *Prehistory of the Eastern Arctic* (Orlando, FL: Academic Press, 1985), Chapters 8 and 9.

64 On the disputes over Peary's claim, see Wally Herbert, *The Noose of Laurels: Robert E. Peary and the Race to the North Pole* (New York: Athenaeum, 1989), Chapters 15-21.

65 Robert E. Peary, *The North Pole* (New York: Frederick A. Stokes, 1910), 43-44, 46, 47.

66 Ibid., 48.

67 Robert Peary, 'Report of R.E. Peary, C.E., U.S.N., on Work Done in the Arctic in 1898-1902,' *Bulletin of the American Geographical Society* 35, 5 (1903): 499.

68 Donald B. MacMillan, 'Food Supply of the Smith Sound Eskimos: A Story of Primitive Life Maintained on the Natural Resources of a Frozen Arctic Land and Sea,' *American Museum of Natural History Journal* 18, 3 (1918): 174. Knud Rasmussen estimated the number of muskoxen killed each year at three hundred. Rasmussen, *Greenland by the Polar Sea: The Story of the Thule Expedition from Melville Bay to Cape Morris Jesup,* trans. Asta and Rowland Kenney (London: Heinemann, [1921]), 17.

69 Comer to Stefansson, 29 May 1919, LAC, RG 85, vol. 1203, file 401-3.

70 J.B. Harkin, *The Origin and Meaning of the National Parks of Canada: Extracts from the Papers of the Late Jas. B. Harkin, First Commissioner of the National Parks of Canada* (Saskatoon: H.R. Larson, 1957), 5, 7, 9.

71 Entry for William Harkin in the *Canadian Parliamentary Companion and Annual Register,* 1880. Other information on the family has been taken from the 1881, 1901, and 1911 census records, available online at the Canadian Genealogy Centre website, and from E.J. Hart, *J.B. Harkin: Father of Canada's National Parks* (Edmonton: University of Alberta Press, 2010), 1-5.

72 Untitled summary of Harkin's career, probably by Harkin himself, LAC, MG 30 E169, James Bernard Harkin papers (JBH), vol. 2, file 'Canadian Sovereignty of Arctic Islands re: V.O. Stefansson.' Some inaccuracies in this account have been corrected by consulting the *Civil Service Lists* and LAC, Department of Indian Affairs, RG 10, vol. 3052, file 242,830.
73 Harkin, *Origin and Meaning of the National Parks,* 5.
74 M.B. Williams, *Guardians of the Wild* (London: Nelson, 1936), 2, 7, 140. On the Conservatives' view of Harkin as a Liberal placeman, see C.J. Taylor, *Negotiating the Past: The Making of Canada's National Historic Parks and Sites* (Montreal/Kingston: McGill-Queen's University Press, 1990), 110.
75 Norman Luxton, quoted in Hart, *J.B. Harkin,* 485.
76 See M.H. Long, 'The Historic Sites and Monuments Board of Canada,' Canadian Historical Association, *Report of the Annual Meeting* 33 (1954): 3-4, and Taylor, *Negotiating the Past,* 25-31.
77 Morris Zaslow, *The Opening of the Canadian North, 1870-1914* (Toronto: McClelland and Stewart, 1971), 243.
78 The very scanty correspondence between Harkin and the RNWMP on the enforcement of the Migratory Birds Convention Act and the Northwest Game Act during 1919 can be found in LAC, Royal Canadian Mounted Police, RG 18, vol. 2172, files 27 and 28. In 1919-20 three infringements of the Northwest Game Act were investigated; in 1920-21 there was only one investigation. During 1921-22, however, the number rose to fifteen. Sandlos, *Hunters at the Margin,* 159.

Chapter 2: The Danish Threat

1 Maurice Pope, *Public Servant: The Memoirs of Sir Joseph Pope, Edited and Completed by Maurice Pope* (Toronto: Oxford University Press, 1960), 285.
2 Quoted in Robert Bothwell, *Loring Christie: The Failure of Bureaucratic Imperialism* (New York: Garland, 1988), 47.
3 Quoted in John Hilliker, *Canada's Department of External Affairs,* vol. 1, *The Early Years, 1909-1946* (Montreal/Kingston: McGill-Queen's University Press, 1990), 80.
4 Harkin to Cory, two memos, both dated 11 July 1919, Library and Archives Canada (LAC), Northern Affairs Program, RG 85, vol. 1203, file 401-3.
5 Cory to McLean, 18 July 1919, and McLean to Cory, 24 July 1919, ibid.
6 *Treaties and Other International Agreements of the United States of America, 1776-1949,* vol. 7, *Denmark-France* (Washington, DC: United States Government Printing Office, 1971), 62. See also *Foreign Relations of the United States: The Lansing Papers, 1914-1920,* vol. 2 (Washington, DC: United States Government Printing Office, 1940), 501-11, and Charles Callan Tansill, *The Purchase of the Danish West Indies* (Gloucester, MA: Peter Smith, 1966).
7 See Philip Wigley, *Canada and the Transition to Commonwealth: British-Canadian Relations, 1917-1926* (Cambridge: Cambridge University Press, 1977), Chapter 4.
8 Curzon to Grevenkop-Castenskiold, 27 August 1919, and Grevenkop-Castenskiold to Curzon, 28 August 1919, National Archives of the United Kingdom: Public Record Office (TNA: PRO), FO 371/4084/9296.
9 Typed transcript of evidence, LAC, Royal Commission on Possibilities of Reindeer and Musk-Ox Industries in the Arctic and Sub-Arctic Regions, RG 33-105, vol. 1.
10 See Stefansson's 1917 memorandum to McBride, printed in Richard Diubaldo, *Stefansson and the Canadian Arctic* (Montreal: McGill-Queen's University Press, 1978), 136-37.

11 Keedick later sued Stefansson for $15,000 in damages. Stefansson replied with his own lawsuit. He alleged that Keedick had continued to make bookings after Stefansson gave him notice that he could not continue, thus damaging Stefansson's reputation when the bookings were not fulfilled. See 'Lecture Manager Sues Stefansson,' *New York Times,* 13 November 1921, E9, and 'Press Agent Too Good,' ibid., 18 January 1922, 9.
12 Stefansson to Harkin, 29 March 1920, Trent University Archives (TUA), Trevor Lloyd papers (TL), file 87-104-9-50.
13 Stefansson to Meighen, 12 March 1920, LAC, Canadian Parks Service, RG 84, vol. 166, file U226.
14 See Stefansson to Harkin, 12 April 1920, and Harkin to Cory, 27 May 1920, LAC, RG 85, vol. 1147, file 270-8-2.
15 Rudyard Kipling, 'The Verdict of Equals,' in Kipling, *A Book of Words* (London: Macmillan, 1928), 71-72.
16 See 'Mr. Stefánsson's Expedition,' *Geographical Journal* 46, 6 (December 1915): 457-59, and 'Letter from Mr. Stefánsson,' ibid., 52, 4 (October 1918): 248-55.
17 See Ernest F. Chafe, 'The Voyage of the "Karluk" and Its Tragic Ending,' *Geographical Journal* 51, 5 (May 1918): 307-16.
18 Hinks to Stefansson, 13 September 1918, Royal Geographical Society Archives (RGS), CB 8/80. All quotations from material in the Royal Geographical Society Archives are by kind permission of the Society.
19 He did, in fact, receive the RGS medal the next year, but he was not able to accept it in person.
20 'Meat from the Arctic: Explorer on Reindeer Farming,' *The Times* (London), 25 March 1920, 13.
21 Hinks to Debenham, 12 April 1920, RGS, CB 8/80.
22 Kathleen Kennet, *Self-Portrait of an Artist: From the Diaries and Memoirs of Lady Kennet (Kathleen, Lady Scott)* (London: John Murray, 1949), 143.
23 'Sir E. Shackleton's 100th Lecture,' *The Times* (London), 24 February 1920, 19.
24 Quoted in the flyer for the lectures. There is a copy in the Rauner Special Collections Library (RSCL), Stef MS-196 (5) S.
25 Vilhjalmur Stefansson, *Discovery: The Autobiography of Vilhjalmur Stefansson* (New York: McGraw-Hill, 1964), 237.
26 Stefansson to Shackleton, 15 April 1920, RSCL, Stef MS-196 (5) S; copy in LAC, RG 85, vol. 583, file 571. All quotations from letters in the Stefansson Collection are by kind permission of the Dartmouth College Library.
27 In his autobiography, Stefansson denied the claim by Shackleton's biographers James and Margery Fisher that he had made a 'cordial offer of help' to Shackleton's projected Arctic expedition. 'I certainly made no "cordial offer of help," because I was entirely unaware of his plans,' Stefansson wrote. Stefansson, *Discovery,* 239.
28 Shackleton to Stefansson, 17 April 1920, RSCL, Stef MS-196 (5) S. The correspondence between the two explorers strongly suggests that Shackleton attempted to put Stefansson in touch with wealthy men who might have been willing to back his reindeer project if the Hudson's Bay Company declined to do so.
29 See Diubaldo, *Stefansson and the Canadian Arctic,* 149-51.
30 See Gavin White, 'Henry Toke Munn (1864-1952),' *Arctic* 37, 1 (March 1984): 74.
31 Stefansson to Rasmussen, 13 April 1920, TUA, TL, 87-104-9-25.
32 Stefansson to Harkin, 26 April 1920, and Harkin to Stefansson, 4 May 1920, LAC, RG 85, vol. 1203, file 401-3; Harkin to Cory, 4 May 1920, LAC, MG 30 E169, James Bernard Harkin papers (JBH), vol. 1, file July 1919-October 1920.

33 Knud Rasmussen, *People of the Polar North* (London: K. Paul, Trench, Trübner, 1908), xix.

34 Richard Vaughan, *Northwest Greenland: A History* (Orono: University of Maine Press, 1991), 132, citing L. Mylius-Erichsen and Harald Moltke, *Grønland* (Copenhagen, 1906), 169, 315. See also Harald Lindow, 'Trade and Administration of Greenland,' in M. Vahl et al., eds., *Greenland,* vol. 3, *The Colonization of Greenland and Its History until 1929* (Copenhagen/London: C.A. Reitzel/Oxford University Press, 1929), 49.

35 See Wally Herbert, *The Noose of Laurels: Robert E. Peary and the Race to the North Pole* (New York: Athenaeum, 1989), 111-17. For a good popular account of Sverdrup's explorations, see Gerard Kenney, *Ships of Wood and Men of Iron: A Norwegian-Canadian Saga of Exploration in the High Arctic* (Toronto: Dundurn Press, 2005).

36 Quoted in Jean Malaurie, *The Last Kings of Thule* (New York: Dutton, 1982), 121, 234.

37 H.P. Steensby, 'The Polar Eskimos and the Polar Expeditions,' *Fortnightly Review,* n.s., 86, 515 (November 1909): 892.

38 Ib Nyeboe, 'Stationen Thule,' 1 November 1935, TUA, TL, 87-014-6-2.

39 Rasmussen to O.D. Skelton, 5 May 1925, LAC, Department of External Affairs, RG 25, vol. 1386, file 1324.

40 Rasmussen, 'Handelstationen Thule,' July 1926, TUA, TL, 87-104-6-2. See also Therkel Mathiassen, 'Knud Rasmussen's Sledge Expeditions and the Founding of the Thule Trading Station,' *Geografisk Tidsskrift* 37, 1-2 (1934): 21; Rolf Gilberg, 'Inughuit, Knud Rasmussen, and Thule,' *Études/Inuit/Studies* 12, 1-2 (1988): 47; and Rudolf Sand, 'Nogle Bemærkinger Om Kap York Stationen Thule Af Overretasagfører Rudolf Sand Stationens Administrator,' 22 April 1934, TUA, TL, 87-104-6-2.

41 Nyeboe, 'Stationen Thule.'

42 Peter Freuchen, *Arctic Adventure: My Life in the Frozen North* (New York: Farrar and Rinehart, 1935), 35, 45.

43 See 'Greenland and the Danish Islands,' *New York Times,* 21 August 1916, 10, and 'Peary View as to U.S. "Rights" in Greenland,' *Christian Science Monitor,* 11 September 1916, clipping attached to Marie Peary Stafford, 'Peary's Ideas about US Rights in Greenland,' draft article for Stefansson's unpublished 'Encyclopedia Arctica' (Ann Arbor: University Microfilms International/Hanover, NH: Dartmouth College Library, 1974), Reel 27. See also Stefansson's articles 'United States in Relation to Greenland' and 'Greenland in International Relations,' ibid.

44 Cory to Pope, 12 May 1920, LAC, RG 85, vol. 1203, file 401-3.

45 Typed transcript, LAC, RG 33-105, vol. 1. In his narrative, published two years previously, MacMillan wrote that in 1916 the United States had 'conced[ed] to Denmark ... the right to control all of Greenland.' Therefore, 'our Eskimos, hitherto free and independent,' were 'henceforth subject to the control of a foreign nation.' Donald B. MacMillan, *Four Years in the White North* (New York: Harper, 1918), 276.

46 Rasmussen to Greenland Administration, 8 March 1920, TNA: PRO, FO 371/4084/9296; copies in LAC, Governor General's Office, RG 7, vol. 596, file 27489, and LAC, MG 30 E169, James Bernard Harkin papers (JBH), vol. 1, file July 1919-October 1920.

47 Grevenkop-Castenskiold to Curzon, 12 April [1920], TNA: PRO, FO 371/4084/9296.

48 Jens Brøsted, 'Danish Accession to the Thule District, 1937,' *Nordic Journal of International Law* 57 (1988): 264.

49 Kai Birket-Smith, 'The Greenlanders of the Present Day,' in M. Vahl et al., eds., *Greenland,* vol. 2, *The Past and Present Population of Greenland* (Copenhagen/London: C.A. Reitzel/Humphrey Milford, 1928), 9. The name was given by the Literary Expedition.

50 See 'Knud Rasmussens Ekspedition til Centraleskimoernes Lande' and accompanying sketch map, *Politiken,* 27 April 1921, clipping in LAC, RG 85, vol. 583, file 571.
51 See Robert E. Peary, *Northward over the 'Great Ice': A Narrative of Life and Work along the Shores and upon the Interior Ice-Cap of Northern Greenland in the Years 1886 and 1891-1897* (New York: Frederick A. Stokes, 1898), 1:479-514.
52 Lyle Dick suggests that the Danish claim was based on Ellesmere's status as an Inughuit hunting ground. However, given that Inughuit hunting on Ellesmere dated only from the time of Peary's expeditions and that the Inughuit were not Danish subjects in 1920, this would not have been a valid argument even if the Danes had wished to use it. See Dick, *Muskox Land: Ellesmere Island in the Age of Contact* (Calgary: University of Calgary Press, 2001), 267.
53 Elmer Ekblaw, 'A Recent Eskimo Migration and Its Forerunner,' *Geographical Review* 9, 2 (February 1920): 141-44.
54 He showed a detailed knowledge of the matter in his 25 September 1920 letter to Loring Christie (see note 68).
55 Stefansson to Harkin, 15 May 1920, LAC, JBH, vol. 1, file July 1919-October 1920.
56 Reported in Rudolph Anderson to Belle Anderson, 18 May 1920, LAC, MG 30 B40, Rudolph Martin Anderson and Mae Belle Allstrand Anderson papers (RMA/MBAA), vol. 8, file 5.
57 *Globe* (Toronto), 11 June 1920, 2.
58 Harkin to Cory, 14 and 16 June 1920, LAC, RG 85, vol. 1203, file 401-3.
59 Harkin to Cory, 29 June 1920, LAC, JBH, vol. 1, file July 1919-October 1920. See also Cory to Pope, 23 June and 3 July 1920, and enclosures, LAC, RG 25, vol. 2669, file 9059-B-40. The name Royal North West Mounted Police had changed to Royal Canadian Mounted Police in February 1920.
60 Devonshire to Milner, 13 July 1920, TNA: PRO, FO 371/4084/9296.
61 LAC, RG 25, vol. 2669, file 9059-B-40.
62 David J. Hall and Donald B. Smith, 'Lougheed, Sir James Alexander,' *Dictionary of Canadian Biography,* vol. 15, http://www.biographi.ca.
63 Vilhjalmur Stefansson, 'The Region of Maximum Inaccessibility in the Arctic,' *Geographical Review* 10, 3 (September 1920): 167-72.
64 See Canada, Royal Commission on Possibilities of Reindeer and Musk-ox Industries in the Arctic and Sub-arctic Regions, *Report of the Royal Commission Appointed by Order-in-Council of Date May 20, 1919, to Investigate the Possibilities of the Reindeer and Musk-ox Industries in the Arctic and Sub-arctic Regions of Canada* (Ottawa: The Commission, 1922). The report was tabled in the House of Commons on 4 May 1921.
65 Rasmussen to Stefansson, 11 May 1920, RSCL, Stef MS-196 (5) R.
66 See Knud Rasmussen, 'Project of a Danish Expedition to the Central Eskimo,' *Geographical Journal* 35, 3 (March 1910): 295-99.
67 See Knud Rasmussen, *Across Arctic America: Narrative of the Fifth Thule Expedition* (New York: G.P. Putnam, 1927).
68 Stefansson to Christie, 25 September 1920 (includes copy of Stefansson to Meighen, 16 September), LAC, MG 30 E44, Loring Cheney Christie papers (LCC), vol. 6, file 19.
69 Ibid.
70 Stefansson to Christie, 25 September 1925 (second letter), ibid.
71 Christie, memo, 12 June 1922, LAC, RG 25, vol. 2667, file 9057-B-40.
72 Bothwell, *Loring Christie,* 45.
73 The Advisory Technical Board was formed in June 1920. Its members were the heads of various branches within the department. See LAC, Surveys and Mapping Branch,

RG 88, vol. 222, file 17290. Several historians have assumed that the board was created to deal with the Arctic sovereignty issue, but this is incorrect. For example, see Morris Zaslow, *The Northward Expansion of Canada, 1914-1967* (Toronto: McClelland and Stewart, 1988), 17; William Barr, *Back from the Brink: The Road to Muskox Conservation in the Northwest Territories* (Calgary: Arctic Institute of North America, 1991), 90; Shelagh Grant, *Arctic Justice: On Trial for Murder, Pond Inlet, 1923* (Montreal/Kingston: McGill-Queen's University Press, 2002), 90; and Dick, *Muskox Land,* 274. The minutes of the meetings show that the board dealt mostly with non-Arctic subjects.

74 Minutes of meeting of Advisory Technical Board, 1 October 1920, LAC, MG 30 B57, John Davidson Craig papers (JDC), vol. 1, file 'Reports and Memoranda.'

75 The Americans had not, in fact, transferred their rights to Denmark; rather, they had renounced any intention of exercising their rights. There was considerable resentment in Norway against Denmark on the matter of northern sovereignty. The two countries were united under one crown from 1380 to 1819. In 1819, when Norway was placed under Swedish rule by the Treaty of Kiel, the Danes retained Greenland, Iceland, and the Faroe Islands by what the Norwegians regarded as trickery. A less likely scenario than the one proposed by Stefansson is, therefore, difficult to imagine. See Knud Berlin, *Denmark's Right to Greenland: A Survey of the Past and Present Status of Greenland, Iceland and the Faroe Islands in Relation to Norway and Denmark,* trans. P.T. Federspiel (London/Copenhagen: Oxford University Press/Arnold Busck, 1932).

76 Minutes of meeting of Advisory Technical Board, 1 October 1920.

77 On this trip, which was undertaken to assist Danish explorer Lauge Koch with the transport of his supplies to Thule, see Freuchen, *Arctic Adventure,* 356-57.

78 Rasmussen to the 'Governor of Canada,' 10 August 1920 and covering letter, 15 September 1920, LAC, RG 25, vol. 2669, file 9059-B-40.

79 See J.B. Harkin, 'Memo Re. Northern Islands. Prepared for Information Technical Advisory Board Meeting, November 10th, 1920,' LAC, JBH, vol. 2, file 'Canadian Sovereignty of Arctic Islands Re: V.O. Stefansson,' and 'Peculiarities in Connection with Danish Action Re Ellesmere Land,' undated memo, LAC, JBH, vol. 1, file November-December 1920.

80 Klotz diary, 13 October 1920, LAC, MG 30 B13, Otto Julius Klotz papers (OJK), vol. 4.

81 Ibid., 15 October 1920.

82 Starnes to Perry, 16 October 1920, and Perry to Starnes, 18 October, LAC, Royal Canadian Mounted Police, RG 18, vol. 3757, file G-516-37.

83 Klotz diary, 20 October 1920.

84 Ibid., 25 October 1920.

85 Harkin to Cory, 1 November 1920, LAC, JBH, vol. 1, file July 1919-October 1920. See also minutes of meeting of Advisory Technical Board, 27 October 1920, LAC, JBH, vol. 1, file November-December 1920.

86 Klotz diary, 26 October 1920.

87 Ibid.

88 'Precis of Minutes of Seventeenth Regular Meeting of the Advisory Technical Board, Held on 27th October, 1920,' LAC, RG 88, vol. 222, file 17290.

89 Klotz diary, 28 October 1920.

90 Deville to Cory, 29 October 1920, LAC, RG 88, vol. 5, file 17235.

91 Cory to Deville, 30 October 1920, ibid.

92 For example, see memo, 13 May 1920, LAC, JBH, vol. 1, file May-December 1921, and Harkin to Cory, 26 May 1921, LAC, RG 85, vol. 583, file 571.

Chapter 3: An Expedition to Ellesmere Land

1 Minutes of meeting, 1 October 1920, Library and Archives Canada (LAC), MG 30 B57, John Davidson Craig papers (JDC), vol. 1, file 'Reports and Memoranda, 1905-1923.'

2 See 'Arbitral Award on the Subject of the Difference Relative to the Sovereignty over Clipperton Island,' *American Journal of International Law* 26, 2 (April 1932): 390-94, and Permanent Court of International Justice, Series A/B (Collection of Judgments, Orders and Advisory Opinions), No. 53, *Legal Status of Eastern Greenland* (Leyden: A.W. Sijothoff, 1933).

3 On Joseph Pope's recommendation, Harkin read Oppenheim's book some time during October. Harkin, 'Memo Re. Northern Islands. Prepared for Information Technical Advisory Board Meeting, November 10th, 1920,' LAC, MG 30 E169, James Bernard Harkin papers (JBH), vol. 1, November-December 1920.

4 William Edward Hall, Preface to the third edition of *A Treatise on International Law,* 4th ed. (Oxford: Clarendon Press, 1895).

5 See Joshua Castellino and Steve Allen, *Title to Territory in International Law: A Temporal Analysis* (Aldershot: Ashgate, 2003), Chapter 2. See also Antony Anghie, *Imperialism, Sovereignty, and the Making of International Law* (Cambridge: Cambridge University Press, 2004), and Ken MacMillan, *Sovereignty and Possession in the English New World: The Legal Foundations of Empire, 1576-1640* (Cambridge: Cambridge University Press, 2006). On contradictory claims about the value of discovery and symbolic acts of possession, see James Simsarian, 'The Acquisition of Legal Title to Terra Nullius,' *Political Science Quarterly* 53, 1 (March 1938): 111-28, especially 113-18, 120. Patricia Seed describes the various acts of possession employed by different nations in 'Taking Possession and Reading Texts: Establishing the Authority of Overseas Empires,' *William and Mary Quarterly,* 3rd ser., 49, 2 (April 1992): 183-209.

6 E. de Vattel, *The Law of Nations or the Principles of Natural Law, Applied to the Conduct and to the Affairs of Nations and of Sovereigns,* trans. Charles G. Fenwick (Washington, DC: Carnegie Institution of Washington, 1916), 3:84-85. On Vattel's influence, see Ian Hunter, 'Natural Law, Historiography, and Aboriginal Sovereignty,' *Legal History* 11, 2 (2007): 142-43, 153-60, 165-66, and Mark Hickford, '"Vague Native Rights to Land": British Imperial Policy on Native Title and Custom in New Zealand, 1837-53,' *Journal of Imperial and Commonwealth History* 38, 2 (June 2010): 180-81, 191.

7 James Brown Scott, ed., *Resolutions of the Institute of International Law Dealing with the Law of Nations, with an Historical Introduction and Explanatory Notes* (New York: Oxford University Press, 1916), 87.

8 See Michael Connor, *The Invention of Terra Nullius: Historical and Legal Fictions on the Foundation of Australia* (Sydney: Macleay Press, 2005); Andrew Fitzmaurice, 'The Great Australian History Wars,' News and Events, University of Sydney, http://www.usyd.edu/news/84.html?newsstoryid=948; Michael Connor, 'The Invention of Territorium Nullius,' Australian Politics, http://www.australian-politics.blogspot.com.

9 L. Oppenheim, *International Law: A Treatise,* 3rd ed., ed. Ronald F. Roxburgh (London: Longmans, Green, 1920), 1:386.

10 William Edward Hall, *A Treatise on International Law,* 7th ed., ed. A. Pearce Higgins (Oxford: Clarendon, 1917), 104-5.

11 A part of the southwestern coast was later designated North Lincoln. See the map contained in *Further Papers Relative to the Recent Arctic Expeditions in Search of Sir John Franklin and the Crews of H.M.S. 'Erebus' and 'Terror'* (London: Eyre and Spottiswoode, 1855).

12 For example, see Franz Boas, 'The Configuration of Grinnell Land and Ellesmere Land,' *Science* 5, 108 (27 February 1885): 170-71. According to Boas, the Inughuit told Emil Bessels of the Hall expedition that Grinnell and Ellesmere Lands were indeed separated by a strait. If true, this story demonstrates their limited knowledge of Ellesmere.
13 C.H. Davis, *Narrative of the North Polar Expedition, U.S. Ship Polaris, Captain Charles Francis Hall Commanding* (Washington, DC: Government Printing Office, 1876), 104.
14 G.S. Nares, *Narrative of a Voyage to the Polar Sea during 1875-6 in H.M. Ships 'Alert' and 'Discovery'* (London: Sampson Low, Marston, Searle, and Rivington, 1878), 1:127, 377, and 2:31.
15 Frederick J. Evans, report dated 23 January 1879 (copy), LAC, Department of the Interior, RG 15, vol. 1, file 'Arctic Islands Documents, 1873-1880.'
16 Robert Hall to Secretary of State for the Colonies, 28 January 1879 (copy), ibid.
17 *Canada Gazette,* 9 October 1880, 389. See also Gordon W. Smith, 'The Transfer of Arctic Territories from Great Britain to Canada in 1880, and Some Related Matters, as Seen in Official Correspondence,' *Arctic* 14, 1 (March 1961): 53-73.
18 Adolphus W. Greely, *Report of the Proceedings of the United States Expedition to Lady Franklin Bay, Grinnell Land* (Washington, DC: Government Printing Office, 1888), 1:207-8, 232, 279-317. W.F. King's report on Canada's title to the northern islands mistakenly states that the American flag was raised on Ellesmere. On the inaccuracies contained in the King report, see note 40, below.
19 PC 2640, 2 October 1895, published in the *Canada Gazette,* 19 October 1895, 683.
20 *Canada Gazette,* 14 May 1898, 2613. For the map, see LAC, Privy Council Office, RG 2, vol. 752.
21 William Wakeham, *Report of the Expedition to Hudson Bay and Cumberland Gulf in the Steamship 'Diana'* (Ottawa: S.E. Dawson, 1898), 24.
22 François Gourdeau (Deputy Minister of Marine and Fisheries) to Wakeham, 23 April 1897, LAC, Department of Marine and Fisheries, RG 42, vol. 338, file 13205A.
23 Wakeham to Gourdeau, 28 September 1897, ibid.
24 Otto Sverdrup, *New Land: Four Years in the Arctic Regions* (London: Longmans, Green, 1901), 2:449-50.
25 White to A.H. Whitcher, 5 May 1904, LAC, Department of Energy, Mines and Resources, RG 21, vol. 153, file 38. The official name was changed to Ellesmere Island in 1905, but Ellesmere Land continued to be used for many years.
26 See 'Ellesmere Land,' *Geographical Journal* 24, 2 (August 1904): 230.
27 'Ellesmere Land,' *Bulletin of the American Geographical Society* 36, 11 (1904): 684.
28 See Hugh S. Spence, 'James White, 1863-1928: A Biographical Sketch,' *Ontario History* 27 (1931): 543-44, and Daniel S.C. Mackay, 'James White/Canada's Chief Geographer, 1899-1909,' *Cartographica* 19, 1 (Spring 1982): 51-61. Neither of these articles mentions Arctic sovereignty.
29 James White, 'Place-Names – Northern Canada,' *Ninth Report of the Geographic Board of Canada,* Part 4 (Ottawa: King's Printer, 1910).
30 Sifton to W.F. King, 11 December 1903, LAC, MG 27 II D15, Clifford Sifton papers (CS), vol. 253, pp. 962-63.
31 Laurier to Senator William C. Edwards, 29 October 1903, LAC, MG 26G, Wilfrid Laurier papers (WL), vol. 288.
32 Gourdeau to Low, 8 August 1903, LAC, Department of Transport, RG 12, vol. 48, file 1654-34.

33 A.P. Low, *Report on the Dominion Government Expedition to Hudson Bay and the Arctic Islands on Board the D.G.S. Neptune, 1903-04* (Ottawa: Government Printing Bureau, 1906), 45-46.
34 Ibid., 48.
35 'Proclamation-Ellesmere Island 1904,' LAC, MG 30 B33, Albert Peter Low papers (APL), vol. 1.
36 Low to Pope, 6 April 1905, LAC, MG 30 E86, Joseph Pope papers (JP), vol. 16, file 59.
37 Frederick A. Cook, *My Attainment of the Pole* (New York: Polar Publishing, 1911), 191-92.
38 See correspondence in LAC, RG 42, vol. 116, file 26152, and RG 21, vol. 153, file 0038. On the American Hydrographic Department's acceptance of Canadian place names on Ellesmere, see also *Congressional Record* 53, 17 (13 January 1916): 1098.
39 Sifton to King, 11 December 1903. On the origins of the Low expedition and the King report, see D.J. Hall, *Clifford Sifton*, vol. 2, *A Lonely Eminence, 1901-1929* (Vancouver: UBC Press, 1985), 125-26.
40 W.F. King, *Report upon the Title of Canada to the Islands North of the Mainland of Canada* (Ottawa: Government Printing Bureau, 1905), 5-8, 24-26. The report, which was drawn up mainly from secondary sources rather than from the original reports and narratives of the explorers, contains a number of errors. For example, it states that the Greely expedition raised the American flag on Ellesmere Island. For a critique of the report, see the comments on it by the hydrographer of the Royal Navy, enclosed in Leo Amery to Lord Byng, 20 October 1925, LAC, Department of External Affairs, RG 25, vol. 4252, file 9057-40.
41 Enclosure in White to Mackenzie King, 8 June 1922, LAC, Department of Justice, RG 13, vol. 934, file 6A. Unfortunately, this file contains only a few pages of White's memorandum, and the full document does not seem to have been preserved in the King papers.
42 Richard Finnie, 'Some Thoughts about Captain J.E. Bernier Apropos "Planting the Flag North" by P.D. Baird,' 26 January 1972, LAC, MG 31 C6, Richard Sterling Finnie papers (RSF), vol. 10, file 13.
43 Oswald Finnie, memo to W.W. Cory, 10 December 1927, LAC, Northern Affairs Program, RG 85, vol. 5, file 20 – Bernier.
44 O.D. Skelton to A.F. Lascelles, 23 November 1933, and Bernier to R.B. Bennett, 21 November 1933, LAC, RG 25, vol. 1673, file 748-33. Bernier's achievements are taken at his own valuation in a number of works, including Yolande Dorion-Robitaille, *Captain J.E. Bernier's Contribution to Canadian Sovereignty in the Arctic* (Ottawa: Department of Indian and Northern Affairs, 1978); Season L. Osborne, 'Closing the Front Door of the Arctic: Capt. Joseph E. Bernier's Role in Canadian Arctic Sovereignty' (master's thesis, Carleton University, 2003); Marjolaine Saint-Pierre, *Joseph-Elzéar Bernier: Capitaine et coureur des mers, 1852-1934* (Sillery, QC: Septentrion, 2004); and David Eric Jessup, 'J.E. Bernier and the Assertion of Canadian Sovereignty in the Arctic,' *American Review of Canadian Studies* 38, 4 (Winter 2008): 409-27. The first three authors interpret all criticisms of Bernier in government files as evidence of bureaucratic indifference on sovereignty matters. For a more critical view, see Alan MacEachern, 'Cool Customer: The Arctic Voyage of J.E. Bernier,' *Beaver* 84, 4 (August-September 2004): 30-35.
45 See Gourdeau to Bernier, 23 June 1906, and Bernier to Gourdeau, 20 July 1906, LAC, RG 42, vol. 142, file 27330. Albert Land is part of Victoria Island. The first page of Bernier's final instructions, dated 24 July 1906, is missing from the copy on this file.

There is a complete copy in LAC, Royal Canadian Mounted Police, RG 18, vol. 323, file 1906-744, and the first page is reproduced from Bernier's own copy in Dorion-Robitaille, *Bernier's Contribution to Canadian Sovereignty*, 55.

46 *Debates of the Senate of the Dominion of Canada, 1906-7* (Ottawa: King's Printer, 1907), 266-74.

47 Earl Grey (Governor General) to Lord Elgin (Colonial Secretary), 22 April 1907, LAC, RG 25, vol. 1095, file 1909-238.

48 Bernier to Brodeur, 3 October 1907, LAC, RG 42, vol. 142, file 27330; J.E. Bernier, *Report on the Dominion Government Expedition to the Arctic Islands and Hudson Strait on Board the C.G.S. 'Arctic,' 1906-1907* (Ottawa: C.H. Parmalee, 1909), 50. See also Janice Cavell, '"As Far as 90 North": Joseph Elzéar Bernier's 1907 and 1909 Sovereignty Claims,' *Polar Record* 46, 239 (October 2010): 372-73.

49 J.E. Bernier, *Report on the Dominion of Canada Government Expedition to the Arctic Islands and Hudson Strait on Board the D.G.S. 'Arctic'* (Ottawa: Government Printing Bureau, 1910), 192.

50 Robert E. Peary, *The North Pole* (New York: Frederick A. Stokes, 1910), 297.

51 Wally Herbert, *The Noose of Laurels: Robert E. Peary and the Race to the North Pole* (New York: Athenaeum, 1989), 282.

52 Pope to Cory, 11 September 1909, LAC, RG 25, vol. 1095, file 1909-238.

53 Quoted in the *Canadian Annual Review* (1909): 205.

54 'Capt. Bernier's Story of Voyage,' *Globe* (Toronto), 6 October 1909, 1.

55 See Joseph E. Bernier, 'Canadian Rights in the Arctic,' in Gerald H. Brown, ed., *Addresses Delivered before the Canadian Club of Ottawa* (Ottawa: Mortimer Press, n.d.), 191-92.

56 Canada, House of Commons, *Debates*, 1909-10, 2:2711-12.

57 *Debates of the Senate of the Dominion of Canada, 1909-10*, 184-85.

58 Morris Zaslow, *The Opening of the Canadian North, 1870-1914* (Toronto: McClelland and Stewart, 1971), 265. See also Harry Whitney, *Hunting with the Eskimos* (New York: Century, 1910), 442-43, and Bernier to Minister of Marine and Fisheries, 20 August 1906, with sample letter addressed to whaling captains, LAC, RG 42, vol. 142, file 27330.

59 Borden to Gilbert H. Grosvenor, 21 February 1913, printed in Vilhjalmur Stefansson, *The Friendly Arctic*, 2nd ed. (New York: Macmillan, 1943), xxii.

60 See the February 1913 correspondence in LAC, RG 42, vol. 475, file 84-2-29, and Henry Fairfield Osborn (President of the American Museum of Natural History) to Borden, 27 February 1913, LAC, RG 42, vol. 463, file 84-2-1.

61 Lewis Harcourt to the Officer Administering the Government of Canada, 10 May 1913, LAC, Governor General's Office, RG 7, vol. 412, file 10045. See also 'Report of the Committee of the Privy Council,' 2 June 1913, in the same file.

62 Minute, 3 September 1920, The National Archives of the United Kingdom: Public Record Office (TNA: PRO), FO 371/4084/9296.

63 Ibid.

64 J.D. Gregory to Grevenkop-Castenskiold, 7 September 1920, ibid. Because few copies of the 7 September note found their way into Canadian files, it has not previously been realized that the supposed protest to Denmark was a very mildly worded document indeed. There is one copy in RG 7 and two on External Affairs files (LAC, RG 7, vol. 412, file 10045, and RG 25, files 9058-F-40 and 9059-B-40).

65 Christie, memo to Meighen, 28 October 1920, LAC, MG 26I, Arthur Meighen papers (AM), I2, vol. 13, file 7.

66 Stefansson to Meighen, 30 October 1920, ibid.

67 Borden to Meighen, 3 November 1920, ibid.
68 Klotz diary, 24 November 1920, LAC, MG 30 B13, Otto Julius Klotz papers (OJK), vol. 4.
69 Pope diary, 24 November 1920, LAC, JP, vol. 47.
70 Pope to Meighen, 25 November 1920, LAC, AM, I2, vol. 13, file 7.
71 Deville to Harkin, 11 November 1920, LAC, JBH, vol. 1, file November-December 1920.
72 Harkin, 'Memo Re. Northern Islands. Prepared for information Technical Advisory Board Meeting, November 10th, 1920,' LAC, JBH, vol. 1, file November-December 1920.
73 He returned two External Affairs files about Greenland on 10 November. Harkin to Pope, LAC, RG 25, vol. 4252, file 9057-40.
74 Harkin, 'Memo Re. Northern Islands.'
75 Ibid.
76 Harkin to Cory, 3 November 1920, LAC, JDC, vol. 1, file 'Correspondence, 1903-1922'; C.C. Ballantyne to Lougheed, 5 November 1920, LAC, JBH, vol. 1, file November-December 1920.
77 Cory to Pickels and Cory to J. McArthur, LAC, JBH, vol. 1, file November-December 1920.
78 See Christine W. Billman, 'Jack Craig and the Alaska Boundary Survey,' *Beaver* 51, 2 (Autumn 1971): 44-49. Following the publication of this article, Billman (who was Craig's niece) struck up a correspondence with Richard Finnie. Her letters are filled with affectionate reminiscences of 'Uncle Jack' and 'Aunt Gert.' LAC, RSF, vol. 10, file 13. Other information about Craig has been gleaned from the *Civil Service List of Canada* and LAC, RG 15, vol. 128.
79 Cory to Deville, 16 December 1920, LAC, Surveys and Mapping Branch, RG 88, vol. 222, file 17290.
80 Cory to Perry, 22 December 1920, LAC, RG 18, vol. 3757, file G-516-37. There seems to have been no consultation with the police other than a request to Cortlandt Starnes for an estimate of the cost of establishing five posts and maintaining them for two years. See unidentified author to Harkin, 8 November 1920, LAC, JBH, vol. 1, file November-December 1920.
81 Rasmussen to Henderson, 17 November 1920, LAC, RG 25, vol. 2669, file 9059-B-40.
82 Harkin to Cory, 4 December 1920, LAC, RG 85, vol. 583, file 571.
83 Harkin to Cory, 6 December 1920, LAC, JBH, vol. 1, file November-December 1920.
84 'Inspection of the Steamer Arctic,' *Ottawa Citizen,* 22 December 1920. Clipping in LAC, RG 85, vol. 583, file 571.
85 Stefansson to Meighen and Stefansson to Lougheed, 8 January 1921, LAC, JBH, vol. 1, file January-March 1921; Stefansson to Borden, 8 January 1921, LAC, AM, I2, vol. 13, file 7.
86 Borden to Meighen, 11 January 1921, LAC, AM, I2, vol. 13, file 7.
87 Craig to Harkin, 11 January 1921, LAC, JBH, vol. 1, file January-March 1921.
88 Stefansson to Rutherford, 6 September 1920; Harkin to Anderson, 14 September 1920; and Anderson to Harkin, 27 September 1920, LAC, RG 85, vol. 1203, file 401-3. See also William Waiser, 'Canada Ox, Ovibos, Woolox ... Anything But Musk-Ox,' in Kenneth Coates and William R. Morrison, eds., *For Purposes of Dominion: Essays in Honour of Morris Zaslow* (Toronto: Captus Press, 1989), 189-200.
89 Stefansson to Harkin, 21 November 1920, LAC, JBH, vol. 2, file 'Canadian Sovereignty of Arctic Islands re: V.O. Stefansson.'
90 Harkin, memo, 13 May 1921, LAC, JBH, vol. 1, file May-December 1921.

91 As part of his application for the Baffin Island lease, in June 1920 Stefansson had made a formal, witnessed declaration that he was a British subject, affirming that his statement was 'of the same force and effect as if made under oath and by virtue of the Canada Evidence Act.' Harkin saw this document, but he must have either suspected or known that Stefansson was lying. In either case, Stefansson's willingness to make the declaration cannot have encouraged Harkin to believe in his integrity. The declaration is in LAC, RG 85, vol. 1147, file 270-8-2.
92 Harkin to Cory, 15 March 1921, LAC, JBH, vol. 1, file January-March 1921.

Chapter 4: A Citizen of the British Empire

1 Harkin to Cory, 2 March 1921, and Cory to Lougheed, 23 March 1921, Library and Archives Canada (LAC), MG 30 E169, James Bernard Harkin papers (JBH), vol. 1, file January-March 1921.
2 Harkin to Stefansson, 21 January 1921, Rauner Special Collections Library, Stefansson Collection (RSCL), Stef MS-196 (6) Harkin. An earlier draft of the letter stated that the Department of the Interior would 'be glad to enter into an arrangement with you by which you shall assume command of the exploration party suggested.' This sentence was deleted. Harkin commented that the final version was 'in a safe form.' Undated draft and Harkin to Gibson, 21 January 1921, LAC, JBH, vol. 1, file January-March 1921.
3 Stefansson to Harkin, 2 February 1921, ibid.
4 'Shackleton May Seek North Pole on Next Journey,' and 'A British Explorer's Visit,' *Ottawa Citizen,* 9 February 1921, 5, 16.
5 Copies of these documents can be found in LAC, Northern Affairs Program, RG 85, vol. 583, file 571.
6 This claim was certainly unwarranted. In a letter written on 4 May 1920, Stefansson had stated that he expected to go ahead with his own northern plans, though there still might be 'difficulty' with the Canadian government. However, he assured Shackleton that he did not see competition as harmful, 'and if you still continue interested I shall be glad to give you any information by correspondence.' RSCL, Stef MS-196 (5) S.
7 Meighen to Lougheed, 5 February 1921, LAC, MG 26I, Arthur Meighen papers (AM), I2, vol. 13, file 7.
8 John King Davis, *High Latitude* (Parkville: Melbourne University Press, 1962), 259.
9 Hugh Robert Mill, *The Life of Sir Ernest Shackleton* (London: Heinemann, 1923), 81, 285-86, 289.
10 Quoted in ibid., 84.
11 Louis Bernacchi, *The Saga of the 'Discovery'* (London: Blackie, 1938), 219, 221.
12 Eric Marshall, quoted in Roland Huntford, *Shackleton* (London: Hodder and Stoughton, 1985), 233.
13 Huntford, *Shackleton,* 234.
14 James Fisher and Margery Fisher, *Shackleton* (London: James Barrie, 1957), 200.
15 Quoted in ibid., 328.
16 Stefansson to Harkin (two telegrams), and Harkin to Stefansson, 7 February 1921, LAC, JBH, vol. 1, file January-March 1921.
17 Stefansson to Harkin, 7 February 1921, ibid.
18 Lougheed to Meighen, 14 February 1921, LAC, AM, I2, vol. 13, file 7.
19 Harkin to Cory, 13 May 1921, LAC, RG 85, vol. 583, file 571.
20 At the time, Canadian citizens were defined as British subjects born in Canada, British subjects domiciled in Canada for at least three years, and naturalized aliens. *An Act*

Respecting Immigration (Ottawa: C.H. Parmalee, [1910]); *Revised Statutes of Canada* 1910, 9-10 Edward VII Ch.27 s. 2(f). White British subjects from other parts of the Empire could migrate to Canada at will, provided that they were in good health and appeared to possess sufficient financial means to support themselves. (In theory this right extended to non-whites from the Caribbean and Asia, but in practice their entry was severely restricted.) Because of his Irish birth, Shackleton could, if he wished, have settled in Canada without going through the naturalization process. After three years' residence, he would have had full privileges of citizenship, including the right to vote in Canadian elections. Stefansson, though born in Canada, ceased to be a British subject once his father opted for American citizenship. See Rieko Karatani, *Defining British Citizenship: Empire, Commonwealth, and Modern Britain* (London: Frank Cass, 2003), 57, 76-80; Janice Cavell and Jeff Noakes, 'Explorer without a Country: The Question of Vilhjalmur Stefansson's Citizenship,' *Polar Record* 45, 234 (July 2009): 237-41.

21 Harkin to Cory, 2 March 1921, LAC, JBH, vol. 1, file January-March 1921.

22 Harkin to Cory, 17 February 1921, with attached memorandum by Stefansson, ibid.

23 Meighen to Stefansson, 19 February 1921, LAC, AM, I2, vol. 13, file 7.

24 Harkin to Cory, 22 March 1921, LAC, JBH, vol. 1, file January-March 1921. Harkin did not indicate the source of this idea, but Christie is by far the most likely candidate.

25 Christie, memo, 12 June 1922, LAC, Department of External Affairs, RG 25, vol. 2667, file 9057-B-40.

26 Borden proposed Stefansson as a member on 19 February (the day of Stefansson's successful meeting with Meighen); the vote was taken on 1 March. Vilhjalmur Stefansson, *Discovery: The Autobiography of Vilhjalmur Stefansson* (New York: McGraw-Hill, 1964), 242. According to Rudolph Anderson, sixty-seven votes were cast against Stefansson. Memo enclosed in Anderson to Bartlett, 1 April 1922, LAC, MG 30 B40, Rudolph Martin Anderson and Mae Belle Allstrand Anderson papers (RMA/MBAA), vol. 3, file 5. In his diary, Borden angrily wrote that the club had made itself ridiculous. LAC, MG 26 H, Robert Laird Borden papers (RLB), vol. 449, 16 March 1921.

27 Belle Anderson, memo book, 26 April 1925, LAC, RMA/MBAA, vol. 12, file 4. See also Vilhjalmur Stefansson, *The Adventure of Wrangel Island* (New York: Macmillan, 1925), 80. That Christie was not aware of the tentative arrangements made between Harkin and Stefansson is shown by a letter he wrote to Borden in 1922. See Christie to Borden, 8 September 1922, LAC, MG 30 E44, Loring Cheney Christie papers (LCC), vol. 3, file 6-6.

28 Christie, memos, 17 and 28 February 1921, LAC, AM, I2, vol. 13, file 7.

29 Armstrong to Stefansson, 1 March 1921, ibid. The claim was false; the HBC decided in favour of a post on Wrangel only after hearing from Stefansson that the government intended to annex the island. See Richard Diubaldo, *Stefansson and the Canadian Arctic* (Montreal: McGill-Queen's University Press, 1978), 167, 245-46n28.

30 Stefansson to Armstrong, 7 March 1921, LAC, AM, I2, vol. 13, file 7.

31 Logan to Richard Finnie, 26 July 1974, Trent University Archives (TUA), Trevor Lloyd papers (TL), file 87-014-10-2.

32 Stefansson was well aware of the conflict between his plans and the sector theory. In later years he boasted to historian Gordon Smith that during his period of influence he had persuaded the Canadian government 'to give up placing reliance upon the sector principle,' to which he was 'very much opposed.' Smith to Richard Finnie, 1 June 1977, LAC, RSF, vol. 10, file 14.

33 Edwards to Craig, 10 January 1921, LAC, RG 85, vol. 583, file 571.

34 Pope to Gibson, 15 January 1921, ibid.

35 Pope to Griffith, 15 January 1921, RG 25, vol. 2668, file 9058-F-40.
36 Perry to Thomson, 18 January 1921, LAC, Royal Canadian Mounted Police, RG 18, vol. 3757, file G-516-37.
37 Perry to Thomson, 2 February 1921, ibid.
38 Craig to Cory, 19 January 1921, LAC, MG 30 B57, John Davidson Craig papers (JDC), vol. 1, file 'Correspondence, 1903-1922.'
39 Craig to Doughty, 21 January 1921, LAC, Department of the Interior, RG 15, vol. 2, file 'Arctic Islands, 1920.'
40 Hinks to Thomson, 18 February 1921, TUA, TL, file 87-104-6-7, and Hinks to Colonial Secretary, LAC, RG 25, vol. 2668, file 9058-F-40.
41 Stefansson to Harkin, 21 January 1921, TUA, TL, file 87-104-9-25. MacMillan, while certainly not a model of integrity, was no more unscrupulous than Stefansson himself.
42 'Call to Adventure in the Dim Lands,' *Montreal Gazette,* 23 February 1921, 5.
43 'An Interesting Announcement,' *Montreal Herald,* 24 February 1921, 4.
44 'Shackleton Tells the Story of his Antarctic Voyage,' *Ottawa Citizen,* 24 February 1921, 4.
45 Mackenzie King diary, 23 February 1921, The Diaries of William Lyon Mackenzie King, LAC, http://www.collectionscanada.gc.ca/databases/king. Unlike the *Citizen*'s reporter, King felt that Shackleton's lecture 'was not well delivered.' However, this seems unlikely given Shackleton's long experience as a public speaker and his desire to impress Meighen. Typically, the self-centred King tormented himself with memories of his own poor performance in moving the vote of thanks. 'I had some good thoughts, but felt very ill at ease in speaking, & much discouraged when I concluded. I did not do myself justice in the least,' he wrote.
46 Marginal note on Stefansson to Lougheed, 26 February 1921, LAC, JBH, vol. 1, file January-March 1921.
47 Harkin to Cory, 2 March 1921, ibid.
48 Flora McCrae Eaton to John Craig Eaton, 2 March 1921, Archives of Ontario (AO), John Craig Eaton and Flora McCrae Eaton papers, F228, B422374.
49 'Shackleton Will Leave in May or June for Arctic,' *Ottawa Citizen,* 1 March 1921, 1.
50 'New Expedition to the Arctic May or June,' *Ottawa Journal,* 1 March 1921, 1.
51 Harkin to Cory, 2 March 1921, LAC, JBH, vol. 1, file January-March 1921.
52 Christie, memo, 12 June 1922. Christie did not specify the date of the encounter, but Stefansson's March 1921 visit to Ottawa is the most probable time.
53 Stefansson to Lougheed, 14 March 1921, LAC, JBH, vol. 1, file January-March 1921.
54 Harkin to Cory, 15 March 1921, ibid.
55 Ibid.
56 Harkin to Cory, 22 March 1921, ibid.
57 Harkin to Cory, 21 March 1921, ibid.
58 Ibid.
59 Harkin to Cory, 22 March 1921. Harkin may have resented Lougheed's determination to abolish the Commission of Conservation, which had done excellent work under the leadership of Harkin's patron, Sir Clifford Sifton. If so, it might have given him particular satisfaction to defy Lougheed over the Stefansson expedition. On the Commission of Conservation, see D.J. Hall, *Clifford Sifton,* vol. 2, *A Lonely Eminence, 1901-1929* (Vancouver: UBC Press, 1985), Chapter 11.
60 Cory to Lougheed, 23 March 1921, LAC, JBH, vol. 1, file January-March 1921.
61 Harkin to Stefansson, 26 March 1921, ibid.
62 Stefansson to Harkin, 11 April 1921, LAC, JBH, vol. 1, file April 1921.

Chapter 5: Rasmussen in London

1 Nyeboe to Richard Turner, 3 February 1921, The National Archives of the United Kingdom: Public Record Office (TNA: PRO), FO 371/6759/9296.
2 Grevenkop-Castenskiold to Curzon, 9 March 1921, ibid.
3 Thomson to Perry, 15 March 1921, Library and Archives Canada (LAC), Royal Canadian Mounted Police, RG 18, vol. 3757, file G-516-37.
4 Perley to Borden, 15 August 1914, doc. 22 in *Documents on Canadian External Relations,* vol. 1, *1909-1918* (Ottawa: Department of External Affairs, 1967).
5 Griffith to Lougheed, 16 March 1921, LAC, MG 30 E169, James Bernard Harkin papers (JBH), vol. 1, file January-March 1921.
6 Memo by Charles V. Sale, enclosed in Griffith to Cory, 30 March 1921, ibid.
7 Therkel Mathiassen, *Report on the Expedition* (Copenhagen: Gyldendalske Boghandel, Nordisk Forlag, 1945), 11.
8 Sale to Griffith, 22 March, LAC, Department of External Affairs, RG 25, vol. 565, file 'Knud Rasmussen Expedition to N.A. Arctic Archipelago.'
9 Lambert to Griffith, 16 March 1921, ibid.
10 Churchill to Devonshire, 29 April 1921, LAC, RG 25, vol. 2667, file 9058-F-40.
11 An order-in-council dated 21 March 1921 provided that the High Commission should report to External Affairs commencing on 1 April. See Lovell C. Clark, ed., *Documents on Canadian External Relations,* vol. 3, *1919-1925* (Ottawa: Department of External Affairs, 1970), doc. 32. Either Griffith was not yet aware of this directive or he continued to correspond with the Department of the Interior simply out of habit.
12 Harkin to Cory, 23 March 1921; Harkin to Cory, 30 March 1921; and Lougheed to High Commission, 31 March 1921, LAC, JBH, vol. 1, file January-March 1921; Cory to Griffith, [4 April 1921], LAC, RG 25, vol. 565, file 'Knud Rasmussen Expedition to N.A. Arctic Archipelago.'
13 [Name illegible] to Perry, 23 March 1921, LAC, RG 18, vol. 3757, file G-516-37.
14 Peter Freuchen, *Arctic Adventure: My Life in the Frozen North* (New York: Farrar and Rinehart, 1935), 57.
15 Rolf Gilberg, 'Inughuit, Knud Rasmussen, and Thule,' *Études/Inuit/Studies* 12, 1-2 (1988): 50.
16 'Refutes the Idea of an Expedition to Frozen North,' *Ottawa Journal,* 4 April 1921, clipping in LAC, Northern Affairs Program, RG 85, vol. 585, file 586.
17 Shackleton to Meighen, 5 April 1921, LAC, MG 26I, Arthur Meighen papers (AM), I2, vol. 13, file 7.
18 Maggie Siggins, *Bassett* (Toronto: James Lorimer, 1979), 11-16, 22-23.
19 Unsigned memo, 14 April, LAC, AM, I2, vol. 13, file 7.
20 Shackleton to Christie, undated, ibid.
21 Unidentified clipping with date stamp 11 April and note by Craig, LAC, RG 85, vol. 583, file 571.
22 Knight to Harkin, 12 April 1921, LAC, JBH, vol. 1, file April 1921.
23 Stefansson to Harkin, 19 April 1921, ibid.
24 Harkin to Cory, 19 April 1921, ibid.
25 Cory to Harkin, 20 April 1921, ibid.
26 Harkin to Cory, 4 May 1921, with marginal note by Cory, LAC, JBH, vol. 1, file May-December 1921.
27 Harkin to Cory, 16 April 1921, with marginal note by Cory, LAC, JBH, vol. 1, file April 1921.
28 Draft telegram, LAC, AM, I2, vol. 13, file 7.

29 Churchill to Devonshire, 29 April 1921, LAC, RG 25, vol. 2668, file 9058-F-40.
30 Foreign Office to Marling, 30 April 1921, TNA: PRO, FO 371/6759/9296.
31 Marling to Foreign Office, 30 April 1921, ibid.
32 Shackleton to Bassett, 20 April 1921, LAC, AM, I2, vol. 13, file 7.
33 Bassett to Meighen, 3 May 1921, ibid.
34 Meighen to Shackleton, 9 May 1921, ibid.
35 Harkin to Cory and Harkin to Christie, 25 April 1921, LAC, JBH, vol. 1, file April 1921.
36 Meighen to Allan, 4 May 1921, Hudson's Bay Company Archives (HBCA), RG 2/4/63.
37 Harkin to Cory, 13 May 1921, LAC, RG 85, vol. 583, file 571.
38 Lougheed subsequently denied that he had ever favoured the Shackleton plan. See a statement on the matter drafted by Craig and signed by Lougheed, enclosed in Cory to Harkin, 15 June 1921, ibid. The statement was prepared on Cory's initiative, possibly in order to secure an official account of his department's actions which glossed over the fact that he had tacitly allowed Harkin to defy the minister's wishes. See Cory to Craig, 18 May 1921, ibid.
39 Vilhjalmur Stefansson, *The Friendly Arctic,* 2nd ed. (New York: Macmillan, 1943), 691.
40 Craig to Pickels, 20 May 1921, LAC, MG 30 B63, Harris Christopher Pickels papers (HCP), vol. 1, file 'Correspondence, 1920-1921.'
41 Christie to Borden, 5 September 1922, LAC, MG 30 E44, Loring Cheney Christie papers (LCC), vol. 3, file 6-6.
42 Nobel to Minister of Marine and Fisheries, 28 April 1921, LAC, JBH, vol. 1, file May-December 1921.
43 Harkin to Craig, 10 May 1921, ibid.
44 Harkin to Christie, 5 May 1921, and enclosed draft report by Holmden, LAC, RG 25, vol. 4252, file 9057-40.
45 Harkin to Cory, 26 May 1921, and attached clipping, 'Reasons for a New Arctic Expedition,' *The World's Work* (May 1921): 17, LAC, RG 85, vol. 583, file 571.
46 See A.H. H[inds] to Cory, 1 June 1921, ibid.
47 Canada, House of Commons, *Debates,* 1921, 4:4106, 4113-14.
48 Harkin to Stefansson, 30 May 1921, LAC, JBH, vol. 1, file May-December 1921.
49 Foreign Office to Marling, 9 May 1921, TNA: PRO, FO 371/6759/9296.
50 Marling to Foreign Office, 11 May 1921, ibid.
51 Grevenkop-Castenskiold to Curzon, 13 May 1921, TNA: PRO, FO 371/6758/9296. The note was received in Ottawa on 13 June. See LAC, Governor General's Office, RG 7, vol. 596, file 27489.
52 See note, 27 May 1921, by R.C.S. Stevenson (Second Secretary) on Lambert to Undersecretary, Foreign Office, 25 May 1921, and Gregory (Assistant Under-secretary) to Marling, 31 May 1921, TNA: PRO, FO 371/6759/9296.
53 HBC Winnipeg to HBC London, 19 and 20 May 1921, HBCA, RG 2/4/63.
54 Brabant to Harkin, 25 May 1921, LAC, JBH, vol. 1, file May-December 1921.
55 Harkin to Cory, 30 May 1921, LAC, RG 85, vol. 583, file 571.
56 Sale to Rasmussen, 24 May 1921, HBCA, A.102/1962, and Sale to Winnipeg Office, 26 May 1921, HBCA, RG 2/4/63.
57 Rasmussen to Sale, 2 June 1921, HBCA, A.92/7/1.
58 Rasmussen to Daugaard-Jensen, 4 June 1921, Trent University Archives (TUA), Trevor Lloyd papers (TL), 87-014-8-21 (translation by Trevor Lloyd).
59 Rasmussen to Griffith, 4 June 1921, LAC, RG 25, vol. 565, file 'Knud Rasmussen Expedition to N.A. Arctic Archipelago.' On Griffith's suspicions, see Griffith to Pope, 21 June 1921, LAC, RG 25, vol. 2668, file 9058-F-40.

60 Rasmussen to Lambert, 4 June 1921, LAC, RG 25, vol. 2668, file 9058-F-40.
61 Rasmussen to Sale, 4 June 1921, HBCA, A.92/7/1.
62 Sale to HBC Winnipeg, 4 June 1921, HBCA, RG 2/4/63, and Fitzgerald to Lougheed, 4 June 1921, LAC, RG 85, vol. 583, file 571.
63 Lougheed to Cory, 5 June 1921, ibid.
64 Craig to Cory, 6 June 1921, ibid.
65 Lougheed to Fitzgerald, 8 June 1921, HBCA, RG 2/4/63.
66 Grevenkop-Castenskiold to Curzon, 8 June 1921, and Gregory to Churchill, 9 June 1921, TNA: PRO, FO 371/6759/9296. This request was, of course, an implicit recognition of Canadian sovereignty over the archipelago. Rather than challenging Canadian sovereignty in 1921, Denmark was actually the first foreign state to acknowledge Canadian jurisdiction.
67 Minute on Colonial Office to Foreign Office, 9 June 1921, ibid.
68 Lougheed to Fitzgerald, 10 June 1921, HBCA, RG 2/4/63.
69 Minute on copy of Churchill to Devonshire, 10 June 1921, TNA: PRO, FO 371/6759/9296.
70 Griffith to Pope, 21 June 1921, LAC, RG 25, vol. 2668, file 9058-F-40.
71 Griffith to External Affairs, 10 June 1921, ibid.
72 Cory to Pope, 10 June 1921, and External Affairs to High Commission, 10 June 1921, ibid.; Pope to Governor General's secretary, 11 June 1921; and Devonshire to Churchill, 11 June 1921, LAC, RG 7, vol. 412, file 10045.
73 Griffith to Rasmussen, 11 June 1921, LAC, RG 25, vol. 565, file 'Knud Rasmussen Expedition to N.A. Arctic Archipelago.'
74 Rasmussen to Griffith, 11 June 1921, ibid.
75 Griffith to Sale, 11 June 1921, ibid.
76 Secretary of the Hudson's Bay Company to Officers in Baffin Land, Nelson River, and Mackenzie River Districts, 11 June 1921, HBCA, RG 2/4/63.
77 HBC London to HBC Winnipeg, 13 June 1921, HBCA, RG 2/4/63.
78 Sale to Rasmussen, 13 June 1921, HBCA, A.102/1962.
79 Sale to Griffith, 28 June 1921, LAC, RG 25, vol. 565, file 'Knud Rasmussen Expedition to N.A. Arctic Archipelago.'
80 Griffith to Pope, 21 June 1921, LAC, RG 25, vol. 2668, file 9058-F-40. Charitably, Griffith did not forward a copy of Rasmussen's 4 June letter to Ottawa.
81 Harkin to Cory, 29 June 1921, LAC, JBH, vol. 1, file May-December 1921.
82 John Idington (deputy Governor General) to Churchill, 14 July 1921, LAC, RG 7, vol. 596, file 27489.
83 Minute by Esmond Ovey (First Secretary) on Colonial Office to Foreign Office, 29 July 1921, TNA: PRO, FO 371/6758/9296.
84 Ovey to Grevenkop-Castenskiold, 15 August 1921, ibid.
85 Lauge Koch, 'Report on the Danish Bicentenary Jubilee Expedition North of Greenland 1920-23,' *Meddelelser om Grønland* 70 (1927): 61, 173-75. See also LAC, RG 25, vol. 1756, file 1935-872.

Chapter 6: Wrangel Island

1 Peter Freuchen, *Arctic Adventure: My Life in the Frozen North* (New York: Farrar and Rinehart, 1935), 376.
2 Knud Rasmussen, *Across Arctic America: Narrative of the Fifth Thule Expedition* (New York: G.P. Putnam, 1927), 41-42. Harwood Steele gives the officer's name as Paquet. Steele, *Policing the Arctic* (London: Jarrolds, 1936), 349.

3 Harkin to Cory, 4 January 1922, and Finnie to Perry, 23 January 1922, Library and Archives Canada (LAC), Royal Canadian Mounted Police, RG 18, vol. 3290, file 1921-HQ-1091-G-1; Perry to Finnie, 27 January 1922, LAC, Northern Affairs Program, RG 85, vol. 583, file 567.

4 Quoted in James Fisher and Margery Fisher, *Shackleton* (London: James Barrie, 1957), 436.

5 'Of Such Is History,' *Montreal Gazette,* 29 September 1921, 10.

6 Quoted in Frank Wild, 'The Voyage of the "Quest,"' *Geographical Journal* 61, 2 (February 1923): 82.

7 Ernest Shackleton, *South* (London: Heinemann, 1919), 206, 209.

8 T.S. Eliot, *The Waste Land,* in *The Complete Poems and Plays, 1909-1950* (New York: Harcourt, Brace and World, 1971), part 5, lines 360-63. Eliot changed the number of men from three to two in order to link the incident with the Biblical episode of the risen Christ's appearance beside his disciples on the road to Emmaus.

9 Wild, 'The Voyage of the "Quest,"' 82. Emphasis in original.

10 On the traders, see Gavin White, 'Scottish Traders to Baffin Island, 1910-1930,' *Maritime History* 5, 1 (Spring 1977): 34-50. Besides the lease to the Hudson's Bay Reindeer Company, there were earlier government actions which Harkin, Oswald Finnie, and Jack Craig likely knew about. In 1903 the Department of the Interior issued a quartz mining license to a Scotsman, Robert Kinnes of Dundee; Kinnes subsequently provided the department with evidence to show that he was complying with Canadian regulations. See P.G. Keyes to Fred White, 12 September 1904, LAC, Department of Transport, RG 12, vol. 49, file 1654-34. In 1910 the government granted 960 acres at Pond Inlet to Joseph Bernier, who named his domain 'Berniera' and carried on extensive commercial activities there. Besides trading for furs, he received a prospector's permit, filed mineral claims, and was given a license to mine coal. See Yolande Dorion-Robitaille, *Captain J.E. Bernier's Contribution to Canadian Sovereignty in the Arctic* (Ottawa: Department of Indian and Northern Affairs, 1978), 93, 97.

11 Perry to the President of the Privy Council, undated, June 1921, LAC, RG 18, vol. 3280, file 1920-HQ-681-G-4.

12 Perry to Cory, 29 June 1921, LAC, RG 85, vol. 587, file 591. Shelagh Grant claims that Perry's decision to send Joy north in 1921 was an attempt to force the government's hand on northern sovereignty. 'A pending trial ... would put pressure on the government to allow his [Perry's] plans for the new detachments to go forward,' she writes. Grant, *Arctic Justice: On Trial for Murder, Pond Inlet, 1923* (Montreal/Kingston: McGill-Queen's University Press, 2002), 127. However, there are a number of flaws in her argument. The plans for new detachments were not made by Perry; he was merely carrying out decisions made by the Department of the Interior. Perry therefore had no particular reason to push the government, and it was not his responsibility to do so. Moreover, sovereignty over Baffin Island was not in doubt, and a trial there could do nothing to strengthen Canada's claim to Ellesmere. It is also clear that much of the impetus for the RCMP's decision to investigate the Janes murder came from Janes's father. See the RCMP file on the case (LAC, RG 18, vol. 3280, file 1920-HQ-681-G-4). There is nothing about sovereignty concerns on this file.

Grant describes the precise reason for Perry's change of heart in June 1921 as 'a mystery' (127). However, his letter to Cory makes the reason clear. Perry explained that he had not known about the HBC's plan to establish posts on Baffin Island until 'a few days ago.' This lack of information shows how far Perry was from the centre of decision making on Arctic issues.

Moreover, important files not consulted by Grant show that the only sovereignty concerns related to the James trial came from the Department of the Interior. On 24 October 1922 Cortlandt Starnes wrote to Finnie enclosing preliminary reports on the case and the original depositions taken by Joy on Baffin Island. He recommended that a trial be held at Pond Inlet for two reasons: first, because it was less expensive to send a judge and jury there than to bring all the witnesses south, and second, because it would be the most effective way to make the Inuit aware that they must obey Canadian laws. An unsigned memo in the Department of the Interior file on the case, dated 24 November 1922, pointed out that 'the holding of a court is an administrative act of importance.' Subsequently, Cory wrote to the deputy minister of justice. He repeated the arguments made by Starnes, then added that '[i]ncidentally' a trial at Pond Inlet could also be 'significant from the standpoint of perfecting our title to [the] Northern Islands.' See Starnes to Finnie, 24 October 1922, LAC, RG 13, vol. 279, file 1923-1158; unsigned memo, 24 November 1922, and Cory to Newcombe, 4 December 1922, LAC, RG 85, vol. 609, file 2687.

13 Richard Finnie, 'O.S. Finnie: A Pioneer Northern Administrator' (unpublished article), LAC, MG 31 C6, Richard Sterling Finnie papers (RSF), vol. 6, file 12. On Finnie's activities during the summer of 1921, see Morris Zaslow, *The Northward Expansion of Canada, 1914-1967* (Toronto: McClelland and Stewart, 1988), 25-26.
14 Richard S. Finnie, 'Stefansson as I Knew Him,' *North/Nord* 25, 3 (May-June 1978): 37, 40.
15 Finnie to Harkin, 18 January 1922, and Harkin to Finnie, 25 January 1922, LAC, RG 85, vol. 583, file 567.
16 Bernier to Perry, 23 January 1922, and enclosed copy of Bernier's letter to King, LAC, RG 18, vol. 3757, file G-516-37.
17 Cory to Finnie, 3 February 1922, LAC, RG 85, vol. 583, file 567.
18 Harkin to Craig, 13 February 1922, ibid.
19 Craig, memo, 16 February 1922, LAC, MG 30 B57, John Davidson Craig papers, JDC, vol. 1, file 'Reports and Memoranda.'
20 Finnie to Craig, 13 February 1922, LAC, RG 85, vol. 583, file 567.
21 Finnie to Gibson, 20 February 1922, ibid.
22 Biographical information from 'The Honourable Charles Stewart, 1917-21,' Legislative Assembly of Alberta, http://www.assembly.ab.ca/lao/library/PREMIERS/stewart.htm.
23 Stefansson, *The Adventure of Wrangel Island* (New York: Macmillan, 1925), 72, 76.
24 Copy in the A.J.T. Taylor Arctic Collection, Special Collections and University Archives, University of British Columbia Library. Norman Amor, *Beyond the Arctic Circle: Materials on Arctic Explorations and Travels since 1750 in the Special Collections and University Archives Division of the University of British Columbia Library* (University of British Columbia Occasional Publication No. 1, 1992), describes the collection, which Taylor assembled with Stefansson's help.
25 Stefansson to Hinks, 21 April 1922, Royal Geographical Society Archives (RGS), CB 9/139.
26 Stefansson, *Adventure of Wrangel Island,* 76-77.
27 Ibid., 77.
28 'Plans New Arctic Trip,' *New York Times,* 24 June 1921, 8.
29 Stefansson to King, 11 March 1922, LAC, MG 26 J, William Lyon Mackenzie King papers (WLMK), J1, vol. 82, file '1922, Stefansson.'
30 Stefansson, *Adventure of Wrangel Island,* 77.

31 The exchange of correspondence between Stefansson and Crawford is printed in *Adventure of Wrangel Island,* 79, 81.
32 Ibid., 82.
33 J.T. Crawford to Stewart, 8 May 1925, printed in Canada, House of Commons, *Debates,* 1925, 4:4094.
34 Allan Crawford to Helen Crawford, 6 August 1921, LAC, MG 30 B40, Rudolph Martin Anderson and Mae Belle Allstrand Anderson papers (RMA/MBAA), vol. 3, file 3.
35 Letters from Allan Crawford to an unidentified friend, 14 July and 25 July 1921, LAC, RMA/MBAA, vol. 11, file 17, and vol. 3, file 3.
36 Letter from Galle to his parents, 15 August 1921, quoted in Jennifer Niven, *Ada Blackjack: A True Story of Survival in the Arctic* (New York: Hyperion, 2003), 40.
37 See J.T. Crawford to Meighen, 21 June 1925, and enclosures: 'Wrangel Island Expedition: Statement from the Parents of Allan Crawford,' 19 June 1925, and copies of two letters from Stefansson to Allan Crawford, 7 August 1921, LAC, MG 26I, Arthur Meighen papers, AM, I3, vol. 59, file 5.
38 'Says Stefansson Is Wrong,' *New York Times,* 11 August 1921, 9.
39 Niven, *Ada Blackjack,* 50.
40 Ibid., 8-11, 59-60.
41 Stefansson to Desbarats, telegram, 19 July 1921, LAC, Department of Marine and Fisheries, RG 42, vol. 471, file 84-2-7.
42 Stefansson to Desbarats, telegram, 25 August 1921, ibid.
43 D.J. Hall, *Clifford Sifton,* vol. 2, *A Lonely Eminence, 1901-1929* (Vancouver: UBC Press, 1985), 243.
44 Quoted in ibid., 262.
45 See 'The Geographical Work of the Canadian Arctic Expedition,' *Geographical Journal* 63, 6 (June 1924): 508-25, and 'The Geographical Work of the Canadian Arctic Expedition,' *Geographical Journal* 65, 4 (April 1925): 340-42.
46 See LAC, Department of Justice, RG 13, vol. 268, file 970-22.
47 Stefansson to Armstrong, 7 September 1921, LAC, AM, I2, vol. 13, file 7.
48 Armstrong to Stefansson, 8 September 1921, ibid.
49 'Stefansson Advance Party Leaves Nome for the Arctic,' *New York Times,* 12 September 1921, 12.
50 'Says Stefansson Is Wrong.'
51 'First Stefansson Party at Wrangell,' *New York Times,* 28 September 1921, 17.
52 Reproduced in Stefansson, *Adventure of Wrangel Island,* opposite 119.
53 'First Stefansson Party at Wrangell.'
54 Vilhjalmur Stefansson, 'Northward the Course of Empire,' *MacLean's Magazine,* 1 December 1921, 15, 50-52. The other articles in the series were 'Our North That Never Was,' 15 December 1921; 'How Habitable Is Northern Canada?' 1 January 1922; 'Canada's Caribou Crop,' 15 January 1922; and 'Far North Really Liveable,' 1 February 1922. The material was later reprinted in Stefansson, *The Northward Course of Empire* (New York: Harcourt, Brace, 1922).
55 In private, Robert Bartlett furiously condemned Stefansson as 'a God damn liar a Poltroon and a sneak' (Bartlett to Rudolph Anderson, 1 February 1922, LAC, RMA/MBAA, vol. 3, file 4), but he made no public statement.
56 Rudolph Anderson to Bartlett, 20 August 1919, LAC, RMA/MBAA, vol. 3, file 1.
57 *Ottawa Journal,* 14 January 1922, 12; *Montreal Star,* 14 January 1922; and *Quebec Chronicle,* 16 January 1922. Clippings in LAC, RG 85, vol. 582, file 565.

58 *Ottawa Journal,* 14 January 1922, 12.
59 'Stefansson Bobs Up Again,' *Montreal Standard,* 21 January 1922, 34.
60 Stefansson to Camsell, 5 October 1920, LAC, MG 30 B8, Charles Camsell papers (CC), vol. 3, file 19. A letter from Stefansson to Desbarats, written on the same day, shows that Stefansson feared Camsell might take Anderson's side. LAC, RG 42, vol. 490, file 84-2-55.
61 Camsell to Stefansson, 23 January 1922, LAC, CC, vol. 3, file 19.
62 See Stefansson to King, 22 January 1922, and King to Stefansson, 31 January 1922, LAC, RG 42, vol. 466, file 84-2-4. King also consulted the minister of marine and fisheries, who agreed that there should be 'No action' (marginal note on King to Stefansson, dated 3 February 1922).
63 O'Neill to Camsell, 24 January 1922, LAC, CC, vol. 3, file 16.
64 Jenness to Stefansson, 30 January 1922, and Stefansson to Jenness, 28 February 1922, LAC, RG 13, vol. 933, file 'Anderson, Dr. R.M.'
65 'Mr. Jenness's Statement,' *Ottawa Citizen,* 16 January 1922, 2.
66 Gordon W. Smith, as reported in William Hunt to Richard Finnie, 12 May 1976, LAC, RSF, vol. 8, file 6.
67 Anderson, Chipman, O'Neill, and Jenness to Stewart, 28 February 1922, LAC, Geological Survey of Canada, RG 45, vol. 67, file 4078C.
68 Stefansson to Camsell, 24 February 1922, LAC, CC, vol. 3, file 19. Stefansson was both free and unscrupulous in his use of Borden's name at this time. Later in 1922, he warned Shackleton's biographer H.R. Mill not to discuss Shackleton's Canadian plans. He claimed that Borden was the only man to know the full story of Shackleton's dealings with the Canadian government. According to Stefansson, a false official version of this story was being circulated in order to protect Shackleton's supporters in the cabinet; in deference to Borden's wishes, Stefansson did not want the official version to be publicly contradicted. In fact, Borden knew nothing about the matter. He asked Loring Christie for information on Stefansson's behalf. Christie replied bluntly that 'in his own interest Mr. Stefansson would be well advised to say nothing' about the entire episode. See Borden to Christie, 20 August 1922, and Christie to Borden, 8 September 1922, LAC, LCC, vol. 3, file 6-6; Stefansson to Mill, 26 August and 2 October 1922, Scott Polar Research Institute (SPRI), MS 100/122/1-2.
69 This explanation is given in a number of letters written by both of the Andersons. For example, see Belle Anderson to Helen Crawford, 4 June 1925, LAC, RMA/MBAA, vol. 9, file 20.
70 Stefansson to Camsell, 19 March 1922, LAC, CC, vol. 3, file 19.
71 See Raymond Pearl, review of *The Friendly Arctic,* by Vilhjalmur Stefansson, *Science,* n.s. 55, 1421 (24 March 1922): 320-21, and Diamond Jenness, 'The Friendly Arctic,' *Science,* n.s. 56, 2436 (7 July 1922): 8-12.
72 Rudolph Anderson to William F. Riley, 16 January 1923, LAC, RMA/MBAA, vol. 3, file 7.
73 H.H. Langton, Review of *The Friendly Arctic,* by Vilhjalmur Stefansson, *Canadian Historical Review* 3, 1 (March 1922): 91. See the exchange of correspondence between Langton and White, LAC, RG 13, vol. 934, file 9A.
74 See 'Silver Wave Returns from Wrangell Island,' *Nome Nugget,* 24 September 1921, typed copy in LAC, RMA/MBAA, vol. 32, file 5-1-130.
75 Stefansson to King, 11 March 1922. There is a note at the top of the first page by Stefansson: 'This letter was written at a time when I thought I could not see the Prime Minister. I am handing it in now as a memorandum of our conversation – at the Prime Minister's request.'

76 Gordon G. Henderson, 'Policy by Default: The Origin and Fate of the Prescott Letter,' *Political Science Quarterly* 79, 1 (March 1964): 78-79, citing State Department file 861.0144.

77 Stefansson seems to have chosen five years simply because it best suited his purposes. In a 1926 survey of international law on the acquisition of territory, British jurist Sir Mark Lindley observed that although a period of twenty-five years had been suggested, 'attempts to give a definite limit to the period have met with little support, and what is a reasonable time will vary in different cases, and will depend upon such considerations as the difficulty of establishing political control and effecting colonization in the regions discovered, the relation of other States to the territory, and the urgency of the need for governmental institutions there.' Lindley, *The Acquisition and Government of Backward Territory in International Law* (London: Longmans, Green, 1926), 137. We have found no reference to a period of five years in any work on international law.

78 Stefansson to King, 11 March 1922.

79 For example, see Richard Diubaldo, *Stefansson and the Canadian Arctic* (Montreal: McGill-Queen's University Press, 1978), 171, and Niven, *Ada Blackjack,* 92-93.

80 Stefansson to Hinks, 21 April 1922, RGS, CB 9/139.

81 Stefansson to Harkin, 7 January 1922, and Harkin to Stefansson, 16 January, LAC, RG 85, vol. 1203, file 401-3.

82 The only record of the meeting is a brief and not very informative entry in King's diary, 17 March 1922.

83 'Stefansson Claims Wrangell Island for Great Britain: The Expedition He Sent Out Last Fall Has Established Possession, Says Explorer,' *New York Times,* 20 March 1922, 1.

84 Pope to King, 21 March 1921, LAC, WLMK, J4, vol. 147, file 1205.

85 Joy Esberey suggests that King's well-known interest in the supernatural served the function of reassurance, convincing him that important figures from his past – most notably, his mother and Sir Wilfrid Laurier – approved of his actions from beyond the grave. Joy E. Esberey, *Knight of the Holy Spirit: A Study of William Lyon Mackenzie King* (Toronto: University of Toronto Press, 1980), 131.

86 J.A. Stevenson to J.W. Dafoe, 26 July 1922, LAC, MG 30 D45, John Wesley Dafoe papers (JWD), vol. 2. See also John Hilliker, *Canada's Department of External Affairs,* vol. 1, *The Early Years, 1909-1946* (Montreal/Kingston: McGill-Queen's University Press, 1990), 89.

87 F.R. Scott, 'W.L.M.K,' in Scott, *The Eye of the Needle: Satire, Sorties, Sundries* (Montreal: Contact Press, 1957), 21-22.

88 'Canada Expects to Hold Wrangell: Basing Claim on Discovery and Occupation – British Government Apparently Indifferent,' *New York Times,* 21 March 1922, 3.

89 'Wrangel Island,' *Globe* (Toronto), 23 March 1922, 4.

90 'Stefansson Claims Wrangell Island for Great Britain.'

91 Stefansson to Hinks, 21 April 1922, RGS, CB 9/139.

92 Cory to Finnie, 25 March 1922, LAC, RG 85, vol. 1142, file 1005-5-1.

93 Craig to Cory, 4 April 1922, ibid.

94 Richard Finnie, 'Stefansson as I Knew Him' (draft article), LAC, RSF, vol. 8, file 6.

95 Finnie to Cory, 3 May 1922, LAC, RG 85, vol. 582, file 565.

96 Stefansson to Finnie, 3 May 1922, LAC, RG 85, vol. 1142, file 1005-5-1.

97 Stefansson to Orville Wright, 24 August 1922, Rauner Special Collections Library (RSCL), Stefansson Collection, Stef MS-196 (9) W.

98 Craig to Finnie, 10 May 1922, LAC, RG 85, vol. 1142, file 1005-5-1.

99 Canada, House of Commons, *Debates,* 1922, 2:1750-51.

100 Stefansson to Meighen, 15 May 1922, LAC, AM, I3, vol. 59, file 5.
101 Belle Anderson to William McKinlay, 29 June 1922, LAC, RMA/MBAA, vol. 8, file 7.
102 Nancy Fogelson, *Arctic Exploration and International Relations, 1900-1932* (Fairbanks: University of Alaska Press, 1992), 67-68; Henderson, 'Policy by Default,' 79-81.
103 Churchill to Byng (telegram), 2 June 1922; Churchill to Byng (letter), 3 June 1922, and enclosed copy of Soviet note, 24 May 1922, LAC, WLMK, J1, vol. 81, file '1922, Secretary of State for the Colonies.'
104 Stefansson to King, 31 May, ibid., vol. 82, file '1922, Stefansson.'
105 Stefansson to Ford, 2 June 1922, LAC, MG 30 B81, Vilhjalmur Stefansson papers (VS), vol. 2.
106 These handwritten copies are now in LAC, Department of the Interior, RG 15, vol. 1, file 'Arctic Islands Documents, 1873-1880.' They are essential to understanding the transfer. In the autumn of 1921 the papers were examined by Hensley Holmden. See Holmden to Craig, 31 October 1921, LAC, JDC, vol. 1, file 'Despatches, 1874-1923.'
107 Edwards to Cory, 4 July 1922, LAC, JDC, vol. 1, file 'Correspondence, 1903-1922.'
108 Finnie and Craig to Cory, 7 June 1923, LAC, RG 85, vol. 268, file 1003-6. For reports on Janes's character, see LAC, RG 18, vol. 3280, file 1920-HQ-681-G-4.
109 Herbert Patrick Lee, *Policing the Top of the World* (Toronto: McClelland and Stewart, 1928), 3-4.
110 See LAC, RG 85, vol. 595, file 758.
111 'Canada Will Send Party to Northland,' *Winnipeg Tribune,* 13 July 1922, clipping in HBCA, RG 2/4/63. See also F.C. Mears, 'Prepare to Push Boundary to Shadow of North Pole,' *Globe* (Toronto), 13 July 1922, 1, and 'Canada to Occupy Islands: Expedition Will Establish Title to Territory Opposite Greenland,' *New York Times,* 14 July 1922, 8.
112 Craig, 1922 draft report, LAC, RG 85, vol. 349, file 203.
113 Lee, *Policing the Top of the World,* 5-6.
114 Craig diary, 5 August 1922, LAC, RG 85, vol. 349, file 203.
115 Craig, 'The 1923 Expedition to the Arctic Islands,' LAC, RG 85, vol. 348, file 201-1-8. See also his diary entry for 30 August 1922.
116 Craig diary, 31 July 1922.
117 Lee, *Policing the Top of the World,* 8, 12.
118 Ibid., 23.
119 Ibid., 20.
120 Craig diary, 13 September 1922.
121 Lee, *Policing the Top of the World,* 28.
122 J.D. Craig, 'Canadian Expedition, 1922,' in J.D. Craig, F.D. Henderson, and George P. Mackenzie, *Canada's Arctic Islands* (Ottawa: F.A. Acland, 1927), 9.
123 Craig to Harkin, 28 August 1922, LAC, MG 30 E169, James Bernard Harkin papers (JBH), vol. 1, file January 1922-February 1959.
124 Lee, *Policing the Top of the World,* 36.
125 Craig, 'Canadian Expedition, 1922,' 11.
126 Craig diary, 13 September 1922.
127 Ibid.
128 Craig, 1922 draft report.
129 Craig, 'The 1923 Expedition.'
130 Craig, 1922 draft report.
131 Stefansson to Finnie, 13 May 1922, LAC, RG 85, vol. 1124, file 1005-5-1.
132 Stefansson to Christie, 9 June 1922, and Christie to Stefansson, 12 June 1922, LAC, Department of External Affairs, RG 25, vol. 2667, file 9057-B-40.

133 T.L. Cory, memo, 15 June 1922, LAC, RG 85, vol. 1124, file 1005-5-1.
134 Stewart to Stefansson, telegram, 24 June 1922, ibid.
135 Stefansson to Stewart, telegram, 9 July 1922, ibid.
136 Belle Anderson to Helen Crawford, 4 November 1923, LAC, RMA/MBAA, vol. 9, file 17.
137 Statement by Joseph Bernard, 6 March 1923, LAC, RG 85, vol. 1124, file 1005-5-1.
138 Stefansson statement, enclosed in Stefansson to Cory, 8 August 1922, ibid.
139 Stefansson did not in fact need the entire $3,000, since he had already received some money from Orville Wright. Stefansson to Wright, 24 August 1922.
140 Christie, memo, 9 August 1922, LAC, RG 25, vol. 2667, file 9057-B-40.
141 'Predicts Arctic Trade,' *New York Times,* 10 August 1922, 21.
142 Stefansson to Finnie, 16 March 1923, LAC, RG 85, vol. 1124, file 1005-5-1.
143 'Stefansson and the Possession of Wrangel Island,' *Ottawa Citizen,* 8 August 1922, 2; 'Wrangel Island of Great Value: Will Be of Immense Benefit to Empire, Declares Stefansson,' *Montreal Gazette,* 8 August 1922, 3; and 'Sees All Canada as Productive of Great Wealth,' *Toronto Star,* 9 August 1922, 1-2.
144 'Says Toronto Men [sic] at Wrangel Are Still Safe,' *Toronto Star,* 23 October 1922, 1.

Chapter 7: Stefansson in London

1 F.C. Mears, 'Farthest North is [sic] Post Office Dominion Opens,' *Globe* (Toronto), 16 October 1922, 3.
2 Churchill to Byng, 15 July 1923, and enclosed copy of hydrographer's memo, Library and Archives Canada (LAC), Governor General's Office, RG 7, vol. 412, file 10045.
3 On this expedition, see William Barr, 'The Voyages of *Taymyr* and *Vaygach* to Ostrov Vrangelya, 1910-15,' *Polar Record* 16, 101 (May 1972): 213-34, and L.M. Starokadomskiy, *Charting the Russian Northern Sea Route: The Arctic Ocean Hydrographic Expedition, 1910-1915,* trans. and ed. William Barr (London/Montreal: Arctic Institute of North America/McGill-Queen's University Press, 1976).
4 Devonshire to Byng, 4 November 1922, and enclosed copy of American note, 27 September 1922, LAC, RG 7, vol. 412, file 10045. Henrietta, Jeannette, and Bennett Islands, all mentioned in the Russian note, were discovered in 1881 by the American explorer George Washington De Long.
5 Gordon G. Henderson, 'Policy by Default: The Origin and Fate of the Prescott Letter,' *Political Science Quarterly* 79, 1 (March 1964): 81.
6 Count Beckendorff (Russian Ambassador) to Viscount Grey (Foreign Secretary), 23 October 1916, LAC, RG 7, vol. 412, file 10045.
7 See Pope to Stewart, 16 October 1922; Stewart to Pope, 1 December 1922; and Stewart to Pope, 5 December 1922, LAC, Department of External Affairs, RG 25, vol. 2667, file 9057-B-40.
8 Craig to Finnie, 16 December 1922, LAC, Northern Affairs Program, RG 85, vol. 1124, file 1005-5-1.
9 Stefansson to Cory, 27 December 1922, ibid.
10 Stefansson to Finnie, telegram, 13 February 1923, ibid.
11 John Anderson to Cory, 28 February 1923, ibid.
12 Finnie to John Anderson, 3 March 1923, ibid.
13 Stefansson to Finnie, 8 March 1923, ibid.
14 Stefansson to Finnie, 14 March 1923, and enclosure, 'Interview with Captain Joseph Bernard,' ibid.
15 Richard Diubaldo, *Stefansson and the Canadian Arctic* (Montreal: McGill-Queen's University Press, 1978), 177-79.

16 Stefansson to Finnie, 16 March 1923, and Finnie to Stefansson, 21 March 1923, LAC, RG 85, vol. 1124, file 1005-5-1.
17 Stefansson to Finnie, 24 March 1923, ibid.
18 Pope to King, 22 March 1923, LAC, MG 26 J, William Lyon Mackenzie King papers (WLMK), J4, vol. 147, file 1205.
19 Bernard to King, 21 March 1923 (emphasis in original). The letter was acknowledged and forwarded to Stewart on 22 March. LAC, RG 85, vol. 1124, file 1005-5-1.
20 Finnie to Stefansson, 27 March 1923, ibid.
21 King diary, 27 March 1923, The Diaries of William Lyon Mackenzie King, LAC, http://www.collectionscanada.gc.ca/databases/king.
22 This account follows Stefansson's own recollections, as given in a letter to Finnie, 8 December 1925, LAC, RG 85, vol. 764, file 5064.
23 Stewart to King, 22 August 1923, LAC, WLMK, J1, vol. 95, file '1923, Stewart.'
24 King diary, 7 April 1923.
25 For example, see Philip Wigley, *Canada and the Transition to Commonwealth: British-Canadian Relations, 1917-1926* (Cambridge: Cambridge University Press, 1977), 160-66, and Norman Hillmer and J.L. Granatstein, *Empire to Umpire: Canada and the World into the Twenty-First Century* (Toronto: Thomson Nelson, 2008), 78-81.
26 'Canada Likely Establish Post on Wrangel Id.,' *Ottawa Citizen,* 9 April 1923, 9.
27 Craig to Finnie, 9 April 1923, LAC, RG 85, vol. 1124, file 1005-5-1.
28 Finnie to Gibson, 9 April 1923, ibid.
29 'Wrangel Island Again,' *Ottawa Journal,* 10 April 1923, 8.
30 'The Explorer-Lecturer Who Is Different,' LAC, Department of Justice, RG 13, vol. 934, file 9A.
31 Belle Anderson, memo, 21 April 1923, LAC, MG 30 B40, Rudolph Martin Anderson and Mae Belle Allstrand Anderson papers (RMA/MBAA), vol. 12, file 5. Stefansson made similar statements in a letter written at about the same time. He told Arthur Hinks: 'My personal inclination is to insist that the investigation be gone through with. I hope it will be thoroughgoing, that the hearings will be secret, and that only the findings of the investigating committee will be announced. My reason for insisting that the matter be gone through with is that I have several important plans which I am urging upon the Canadian Government. It is more difficult for me to get them to carry through these plans while there is in the minds of any member of the Government the feeling that there may be even partial justification for the charges that have been made.' Stefansson to Hinks, 22 April 1923, Royal Geographical Society Archives (RGS), CB 9/139.
32 Stefansson to White, 18 April 1923, LAC, RG 13, vol. 934, file 9A.
33 Stefansson to Harkin, 23 April 1923, and enclosure; Harkin to Stefansson, 27 April 1923, LAC, MG 30 E169, James Bernard Harkin papers (JBH), vol. 2, file 'Canadian Sovereignty of Arctic Islands re: V.O. Stefansson.'
34 White to Stefansson, 23 April 1923, LAC, RG 13, vol. 934, file 9A.
35 Pope to White, 25 March 1922, and Thomas Hall to White, 2 April 1923, ibid.
36 'As to Stefansson,' *Ottawa Journal,* 1 May 1923, 6.
37 There is a transcript of the lecture in LAC, RMA/MBAA, vol. 11, file 8. Stefansson had distributed many free tickets but, according to George Desbarats' diary, the size of the audience was only 'fair.' 1 May 1923, LAC, MG 30 E89, George Desbarats papers (GD), vol. 6.
38 Stefansson to Harkin, 9 May 1923, and Harkin to Stefansson, 14 May 1923, LAC, JBH, vol. 2, file 'Canadian Sovereignty of Arctic Islands re: V.O. Stefansson.'

39 Stefansson to Camsell, 20 November 1922, LAC, MG 30 B8, Charles Camsell papers (CC), vol. 3, file 19.
40 Vilhjalmur Stefansson, Burt M. McConnell, and Harold Noice, 'The Friendly Arctic,' *Science,* n.s. 57, 1474 (30 March 1923): 368-69.
41 Ibid., 369-73.
42 Camsell to R.W. Brock, LAC, Geological Survey of Canada, RG 45, vol. 67, file 4078C.
43 Rudolph Anderson to Bartlett, 13 June 1923, LAC, RMA/MBAA, vol. 3, file 7.
44 Rudolph Anderson to Borden, 14 April 1923, and Borden to Anderson, 16 April 1923, LAC, MG 26 H, Robert Laird Borden papers (RLB), vol. 259, file 7.
45 Anderson to Borden, 21 April 1923, and Borden to Anderson, 11 May 1923, ibid.
46 See Belle Anderson to Rudolph Anderson, 14 May 1923, LAC, RMA/MBAA, vol. 3, file 7.
47 Charles Camsell, 'The Friendly Arctic,' *Science,* n.s. 57, 1484 (8 June 1923): 665-66.
48 Bernard to Starnes, 14 May 1923, LAC, Royal Canadian Mounted Police, RG 18, vol. 3183, file G-804-5-25.
49 Starnes to Pope, 15 May 1923, and Pope to Starnes, 17 May 1923, ibid.
50 Byng to Devonshire, 23 April 1923, LAC, RG 7, vol. 411, file 10045.
51 Vilhjalmur Stefansson, *The Adventure of Wrangel Island* (New York: Macmillan, 1925), 145.
52 Ibid., 135.
53 See L.S. Amery, 'Hudson Bay: Its Conditions and Problems,' in J. Castell Hopkins, ed., *Empire Club Speeches, 1910-1911* (Toronto: Saturday Night Press, n.d.), 27-36, and Amery's comments on W.L. Grant's address to the Royal Geographical Society: 'Geographical Conditions Affecting the Development of Canada,' *Geographical Journal* 38, 4 (October 1911): 376-80.
54 Diary of Leo Amery, 1 May, 29 May, 8 July, and 22 July 1923, Churchill College Archives (CCA), AMEL 7/17.
55 Stefansson, *Adventure of Wrangel Island,* 152.
56 See report of the Committee of the Privy Council, 22 February 1913, and Desbarats' instructions to Stefansson, 29 May 1913, LAC, RG 45, vol. 67, file 4078C. The report stated that Stefansson's expedition would 'conduct its explorations in waters and on lands under Canadian jurisdiction or included in the northern zone contiguous to the Canadian territory'; the instructions referred to 'the Northern waters of Canada.' Clearly, there was no official intention that Stefansson should make claims beyond the 141st meridian.
57 Vilhjalmur Stefansson, 'The History and Importance of Wrangell Island,' *Spectator* 4934 (9 June 1923): 958-59, and 4935 (16 June 1923): 998-1000.
58 Vilhjalmur Stefansson, 'Wrangell Island' (letter to the editor), *Spectator* 4937 (30 June 1923): 1079.
59 Foreign Office memo, 9 July 1923, LAC, RG 25, vol. 2667, file 9057-B-40.
60 Stefansson to King, 10 July 1923, LAC, WLMK, J1, vol. 95, file '1923, Stefansson.'
61 Finnie to Gibson, 7 July 1923, LAC, RG 85, vol. 1124, file 1005-5-1.
62 Stefansson, *Adventure of Wrangel Island,* 159.
63 Minutes of Cabinet meeting, 25 July 1923, CAB/23/46, Documents Online, Cabinet Papers, National Archives, http://www.nationalarchives.gov.uk./cabinetpapers. See also Amery diary, 25 July 1923.
64 See correspondence in LAC, RG 7, vol. 411, file 10045.
65 Soviet note, 25 August 1923, LAC, RG 25, vol. 2667, file 9057-B-40.

66 Devonshire to Byng, 7 September 1923, and enclosed copy of code telegram to Moscow, 1 September 1923, LAC, RG 7, vol. 411, file 10045.
67 Stefansson to King, 30 August 1923 (two letters), LAC, MG 26 J, William Lyon Mackenzie King papers (WLMK), J1, vol. 95, file '1923, Stefansson.'
68 Quoted in Jennifer Niven, *Ada Blackjack: A True Story of Survival in the Arctic* (New York: Hyperion, 2003), 335.
69 'Tragedy of North Grieves Explorer: Stefansson at Loss to Explain How Men Perished on Wrangell Island,' *New York Times,* 2 September 1923, 15.
70 'Lack of Experience Cost Explorers' Lives,' *Toronto Star,* 26 September 1923, 3.
71 'Farewell Letter of Lost Explorer Reaches His Wife,' *Toronto Star,* 27 September 1923, 21. Stefansson later publicly blamed Noice for the widespread belief that the men had left Wrangel Island because of lack of food. See *Adventure of Wrangel Island,* 166. He was even able to procure a signed statement from Noice, in which Noice apologized for misleading the public (293-96). Noice told Crawford's father that Stefansson had 'double-crossed' him with a false promise that, if he signed the statement, he would not be criticized in the book. Noice to J.T. Crawford, 1 May 1925, LAC, RG 13, vol. 933, file 'Anderson, Dr. R.M.' See also Chapter 8. For a more detailed account of the conflicts between Stefansson and Noice, see Niven, *Ada Blackjack,* 267-69, 271-73, 284-85, 301-3, 308-10, 326-27, and 348-51.
72 'University Is Opened: Tribute to A. Crawford,' *Toronto Telegram,* 26 September 1923, 20.
73 'Gallant Allan Crawford,' *Globe* (Toronto), 3 September 1923, 4; 'Mr. Stefansson Shows Results,' *Montreal Standard,* 9 September 1923, 26; 'Have Furled Flag on Wrangel Island,' *Mail and Empire* (Toronto), 8 September 1923, 1; and 'The Un-Friendly Arctic,' *Saturday Night,* 15 September 1923, 1.
74 'Wrangel Island Not Ours,' *Toronto Star,* 1 September 1923, 1.
75 'Wrangel Island Tragedy: Government Not Concerned,' *Toronto Evening Telegram,* 1 September 1923, 19.
76 John Anderson, telegram to Taylor relaying Stefansson's message, LAC, RG 13, vol. 924, file 6107. Taylor provided a copy of this message to the Crawfords, who later sent it to James White.
77 Stefansson to King, 5 October 1923, LAC, WLMK, J1, vol. 95, file '1923, Stefansson.'
78 Belle Anderson to Helen Crawford, 29 December 1923, LAC, RMA/MBAA, vol. 9, file 17. Anderson's unidentified informant was likely either Charles Camsell or Arthur Doughty.
79 O.D. Skelton to Isabel Skelton, 14 October 1923, LAC, MG 30 D33, Oscar Douglas Skelton papers (ODS), vol. 2, file 5.
80 'Wrangel Island Tragedy,' *The Times* (London), 9 October 1923, 11.
81 Starnes to Finnie, 7 March 1923, LAC, RG 85, vol. 268, file 1003-6.
82 Craig to Finnie, 10 March 1923, ibid.
83 Bernier to Craig, 12 March 1923, ibid.
84 Cory to Starnes, 26 May 1923, ibid.
85 Starnes to Cory, 30 May 1923, ibid.
86 Craig to Finnie, 12 June 1923, ibid. See also Finnie and Craig to Cory, 7 June 1923, ibid.
87 Finnie to Cory, 30 June 1923, ibid.
88 J. Dewey Soper, *Canadian Arctic Recollections: Baffin Island, 1923-1931,* ed. Shirley Milligan (Saskatoon: Institute for Northern Studies, University of Saskatchewan, 1981), xii.
89 See Dudley Copland, *Livingstone of the Arctic* (Lancaster, ON: Canadian Century Publishers, 1978 [1967]).

90 See L.T. Burwash, *Canada's Western Arctic: Report on Investigations in 1925-26, 1928-29, and 1930* (Ottawa: F.A. Acland, 1931).
91 Craig, 'The 1923 Expedition to the Arctic Islands,' LAC, RG 85, vol. 348, file 201-1-8.
92 Ibid.
93 Lauge Koch, 'North of Greenland,' *Geographical Journal* 64, 1 (July 1924): 17.
94 Craig, 'The 1923 Expedition.'
95 Caption to the frontispiece of Harwood Steele, *Policing the Arctic* (London: Jarrolds, 1936).
96 Craig, 'The 1923 Expedition.'
97 Finnie to Fitzgerald, 11 October 1923, Hudson's Bay Company Archives (HBCA), RG 2/4/63.
98 Douglas report, 2 October 1922, LAC, RG 18, vol. 3290, file 1921-HQ-1091-G-1.
99 Joy report, 7 September 1922, ibid.
100 Munn to Ralph Parsons, 15 March 1923, HBCA, RG 2/4/63. Munn sent a blistering seven-page attack on Rasmussen to the HBC. 'Explorers Ethics Past and Present,' HBCA, A.102/1962. He also suggested to Finnie that a police officer should board Rasmussen's supply ship to search for furs. Brabant to Fitzgerald, 24 March 1923, HBCA, RG 2/4/63.
101 Munn to Sale, 28 September 1923, HBCA, RG 2/4/63. The origin of the rumours proved to be a visit to Rasmussen's camp by trader Jean Berthie. Berthie, who did some trading on this occasion, was mistaken for a member of the Danish party by the Inuit.
102 Peter Freuchen, *Vagrant Viking: My Life and Adventures* (New York: Julian Messner, 1953), 190-91.
103 Craig, 'The 1923 Expedition.'
104 Soper's photo collection is now in the University of Alberta Archives, J. Dewey Soper fonds, Series 3. It can be viewed online at http://www.ualberta.ca/ARCHIVES. For a detailed account of the trial, see Shelagh Grant, *Arctic Justice: On Trial for Murder, Pond Inlet, 1923* (Montreal/Kingston: McGill-Queen's University Press, 2002).
105 Rudolph Anderson to Harry Allstrand, 25 October 1923, LAC, RMA/MBAA, vol. 3, file 7.
106 Stefansson to King, 2 January 1924, LAC, RG 25, vol. 2667, file 9057-B-40.
107 F.A. McGregor to Stefansson, 5 January 1924, ibid.
108 King diary, 3 December 1923.
109 Belle Anderson, memo book, LAC, RMA/MBAA, vol. 12, file 4.
110 Desbarats diary, 8 April 1923, LAC, GD, vol. 6.
111 Canada, House of Commons, *Debates,* 1924, 2:1110.
112 Belle Anderson to Helen Crawford, 11 April 1924, LAC, RMA/MBAA, vol. 9, file 17.
113 See 'Canada Foregoes Claim on Wrangel, Declares Stewart,' *Globe* (Toronto), 8 April 1924, 2.
114 Stefansson to Stewart, 2 June 1924, LAC, RG 85, vol. 1124, file 1005-5-1.
115 Donat LeBourdais, 'Staking Wrangel Island,' in Vilhjalmur Stefansson, *The Adventure of Wrangel Island* (London: Jonathan Cape, 1926), 395-96, 401. The impression given to Lomen by Stefansson that he might eventually persuade the State Department to back an American claim was highly misleading. Six months earlier, Robert Bartlett had reported from Washington that 'Windjammer's stock is damn low let me tell you. What a change ... That Wrangel Island thing clinched it.' Bartlett to Rudolph Anderson, 1 November 1923, LAC, RMA/MBAA, vol. 3, file 9.
116 Stefansson to Ramsay MacDonald, 2 June 1924, LAC, RG 7, vol. 411, file 10045.
117 Stefansson to Stewart, 2 June 1924, LAC, RG 85, vol. 1124, file 1005-5-1.

118 J.H. Thomas to Byng, 18 June 1924, LAC, RG 25, vol. 2667, file 9057-B-40.
119 Stewart to King, 14 July 1924, and enclosed draft, ibid. See also Byng to Thomas, 24 July 1924, LAC, RG 7, vol. 411, file 10045.
120 G.R. Warner to Stefansson, 8 August 1924, LAC, RG 7, vol. 411, file 10045.
121 Marginal note on copy of Cory to Pope, 6 September 1924, LAC, RG 85, vol. 1124, file 1005-5-1.
122 LeBourdais, 'Staking Wrangel Island,' 406-7.
123 J.C. Hill (acting Assistant Agent, British commercial mission, Vladivostok) to P.H. Hodgson (Chargé d'Affaires, Moscow), 31 October 1924, LAC, RG 7, vol. 411, file 10045.
124 Melody Webb, 'Arctic Saga: Vilhjalmur Stefansson's Attempt to Colonize Wrangel Island,' *Pacific Historical Review* 61, 2 (May 1992): 237.
125 Stefansson to Lomen, 20 July 1925, quoted in ibid., 238.
126 Knud Rasmussen, *The Netsilik Eskimos: Social Life and Spiritual Culture.* Report of the Fifth Thule Expedition, 1921-24, vol. 8, nos. 1-2 (Copenhagen: Gyldendalske Boghandel, Nordisk Forlag, 1931), 134, 79-80, 131.
127 Starnes to Pope, 4 October 1924, LAC, RG 25, vol. 2668, file 9058-F-40.
128 H. Ostermann and E. Holtved, *The Alaskan Eskimos: As Described in the Posthumous Notes of Dr. Knud Rasmussen.* Report of the Fifth Thule Expedition, 1921-24, vol. 10, no. 3 (Copenhagen: Gyldendalske Boghandel, Nordisk Forlag, 1952), 9, 58-59.
129 Ibid., 89-91. Nicholas II's mother, Empress Marya Feodorovna, was born Princess Dagmar of Denmark.
130 Account of a conversation with Bernard, Belle Anderson memo book, 4 March 1931, LAC, RMA/MBAA, vol. 12, file 4.
131 Ostermann and Holtved, *The Alaskan Eskimos,* 266.

Chapter 8: The Sector Claim

1 Finnie to Nobel, 30 October 1924, Library and Archives Canada (LAC), Northern Affairs Program, RG 85, vol. 755, file 4691.
2 Boggild to Finnie, 17 November 1924, ibid.
3 Boggild to Pope, 24 November 1924, LAC, Department of External Affairs, RG 25, vol. 2668, file 9058-F-40.
4 See photo in the *Washington Post,* 16 November 1924.
5 Daugaard-Jensen to Finnie, 15 December 1924, LAC, RG 85, vol. 755, file 4691.
6 Draft memo, 2 December 1924, Trent University Archives (TUA), Trevor Lloyd papers (TL), 87-104-6-2. Translation by Trevor Lloyd.
7 See Jens Brøsted, 'Danish Accession to the Thule District, 1937,' *Nordic Journal of International Law* 57 (1988): 259.
8 F.D. Henderson, 'Canadian Expedition, 1924,' in J.D. Craig, F.D. Henderson, and George P. Mackenzie, *Canada's Arctic Islands* (Ottawa: F.A. Acland, 1927), 36.
9 Craig to Finnie, 15 April 1925, LAC, RG 85, vol. 268, file 1003-6.
10 'Expect MacMillan to Find a Continent,' *New York Times,* 11 April 1925, 15.
11 See LAC, RG 85, vol. 14, file 20 – MacMillan, especially Finnie to MacMillan, 22 October 1924, and MacMillan to Finnie, 27 October 1924. Rasmussen later passed on a report from Thule that MacMillan's party had killed twelve muskoxen and remarked: '[N]ow you are able to see that you cannot trust declarations from Mc.Millan [sic].' Rasmussen to Finnie, 6 June 1925, LAC, RG 85, vol. 584, file 573.
12 Craig to Finnie, 15 April 1925, LAC, RG 85, vol. 268, file 1003-6.

13 See LAC, RG 85, vol. 755, file 4676.
14 Finnie to Gibson, 20 April 1925, LAC, RG 85, vol. 268, file 1003-6.
15 Stewart to the Governor General in Council, 20 April 1925, and Skelton to Harkin, 23 April 1925, LAC, MG 30 E169, James Bernard Harkin papers (JBH), vol. 1, file January 1922-February 1959.
16 With the creation of the Department of National Defence, the Department of the Naval Service (of which Desbarats had formerly been deputy minister) had ceased to exist. Responsibility for matters connected to the Canadian Arctic Expedition (for example, the publication of scientific reports) was transferred to the Department of Marine and Fisheries.
17 See Craig to Cory, 21 June 1921; Cory to Craig, 24 June 1921; and Craig to Cory, 7 July 1921, LAC, RG 85, vol. 14, file 20 – MacMillan.
18 Finnie to Craig, 13 March 1924, and Craig to Finnie, 14 March 1924, LAC, RG 85, vol. 85, file 202-2-1.
19 Daly to Finnie, 10 March 1925, and Daly to Cory, 27 April 1925, ibid.
20 Minutes of first meeting of Northern Advisory Board, 24 April 1925, LAC, RG 25, vol. 2669, file 9062-C-40.
21 Vilhjalmur Stefansson, *The Adventure of Wrangel Island* (New York: Macmillan, 1925), 25, 71-72, 75-80.
22 Belle Anderson, memo book, LAC, MG 30 B40, Rudolph Martin Anderson and Mae Belle Allstrand Anderson papers (RMA/MBAA), vol. 12, file 4.
23 Belle Anderson to W.L. McKinlay, 5 March 1926, RMA/MBAA, vol. 8, file 14.
24 Rudolph Anderson to Bartlett, 23 April 1925, RMA/MBAA, vol. 3, file 12.
25 Belle Anderson to Helen Crawford, 28 December 1925, RMA/MBAA, vol. 10, file 1.
26 Belle Anderson to Helen Crawford, 31 December 1926, ibid.
27 H.M.U., 'Review of "The Adventure of Wrangel Island," by V. Stefansson,' LAC, MG 26 J, William Lyon Mackenzie King papers (WLMK), J4, vol. 147, file 1205.
28 Borden's diary – a maddeningly laconic document – records a visit from Stefansson on Christmas Day, 1921. Borden was then in Washington, serving as the Canadian representative at the International Conference on Naval Limitation. Stefansson gave Borden a copy of *The Friendly Arctic* as a Christmas gift. The two men discussed Wrangel, but the diary provides no details of what was said. Stefansson returned the next day for another conversation on the same subject. Borden passed along some unspecified information about Stefansson's plans to the British representative at the disarmament conference, Sir Maurice Hankey. Stefansson again discussed Wrangel Island with Borden on the day after his first meeting with Mackenzie King. Borden diary, 25 December 1921, 26 December 1921, and 18 March 1922, LAC, MG 26 H, Robert Laird Borden papers (RLB), vol. 449.
29 Finnie to Cory, 5 May 1925, LAC, RG 85, vol. 584, file 573.
30 Belle Anderson to Helen Crawford, 4 May 1925, LAC, RMA/MBAA, vol. 9, file 19.
31 Minutes of the meeting of the Advisory Board on Wildlife Protection, 5 May 1925, LAC, RG 85, vol. 145, file 400-6.
32 Belle Anderson to Helen Crawford, 4 May.
33 See TUA, TL, 87-014-9-19.
34 Belle Anderson to Joseph Bernard, 1 May 1925, LAC, RMA/MBAA, vol. 8, file 13.
35 See LAC, RG 25, vol. 1442, file 417-D.
36 Rasmussen to Skelton, 5 May 1925, LAC, RG 25, vol. 1386, file 1324.
37 Rasmussen to Finnie, 11 May 1925, LAC, RG 85, vol. 584, file 573.

38 Knud Rasmussen, 'The Fifth Thule Expedition, 1921-24: The Danish Ethnographical and Geographical Expedition from Greenland to the Pacific,' *Geographical Journal* 67, 2 (February 1926): 124, 126.
39 Cory to Byrd, 12 May 1925, LAC, RG 85, vol. 14, file 20 – MacMillan, and minutes of second meeting of Northern Advisory Board, 13 May 1925, LAC, RG 25, vol. 2669, file 9062-C-40.
40 Byrd to Cory, 8 May 1925, LAC, RG 85, vol. 14, file 20 – MacMillan.
41 Byrd to Cory, 28 May 1925, ibid.
42 T.L. Cory to W.W. Cory, 20 May 1925, and attached draft bill, LAC, RG 85, vol. 85, file 202-2-1.
43 Richard Finnie to Gordon W. Smith, 3 May and 10 June 1977, LAC, MG 31 C6, Richard Sterling Finnie papers (RSF), vol. 10, file 14. Oswald Finnie recounted this incident to his son in 1937.
44 White to Skelton, 25 May 1925, and enclosed memo, LAC, RG 25, vol. 4252, file 9057-40. Both White and Skelton were vaguely aware of the British sector claim in the Antarctic, but they were not well informed about it and did not use it as a model. See White to Skelton, 10 July 1925, LAC, Department of Justice, RG 13, vol. 924, file 6100B; Skelton to White, 13 July 1923, and enclosed copy of Skelton to Lucien Pacaud, 4 July 1923, LAC, RG 13, vol. 930, file 6303B. On Canada's claim to the Sverdrup Islands, see also White, 'Memorandum re Sverdrup Expedition, 1989-1902,' LAC, RG 13, vol. 934, file 10.
45 Canada, House of Commons, *Debates,* 1925, 4:3772-73.
46 'Our Arctic Heritage,' *Montreal Gazette,* 4 June 1925, 10.
47 See John H. Bryant and Harold N. Cones, *Dangerous Crossings: The First Modern Polar Expedition, 1925* (Annapolis, MD: Naval Institute Press, 2000), 30-38.
48 E.F. McDonald to Wilbur, 28 February 1925, printed in ibid., 177-81.
49 'Disputes Claim of Canada to Islands,' *Ottawa Citizen,* 3 June 1925, 8.
50 MacMillan to Wilbur, 5 June 1925, quoted in Nancy Fogelson, *Arctic Exploration and International Relations, 1900-1932* (Fairbanks: University of Alaska Press, 1992), 93.
51 'Title to Arctic Lands,' *Washington Post,* 9 June 1925, 6.
52 'Canada's Arctic Land Claim May Be Valid on Fine Point,' *Washington Star,* 7 June 1925, clipping in LAC, RG 25, vol. 4252, file 9057-40. In contrast, the *Post's* report on the 1925 Eastern Arctic Patrol gave the impression that it was the first Canadian expedition to the archipelago. See 'Canada to Take Lands in Arctic Ahead of U.S.,' 9 June 1925, 2.
53 Secretary of State to Norwegian Minister, 2 April 1924, quoted in Gordon G. Henderson, 'Policy by Default: The Origin and Fate of the Prescott Letter,' *Political Science Quarterly* 79, 1 (March 1964): 88.
54 Secretary of State to Prescott, 13 May 1924, quoted in ibid., 90.
55 See 'Won't Claim Title to Lands in Arctic: Washington Will Ask MacMillan Expedition to Leave Sovereignty for Later Decision: Norway's Rights Asserted,' *New York Times,* 12 June 1925, 21; 'Wilbur Not Pushing Arctic Land Claims,' *Washington Post,* 12 June 1925, 1; 'MacMillan Expedition without Instructions: Not Officially Advised on Course to Be Followed in Flying Over Disputed Territory,' *Washington Star,* 12 June 1925, clipping in LAC, RG 85, vol. 14, file 20 – MacMillan.
56 Craig to Finnie, 1 May 1925, LAC, RG 85, vol. 268, file 1003-6.
57 Finnie to Gibson, 2 May 1925, ibid.
58 Minutes of second meeting, LAC, RG 25, vol. 2669, file 9062-C-40. D.H. Dinwoodie mistakenly states that this was a new plan devised to meet the American challenge.

Dinwoodie, 'Arctic Controversy: The 1925 Byrd-MacMillan Expedition Example,' *Canadian Historical Review* 53, 1 (March 1972): 59.

59 Finnie to Starnes, 27 May 1925, LAC, RG 85, vol. 268, file 1003-6.

60 Starnes to Finnie, 29 May 1925, ibid.

61 Canada, House of Commons, *Debates,* 1925, 4:4084, 4093.

62 Stefansson, *Adventure of Wrangel Island,* 105-7. In this case, the blame was laid mainly on Knight, who, Stefansson claimed, had had a faith in the 'friendly Arctic' that was 'even greater than my own.'

63 See Chapter 7, note 71, above.

64 Noice to J.T. Crawford, 1 May 1925, LAC, RG 13, vol. 933, file 9A.

65 See 'Food Shortage Led to Wrangel Tragedy: Professor and Mrs. Crawford Discuss the Views of Stefansson,' *Toronto Star,* 29 April 1925, 27, and 'Parents of Dead Youth Say Stefansson to Blame,' unidentified clipping, LAC, JBH, vol. 1, file January 1922-February 1959.

66 Printed in *Debates,* 1925, 4:4094.

67 Stefansson to Finnie, 19 September 1925, LAC, RG 85, vol. 764, file 5064. Stefansson repeatedly named $30,000 as the amount he had spent from his personal savings and borrowed funds, but it seems highly unlikely that any such sum had in fact been expended.

68 Stefansson to Finnie, 8 December 1925, LAC, RG 85, vol. 764, file 5064.

69 Gibson to Finnie, 21 January 1926, ibid.

70 Stefansson to Finnie, 6 May 1926, ibid.

71 Finnie to Stefansson, 17 May 1926, ibid.

72 Vilhjalmur Stefansson, *Discovery: The Autobiography of Vilhjalmur Stefansson* (New York: McGraw-Hill, 1964), 296-97.

73 Gibson to Finnie, 4 May 1925, and attached clipping ('That McMillan Expedition,' *Ottawa Journal,* 2 May 1925, 6), LAC, RG 85, vol. 14, file 20 – MacMillan. It is entirely possible that Finnie and Craig were behind the article, since the views expressed coincide with those in their correspondence with one another.

74 'Canada's Claims Are Explained,' *New York Times,* 13 June 1925, 17.

75 Byng to Chilton, 12 June 1925, LAC, RG 85, vol. 14, file 20 –MacMillan.

76 Chilton to Byng, 12 June 1925, ibid.

77 Chilton to Kellogg, 15 June 1925, LAC, RG 25, vol. 2668, file 9058-B-40.

78 Kellogg to Chilton, 19 June 1925, ibid.

79 Harwood Steele, *Policing the Arctic* (London: Jarrolds, 1936), 260.

80 Cory to Starnes, 29 June 1925, LAC, RG 85, vol. 14, file 20 –MacMillan.

81 Minutes of fourth meeting of Northern Advisory Board, 11 June 1925, LAC, RG 25, vol. 2669, file 9062-C-40.

82 Finnie to Gibson, 28 May 1925, LAC, RG 85, vol. 759, file 4834.

83 Richard Finnie, 'Farewell Voyages: Bernier and the "Arctic,"' *Beaver* 54, 1 (Summer 1974): 47, 50-51.

84 See documents from April-June 1925 in LAC, RG 85, vol. 5, file 20 – Bernier. In January 1926 Finnie told Cory resignedly, 'We have written [Bernier] many times regarding newspaper interviews and of course he always tells us that he never gave them or he will not do it again. I do not suppose there is any use in writing him again about it.' See Finnie to Cory, 30 January 1926, in the same file.

85 Richard E. Byrd, *To the Pole: The Diary and Notebook of Richard E. Byrd, 1925-1927,* ed. Raimund E. Goerler (Columbus: Ohio State University Press, 1998), 37.

86 Affidavit of George Patton Mackenzie, 6 November 1925, and affidavit of Lazare Morin, 3 November 1925, LAC, RG 25, vol. 2668, file 9058-B-40. See also Mackenzie's diary, in LAC, MG 30 E529, George Patton Mackenzie papers (GPM), vol. 1. In his entry for 20 August, Mackenzie recorded that Byrd had excused his failure to apply for a permit in good time by saying he had been told to 'leave the diplomatic end to others.'
87 OPNAV to Byrd, 20 August 1925, reproduced in Season L. Osborne, 'Closing the Front Door of the Arctic: Capt. Joseph E. Bernier's Role in Canadian Arctic Sovereignty' (master's thesis, Carleton University, 2003), 216.
88 It seems highly likely that the message was not received. The American radio operator was negligent in his duty, preferring to chat over the airwaves with amateur radio friends. See Bryant and Cones, *Dangerous Crossings,* 123-24, 161-62.
89 Quoted in Fogelson, *Arctic Exploration and International Relations,* 96.
90 Wilfrid Bovey to J.H. MacBrien, 25 January 1927, LAC, RG 25, vol. 2668, file 9058-B-40. MacMillan also told Bovey that, having been born in Nova Scotia, 'he did not want to be party to anything unfriendly to Canada.' MacMillan was born in Provincetown, Massachusetts.
91 George P. Mackenzie, 'Canadian Expedition, 1925,' in J.D. Craig, F.D. Henderson, and George P. Mackenzie, *Canada's Arctic Islands* (Ottawa: F.A. Acland, 1927), 45-46.
92 See correspondence in LAC, RG 85, vol. 15, file 20 – MacMillan.
93 Starnes to Finnie, 22 May 1926, ibid.
94 Memo by Irving N. Linnell, 16 September 1926, quoted in Dinwoodie, 'Arctic Controversy,' 63.
95 See Henderson, 'Policy by Default,' 90.
96 Chilton to Chamberlain, 28 April 1926, enclosed in Chilton to Byng, 28 April 1926, LAC, RG 25, vol. 1422, file 417-B.
97 Quoted in Thomas M. Tynan, 'Canadian-American Relations in the Arctic: The Effect of Environmental Influences upon Territorial Claims,' *Review of Politics* 41, 3 (July 1979): 407.
98 Henderson, 'Policy by Default,' 94.
99 Skelton to H. Chandler, 13 July 1925, LAC, RG 25, vol. 4252, file 9057-40.
100 See letters in LAC, RG 13, vol. 924, file 6100B, and vol. 931, file 6312.
101 David Hunter Miller, 'Political Rights in the Arctic,' *Foreign Affairs* 4, 1 (October 1925): 51-52. The idea of an 'ownership by everybody' had been put forward by the American writer Thomas Willing Balch. Balch proposed that the polar regions should be considered the 'common possessions of all the family of nations.' 'The Arctic and Antarctic Regions and the Law of Nations,' *American Journal of International Law* 4, 2 (April 1910): 275. See also James Brown Scott, 'Arctic Exploration and International Law,' *American Journal of International Law* 3, 4 (October 1909): 941.
102 Miller, 'Political Rights,' 57-60.
103 Skelton to Starnes, 17 September 1925, LAC, Royal Canadian Mounted Police, RG 18, vol. 3757, file G-516-37.
104 Miller, 'Political Rights,' 50.
105 Lincoln Ellsworth, *Beyond Horizons* (New York: Doran, Doubleday, 1940), 217-18.
106 Quoted in Dinwoodie, 'Arctic Controversy,' 64.

Conclusion: Canada of Itself

1 Unhappy with his pension of $2,400 a year, the indefatigable old explorer informed Prime Minister R.B. Bennett in 1933 that 'the recognition hitherto accorded to my

achievement is incompatible with its significance to generations still to come.' Bernier to Bennett, 21 November 1933, Library and Archives Canada (LAC), Department of External Affairs, RG 25, vol. 1673, file 748-33.

2 George P. Mackenzie, 'Canadian Expedition, 1926,' in J.D. Craig, F.D. Henderson, and George P. Mackenzie, *Canada's Arctic Islands* (Ottawa: F.A. Acland, 1927), 51.

3 Finnie to White, 29 December 1925, LAC, Department of Justice, RG 13, vol. 922, file 'O. S. Finnie.'

4 Minutes of meeting held 21 December 1925, ibid.

5 Minutes of meeting of the Advisory Board on Wildlife Protection, 29 April 1925, LAC, Northern Affairs Program, RG 85, vol. 145, file 400-6.

6 Finnie to Cory, 24 December 1925, LAC, RG 85, vol. 68, file 201-1. Finnie chose to charter the *Beothic* rather than one of the HBC's ships for the 1926 patrol because he did not want to add to the company's prestige in the eyes of the Inuit.

7 Finnie to Rasmussen, 14 February 1928, LAC, RG 85, vol. 584, file 573.

8 PC 1146, 19 July 1926, published in the *Canada Gazette,* 31 July 1926, 382-83.

9 Finnie to Rasmussen, 23 August 1928, LAC, RG 85, vol. 584, file 573.

10 Translation of an article in *Krassnoe Znami,* enclosed in W.G. Ormsby Gore to Byng, 25 March 1925, LAC, MG 26 J, William Lyon Mackenzie King papers (WLMK), J1, vol. 123, file '1925, Secretary of State for the Colonies.' The two dates given for the Washington Convention reflect the difference between the western and Russian calendars.

11 Translation of decree, enclosed in Amery to Byng, 11 August 1926, LAC, RG 25, vol. 2667, file 9057-B-40. See also W. Lakhtine, 'Rights over the Arctic,' *American Journal of International Law* 24, 4 (October 1930): 703-17, and Leonid Timtchenko, 'The Russian Arctic Sectoral Concept: Past and Present,' *Arctic* 50, 1 (March 1997): 29-35.

12 See William Barr, 'Eskimo Relocation: The Soviet Experience on Ostrov Vrangelya,' *Musk-Ox* 20 (1977): 9-20.

13 Christie to Newton Rowell, 22 January 1926, LAC, RG 25, vol. 1281, file 1576-20.

14 Ludwig Aubert to Skelton, 26 March 1928, in Alex I. Inglis, ed., *Documents on Canadian External Relations,* vol. 4, *1926-1930* (Ottawa: Department of External Affairs, 1971), doc. 902.

15 Carl Hambro, quoted in Thorleif Tobias Thorleifsson, 'Norway "Must Really Drop Their Absurd Claims Such As That to the Otto Sverdrup Islands." Bi-Polar International Diplomacy: The Sverdrup Islands Question, 1902-1930' (master's thesis, Simon Fraser University, 2006), 64. This thesis offers a good overview of the issue, and it makes excellent use of Norwegian primary sources. However, it is flawed by lack of research in Canadian archives. For the Canadian perspective, the author relies solely on the documents published in the 1926-30 volume of *Documents on Canadian External Relations.* As a result, he gives too much credit to Skelton, largely ignoring the essential preliminary work done by other members of the Department of External Affairs and by the Northern Advisory Board.

16 Ferguson to King, 9 May 1929, LAC, RG 25, vol. 2960, file 1. Ferguson was subsequently appointed high commissioner in London.

17 L.B. Pearson, 'The Question of Ownership of the Sverdrup Islands,' 28 October 1929, LAC, RG 25, vol. 2667, file 9057-A-40. Significantly, neither Pearson nor anyone else involved in the negotiations cited Joseph Bernier's 1 July 1909 declaration as a possible basis for Canadian sovereignty over the Sverdrup Islands. By 'the sector claim,' they meant the announcement made by Stewart in 1925.

18 *Globe* (Toronto), 21 November 1929, 5.
19 *Canada Treaty Series,* 1930, No. 17.
20 'Canada Gets Arctic Islands: Makes History in Process,' *Toronto Star Weekly,* 22 November 1930.
21 Permanent Court of International Justice, Series A/B (Collection of Judgments, Orders and Advisory Opinions), No. 53, *Legal Status of Eastern Greenland* (Leyden: A.W. Sijothoff, 1933), 50-51, 46.
22 Gillian Triggs, *International Law and Australian Sovereignty in Antarctica* (Sydney: Legal Books, 1986), 30. See also C.H.M. Waldock, 'Disputed Sovereignty in the Falkland Islands Dependencies,' *British Yearbook of International Law* (1948): 315-16, 334-37, and Donald Rothwell, *The Polar Regions and the Development of International Law* (Cambridge: Cambridge University Press, 1996), 59-62. For a contemporary analysis of the significance of the Eastern Greenland decision, see Friedrich August Freiherr von der Heydte, 'Discovery, Symbolic Annexation and Virtual Effectiveness in International Law,' *American Journal of International Law* 29, 3 (July 1935): 448-71.
23 Permanent Court of International Justice, Series C (Pleadings, Oral Statements and Documents), No. 64, *Legal Status of Eastern Greenland* (Leyden: A.W. Sijothoff, 1933), 1865.
24 Thomas M. Tynan, 'Canadian-American Relations in the Arctic: The Effect of Environmental Influences upon Territorial Claims,' *Review of Politics* 41, 3 (July 1979): 407-8, citing a memo from Sumner Wells to Roosevelt, 6 January 1939. The policy was reconsidered at this time because Richard Byrd was urging the American government to make a territorial claim in the Antarctic. See Jason Kendall Moore, 'Bungled Publicity: Little America, Big America, and the Rationale for Non-Claimancy, 1946-61,' *Polar Record* 40, 212 (January 2004): 19.
25 United States Army Air Forces Study on Problems of Joint Defense in the Arctic, 29 October 1946, printed in Shelagh Grant, *Sovereignty or Security? Government Policy in the Canadian North, 1936-1950* (Vancouver: UBC Press, 1988), Appendix G, 308. Grant does not seem to have grasped the importance of these statements, perhaps because of her entirely inadequate understanding of the Eastern Greenland decision and its consequences for international law. She mistakenly states that in 1931, the International Court of Justice 'ruled that occupation superseded claims of discovery or contiguity.' Ibid., 315n31. Not only is Grant mistaken about the nature of the decision, she also makes important factual errors. The decision was made in 1933, not 1931, by the Permanent Court of International Justice, not the International Court of Justice. The International Court of Justice did not then exist.

In her most recent book, *Polar Imperative,* Grant states that the court (this time she calls it the Permanent International Court of Justice) awarded eastern Greenland to Denmark on the basis of contiguity. Again, she misses the true import of the decision. Shelagh Grant, *Polar Imperative: A History of Arctic Sovereignty in North America* (Vancouver: Douglas and McIntyre, 2010), 241.

26 N.D. Bankes, 'Forty Years of Canadian Sovereignty Assertion in the Arctic, 1947-87,' *Arctic* 40, 4 (December 1987): 287.
27 Rolf Gilberg, 'Thule,' *Arctic* 29, 2 (June 1976): 86.
28 In August 2007 President George W. Bush stated that 'the United States does not question Canada's sovereignty over its Arctic islands.' However, he contended that the Northwest Passage is an international strait. See 'Bush Agrees Arctic Is Ours, But Waterway International,' *Hamilton Spectator,* 22 August 2007, A3. Many news stories

emphasized the disagreement over the Northwest Passage rather than the agreement over the islands. For example, see 'The U.S. Interest in Canada's Claim,' *Globe and Mail,* 22 August 2007, A14.

29 See Rob Huebert, 'The Return of the Vikings: New Challenges for the Control of the Canadian North,' Naval Officers' Association of Canada, http://www.noac-national.ca/article/Heubert/The_Return_of_the_Vikings.html; Rob Huebert, 'Return of the Vikings,' *Globe and Mail,* 28 December 2002, A17; Dianne DeMille, 'Denmark "Goes Viking" in Canada's Arctic Islands – Strategic Resources of the High Arctic Entice the Danes,' Canadian American Strategic Review, http://www.sfu.ca.casr/id-arcticviking1.htm; Ken Coates and Bill Morrison, 'Saving our Sovereignty,' *National Post,* 23 August 2006, A13.

30 If the Northwest Passage is deemed to be internal waters, then the Canadian government can exercise the same degree of control over the passage as it does over the mainland. If the northern waters are classified as an international strait, Ottawa cannot deny foreign ships the right of innocent passage. For an excellent overview of the issues, see Donald McRae, 'Arctic Sovereignty? What Is at Stake?' *Behind the Headlines* 64, 1 (January 2007).

31 See John Sandlos, 'Where the Reindeer and Inuit Should Play: Animal Husbandry and Ecological Imperialism in Canada's North' (paper presented at the Canadian Historical Association annual meeting, University of Saskatchewan, May 2007).

32 Craig to Finnie, 9 May 1924, and Cory to Finnie, 12 May 1924, LAC, RG 85, vol. 60, file 160-9.

33 See Craig to McKeand, 25 July 1925, and Finnie to Starnes, 3 October 1925, LAC, RG 85, vol. 759, file 4834. Steele did manage to publish a series of articles in the *Montreal Star* after his return from the North. Steele to Richard Finnie, 1 September 1975, LAC, MG 31 C6, Richard Sterling Finnie papers (RSF), vol. 8, file 12.

34 John O. LaGorce to White, 4 January 1926, LAC, RG 13, vol. 930, file 6298.

35 'The Canadian Government Arctic Expedition of 1926,' LAC, MG 30 E529, George Patton Mackenzie papers (GPM), vol. 1, file 'Eastern Arctic Expedition, 1926.'

36 Peter Geller, *Northern Exposures: Photographing and Filming the Canadian North, 1920-45* (Vancouver: UBC Press, 2004), 44.

37 Craig to Finnie, 8 May 1924, LAC, RG 85, vol. 60, file 160-9.

38 Camsell to Canadian Press, 9 September 1926, LAC, Geological Survey of Canada, RG 45, vol. 62, file 3483N3.

39 'Ottawa Official Conquers Wastes of Baffin Land' (nine photographs) and 'First White Man to Survive Trip across Baffin Land,' *Toronto Star,* 11 October 1926, 19.

40 See A.Y. Jackson, foreword to *Livingstone of the Arctic,* by Dudley Copland (Lancaster, ON: Canadian Century Publishers, 1978 [1967]).

41 Brown to Stewart, 30 January 1903, LAC, RG 85, vol. 12, file 20 – A.Y. Jackson.

42 For example, see C.R. Greenaway, 'Canada's Arctic Declared Rich in Wealth: Painter Says Canada Does Not Appreciate Arctic Possessions,' *Toronto Star,* 10 September 1927, 1, 15, and 'Arctic Fine Material for Canadian Artist,' *Toronto Star,* 17 September 1927, 27. The *Star* also ran a lengthy feature about Joy's patrol to the Sverdrup Islands. Thomas Wayling, 'Thirteen Hundred-Mile Fight against Snowstorm, Blindness, Fogs, Starvation Is Routine Report of R.C.M.P. Officer,' *Toronto Star,* 10 September 1927, 14.

43 A.Y. Jackson, 'Up North,' *Canadian Forum* 8, 87 (December 1927): 478-80. Jackson's diary of the trip was later published. See *The Arctic 1927* (Manotick, ON: Penumbra Press, 1982).

44 Banting to Finnie, 3 June 1930, LAC, RG 85, vol. 69, file 201-1.
45 Finnie to Jackson, 31 December 1930, LAC, RG 85, vol. 12, file 20 – A.Y. Jackson.
46 Peter Geller's *Northern Exposures* provides an excellent account of the production of photographs on the northern patrols; however, his analysis of the dissemination of these images is less satisfactory. Geller discusses only the use of photographs in the official reports, which were read by a small group of Canadians with a particular interest in the Far North. Periodicals such as the *Canadian Geographical Journal*, on the other hand, brought the photographs to a wide audience.
47 Rasmussen to Finnie, undated [November 1927], and Finnie to Rasmussen, 14 February 1928, LAC, RG 85, vol. 584, file 573. On Finnie's goals with regard to the Inuit, see also Diamond Jenness, *Eskimo Administration,* part 2, *Canada,* Arctic Institute of North America Technical Paper No. 14 (May 1964), 30-32.
48 Richard Finnie, 'O.S. Finnie: A Pioneer Northern Administrator,' unpublished manuscript, LAC, RSF, vol. 6, file 12.
49 See the comments quoted in Grant, *Sovereignty or Security?* 27. On Gibson and northern administration, see Morris Zaslow, *The Northward Expansion of Canada, 1914-1967* (Toronto: McClelland and Stewart, 1988), 300-1, and Grant, *Sovereignty and Security?* 28-30. On his administration of the national parks system, see W.F. Lothian, *A History of Canada's National Parks, Vol. 2* (Ottawa: Parks Canada, 1977), 19-20.
50 See C.J. Taylor, *Negotiating the Past: The Making of Canada's National Historic Parks and Sites* (Montreal/Kingston: McGill-Queen's University Press, 1990), 110. According to one of his former subordinates, W.F. Lothian, Harkin had long felt that his career was stagnating. He was on poor terms with Conservative prime minister R.B. Bennett. Not only did Bennett pass him over for promotion to assistant deputy minister, he also 'regularly telephoned Harkin to request his resignation.' Ibid., 223n12.
51 Prentice G. Downes, 'Prentice G. Downes's Eastern Arctic Journal, 1936,' *Arctic* 36, 3 (September 1983): 235, 245. McKeand did at least have a certain flair for publicity. He was fond of giving talks on the Far North, and in November 1933 he made a national radio broadcast in which he told the story of the previous summer's patrol, suggesting to listeners 'that, in your imagination, you stand with us of the Government party, on the deck of the Hudson's Bay Company icebreaker "Nascopie."' LAC, RG 85, vol. 55, file 160-1-1.
52 Mackenzie to Finnie, 26 February 1931, and Finnie to Gibson, 3 March 1931, LAC, RG 85, vol. 70, file 201-1. The Bache post was visited by sledge patrols throughout the 1930s; for example, when Danish scientists passed by the post in 1940, they found a note recording a Canadian visit in 1939. Notes by Christian Vibe on the Danish Thule and Ellesmere Island Expedition, 1939-20, 19 April 1945, LAC, RG 85, vol. 916, file 00136.
53 'Krüger Search Expeditions, 1932,' *Polar Record* 8 (July 1934): 127, 129. See also William Barr, 'The Career and Disappearance of Hans K.E. Krüger, Arctic Geologist, 1886-1930,' *Polar Record* 29, 171 (October 1993): 277-304, and *Red Serge and Polar Bear Pants: The Biography of Harry Stallworthy, RCMP* (Edmonton: University of Alberta Press, 2004), 141-63.
54 John Q. Adams, 'Settlements of the Northeastern Canadian Arctic,' *Geographical Review* 31, 1 (January 1941): 124, 112.
55 L.B. Pearson, 'Canada Looks "Down North,"' *Foreign Affairs* 24, 4 (July 1946): 638, 641.
56 Charles Camsell, 'The New North,' *Canadian Geographical Journal* 33, 6 (December 1946): 265.
57 Hugh L. Keenleyside, *Memoirs of Hugh L. Keenleyside,* vol. 2, *On the Bridge of Time* (Toronto: McClelland and Stewart, 1982), 269, 280-82.

58 Hugh L. Keenleyside, 'Recent Developments in the Canadian North,' *Canadian Geographical Journal* 39, 4 (October 1949): 171.

59 A few years later, the question of northern sovereignty was raised in the House of Commons. The minister of northern affairs and natural resources, Jean Lesage, was initially under the impression that Canada 'took Ellesmere Island over when the Norwegians renounced their claim in 1923.' Herbert W. Herridge to Lorris Borden, 27 January 1956, LAC, MG 30 B46, Lorris Elijah Borden papers (LEB), vol. 1, file 'Correspondence, 1954-59.'

60 Both Douglas Robertson and John Adams followed Miller's account, though they may have received it from government officials rather than from the *Foreign Affairs* article. This version of events had obviously become common currency in Ottawa by the 1930s. See Douglas Robertson, *To the Arctic with the Mounties* (Toronto: Macmillan, 1934), 285-86, and John Adams, 'Settlements of the Northeastern Canadian Arctic,' 114. Adams listed the officials with whom he spoke or corresponded in a note on 112.

61 In 1970, after the *Manhattan* voyage had roused great public interest in Arctic sovereignty, Mabel Williams wrote to Jean Chrétien, then the minister of Indian affairs and northern development, about Harkin's important role during the 1920s. It is evident from Chrétien's reply that Williams had suggested the Harkin papers as a useful source on the history of Canada's claims. Chrétien to Williams, 11 June 1970, LAC, R12219-0-3-E, Mabel Berta Williams papers (MBW), vol. 1, file 31. Williams knew of Harkin's desire for eventual recognition. See a letter from Dorothy Barber to Williams, quoted in Alan MacEachern, *Natural Selections: National Parks in Atlantic Canada, 1935-1970* (Montreal/Kingston: McGill-Queen's University Press, 2001), 249n7.

62 D.M. LeBourdais, 'Everything Points North,' *Canadian Forum* 25, 300 (January 1946): 240-41.

63 Richard Finnie to Robert Logan, 13 May 1972, LAC, RSF, vol. 10, file 13.

64 Richard Finnie attributed the decisions made in 1931 to bureaucratic rivalry and malice rather than to economic necessity. He wrote: 'The shibboleth of economy was made the excuse for political juggling. The Northwest Territories and Yukon Branch, built up in ten years to an efficient and productive organization, just hitting its stride, was dissolved. The Director, highly respected everywhere ... was superannuated while still in his prime ... All those years of experience and accumulated knowledge were sacrificed.' Richard S. Finnie, *Canada Moves North* (New York: Macmillan, 1942), 68. Decades later, he suggested that his father had been ousted as the result of 'skullduggery' on the part of Gibson, whom he described as a 'sly politician.' Richard Finnie to Erling Porsild, 1 September 1974, Trent University Archives (TUA), Trevor Lloyd papers (TL), 87-104-10-2.

65 'Reveals Canada Once Owned Northern Island,' *Ottawa Citizen,* 27 March 1952, 2.

66 Vilhjalmur Stefansson, *Discovery: The Autobiography of Vilhjalmur Stefansson* (New York: McGraw-Hill, 1964), 229-32, 237-39.

67 For example, see David Judd, 'Canada's Northern Policy: Retrospect and Prospect,' *Polar Record* 14, 92 (May 1969): 593; Grant, *Sovereignty or Security?* xvii, 239; Lyle Dick, *Muskox Land: Ellesmere Island in the Age of Contact* (Calgary: University of Calgary Press, 2001), 304.

68 Logan to T.L. Cory, 19 June 1925, LAC, MG 30 B68, Robert Archibald Logan papers (RAL), vol. 1, file 'Licenses for Air Habours in Arctic, 1925.'

69 For a detailed discussion of the Eastern Greenland case in relation to Australia's Antarctic sector claim, see Triggs, *International Law and Australian Sovereignty,* Chapter 2. As Triggs points out, the fact that Australia 'defines the extent of its Antarctic claim

by a sector' does 'no more than put other States on notice of its intent and will to act as sovereign within the area' (95). Like Canada, Australia ultimately rests its claim on acts of occupation within the sector. Because of the Eastern Greenland decision, there is no need for occupation to be carried into every corner of the territory claimed. See J.O. Parry, 'Title by Effective Occupation,' 15 May 1954, and 'Sector Theory and Floating Ice Islands,' 30 August 1954, LAC, RG 25, vol. 6298, file 9057-40 (material declassified in 2007).

Bibliography

Manuscript Sources

Archives of Ontario, Toronto (AO)
F228, John Craig Eaton and Flora McCrae Eaton papers

Churchill College Archives, Cambridge (CCA)
AMEL 7/17, Diary of Leo Amery

Hudson's Bay Company Archives, Winnipeg (HBCA)
A.92/7/1, Correspondence, 5th Thule Expedition
A.102/1962, Rasmussen Expedition (London)
RG 2/4/63, Knud Rasmussen (Winnipeg)

Library and Archives Canada, Ottawa (LAC)
Government Departments
RG 2, Privy Council Office
RG 7, Governor General's Office
RG 10, Department of Indian Affairs
RG 12, Department of Transport
RG 13, Department of Justice
RG 15, Department of the Interior
RG 18, Royal Canadian Mounted Police
RG 21, Department of Energy, Mines and Resources
RG 25, Department of External Affairs
RG 33-105, Royal Commission on Possibilities of Reindeer and Musk-ox Industries in the Arctic and Sub-arctic Regions
RG 37, Public Archives of Canada
RG 42, Department of Marine and Fisheries
RG 45, Geological Survey of Canada
RG 84, Canadian Parks Service
RG 85, Northern Affairs Program
RG 88, Surveys and Mapping Branch

Personal Papers
MG 30 B40, Rudolph Martin Anderson and Mae Belle Allstrand Anderson (RMA/MBAA)
MG 30 B46, Lorris Elijah Borden (LEB)

MG 26 H, Robert Laird Borden (RLB)
MG 30 B8, Charles Camsell (CC)
MG 30 B66, Kenneth Gordon Chipman (KGC)
MG 30 E44, Loring Cheney Christie (LCC)
MG 30 B57, John Davidson Craig (JDC)
MG 30 D45, John Wesley Dafoe (JWD)
MG 30 E89, George Desbarats (GD)
MG 31 C6, Richard Sterling Finnie (RSF)
MG 27 II D7, George Foster (GF)
MG 30 B129, William Harold Grant (WHG)
MG 30 E169, James Bernard Harkin (JBH)
MG 26 J, William Lyon Mackenzie King (WLMK)
MG 26 J-13, Diaries of Prime Minister William Lyon Mackenzie King.
http://www.collectionscanada.gc.ca/databases/king.
MG 30 B13, Otto Julius Klotz (OJK)
MG 26G, Wilfrid Laurier (WL)
MG 30 B68, Robert Archibald Logan (RAL)
MG 30 B33, Albert Peter Low (APL)
MG 30 E529, George Patton Mackenzie (GPM)
MG 26I, Arthur Meighen (AM)
MG 30 B63, Harris Christopher Pickels (HCP)
MG 30 E86, Joseph Pope (JP)
MG 27 II D15, Clifford Sifton (CS)
MG 30 D33, Oscar Douglas Skelton (ODS)
MG 30 B81, Vilhjalmur Stefansson (VS)
R12219-0-3-E, Mabel Berta Williams (MBW)

Census Records
Canadian Genealogy Centre. http://www.collectionscanada.gc.ca/genealogy/index-e.html.

The National Archives of the United Kingdom: Public Record Office, Kew (TNA: PRO)
FO/371/4084, 5386, 6758, 6759
Documents Online. Cabinet Papers. http://www.nationalarchives.gov.uk/cabinetpapers.

Rauner Special Collections Library, Dartmouth College, Hanover, New Hampshire (RSCL)
Stefansson Collection

Royal Geographical Society Archives, London (RGS)
RGS/CB 8/80, Correspondence with Stefansson, 1915-1920
RGS/CB 9/139, Correspondence with Stefansson, 1921-1930

Scott Polar Research Institute, Cambridge (SPRI)
MS 100/122/1-2, Stefansson letters to H.R. Mill, 1922

Special Collections and University Archives, University of British Columbia Library, Vancouver
A.J.T. Taylor Arctic Collection

Trent University Archives, Peterborough, Ontario (TUA)
87-014, Trevor Lloyd papers (TL)

University of Alberta Archives, Edmonton
J. Dewey Soper fonds, Series 3, Photographs and Sketches. http://www.ualberta.ca/ARCHIVES.

Published Primary Sources

Newspapers
Globe (Toronto)
Mail and Empire (Toronto)
Montreal Gazette
Montreal Herald
Montreal Standard
Montreal Star
New York Times
Ottawa Citizen
Ottawa Journal
The Times (London)
Toronto Star
Toronto Telegram
Washington Post

Books and Articles
Amery, L.S. 'Hudson Bay: Its Conditions and Problems.' In J. Castell Hopkins, ed., *Empire Club Speeches 1910-1911,* 27-36. Toronto: Saturday Night Press, n.d.
'Arbitral Award on the Subject of the Difference Relative to the Sovereignty over Clipperton Island.' *American Journal of International Law* 26, 2 (April 1932): 390-94.
Bartlett, Robert. *The Last Voyage of the* Karluk. Boston: Small, Maynard, 1916.
Bernacchi, Louis. *The Saga of the 'Discovery.'* London: Blackie, 1938.
Bernier, Joseph E. 'Canadian Rights in the Arctic.' In Gerald H. Brown, ed., *Addresses Delivered before the Canadian Club of Ottawa (1909),* 190-92 (Ottawa: Mortimer Press, n.d.).
–. *Report on the Dominion of Canada Government Expedition to the Arctic Islands and Hudson Strait on Board the D.G.S. 'Arctic.'* Ottawa: Government Printing Bureau, 1910.
–. *Report on the Dominion Government Expedition to the Arctic Islands and Hudson Strait on Board the C.G.S. 'Arctic,' 1906-1907.* Ottawa: C.H. Parmalee, 1909.
Boas, Franz. 'The Configuration of Grinnell Land and Ellesmere Land.' *Science* 5, 108 (27 February 1885): 170-71.
Borden, Robert. Introduction to *The Friendly Arctic,* by Vilhajalmur Stefansson, 2nd ed., xxvii-xxxi. New York: Macmillan, 1943.
Burwash, L.T. *Canada's Western Arctic: Report on Investigations in 1925-26, 1928-29, and 1930.* Ottawa: F.A. Acland, 1931.

Byrd, Richard E. *To the Pole: The Diary and Notebook of Richard E. Byrd, 1925-1927.* Edited by Raimund E. Goerler. Columbus: Ohio State University Press, 1998.

Camsell, Charles. 'The Friendly Arctic' (letter). *Science* n.s. 57, 1484 (8 June 1923): 665-66.

–. 'The New North.' *Canadian Geographical Journal* 33, 6 (December 1946): 265-77.

Canada. Royal Commission on Possibilities of Reindeer and Musk-Ox Industries in the Arctic and Sub-Arctic Regions. *Report of the Royal Commission Appointed by Order-in-Council of Date May 20, 1919, to Investigate the Possibilities of the Reindeer and Musk-Ox Industries in the Arctic and Sub-Arctic Regions of Canada.* Ottawa: The Commission, 1922.

Chafe, Ernest F. 'The Voyage of the "Karluk" and its Tragic Ending.' *Geographical Journal* 51, 5 (May 1918): 307-16.

Clark, Lovell C., ed. *Documents on Canadian External Relations.* Vol. 3, *1919-1925.* Ottawa: Department of External Affairs, 1970.

Cook, Frederick A. *My Attainment of the Pole.* New York: Polar Publishing, 1911.

Craig, J.D. 'Canadian Expedition, 1922.' In J.D. Craig, F.D. Henderson, and George P. Mackenzie, *Canada's Arctic Islands,* 5-12. Ottawa: F.A. Acland, 1927.

Davis, C.H. *Narrative of the North Polar Expedition, U.S. Ship Polaris, Captain Charles Francis Hall Commanding.* Washington, DC: Government Printing Office, 1876.

Davis, John King. *High Latitude.* Parkville: Melbourne University Press, 1962.

Documents on Canadian External Relations. Vol. 1, *1909-1918.* Ottawa: Department of External Affairs, 1967.

Downes, Prentice G. 'Prentice G. Downes's Eastern Arctic Journal, 1936.' *Arctic* 36, 3 (September 1983): 232-50.

Ekblaw, Elmer. 'A Recent Eskimo Migration and Its Forerunner.' *Geographical Review* 9, 2 (February 1920): 141-44.

Eliot, T.S. *The Complete Poems and Plays, 1909-1950.* New York: Harcourt, Brace and World, 1971.

Ellsworth, Lincoln. *Beyond Horizons.* New York: Doran, Doubleday, 1940.

Finnie, Richard S. *Canada Moves North.* New York: Macmillan, 1942.

–. 'Farewell Voyages: Bernier and the "Arctic."' *Beaver* 54, 1 (Summer 1974): 44-54.

–. 'Stefansson as I Knew Him.' *North/Nord* 25, 3 (May-June 1978): 36-43, and 25, 4 (July-August 1978): 12-19.

–. 'Stefansson's Unsolved Mystery.' *North/Nord* 25, 6 (November-December 1978): 2-7.

Foreign Relations of the United States: 1925. Vol. 1. Washington, DC: United States Government Printing Office, 1940.

Foreign Relations of the United States: The Lansing Papers, 1914-1920. Vol. 2. Washington, DC: United States Government Printing Office, 1940.

Freuchen, Peter. *Arctic Adventure: My Life in the Frozen North.* New York: Farrar and Rinehart, 1935.

–. *Vagrant Viking: My Life and Adventures.* New York: Julian Messner, 1953.

Further Papers Relative to the Recent Arctic Expeditions in Search of Sir John Franklin and the Crews of H.M.S. 'Erebus' and 'Terror.' London: Eyre and Spottiswoode, 1855.

'The Geographical Work of the Canadian Arctic Expedition.' *Geographical Journal* 63, 6 (June 1924): 508-25.

'The Geographical Work of the Canadian Arctic Expedition.' *Geographical Journal* 65, 4 (April 1925): 340-42.

Grant, W.L. 'Geographical Conditions Affecting the Development of Canada.' *Geographical Journal* 38, 4 (October 1911): 362-81.

Greely, Adolphus W. *Report of the Proceedings of the United States Expedition to Lady Franklin Bay, Grinnell Land.* 2 vols. Washington, DC: Government Printing Office, 1888.

Hall, William Edward. *A Treatise on International Law.* 4th ed. Oxford: Clarendon Press, 1895.

–. *A Treatise on International Law.* 7th ed. Edited by A. Pearce Higgins. Oxford: Clarendon, 1917.

Harkin, J.B. *The Origin and Meaning of the National Parks of Canada: Extracts from the Papers of the Late Jas. B. Harkin, First Commissioner of the National Parks of Canada.* Saskatoon: H.R. Larson, 1957.

Henderson, F.D. 'Canadian Expedition, 1924.' In J.D. Craig, F.D. Henderson, and George P. Mackenzie, *Canada's Arctic Islands,* 29-41. Ottawa: F.A. Acland, 1927.

Hewitt, C. Gordon. *The Conservation of the Wildlife of Canada.* New York: Scribner's, 1921.

[Hinks, Arthur]. 'Wrangel Island.' *Geographical Journal* 62, 2 (December 1923): 440-44.

Hooper, C.L. *Report of the Cruise of the U.S. Revenue Steamer Thomas Corwin in the Arctic Ocean, 1881.* Washington, DC: Government Printing Office, 1884.

Inglis, Alex I., ed. *Documents on Canadian External Relations.* Vol. 4, *1926-1930.* Ottawa: Department of External Affairs, 1971.

Jackson, A.Y. *The Arctic 1927.* Manotick, ON: Penumbra Press, 1982.

–. Foreword to *Livingstone of the Arctic,* by Dudley Copland. Lancaster, ON: Canadian Century Publishers, 1978 [1967].

–. 'Up North.' *Canadian Forum* 8, 87 (December 1927): 478-80.

Jenness, Diamond. *Arctic Odyssey: The Diary of Diamond Jenness, 1913-1916.* Edited by Stuart E. Jenness. Ottawa: Canadian Museum of Civilization, 1991.

–. *Eskimo Administration.* Part 2, *Canada.* Arctic Institute of North America Technical Paper No. 14, May 1964.

–. 'The Friendly Arctic' (letter). *Science* n.s. 56, 2436 (7 July 1922): 8-12.

Keenleyside, Hugh L. *Memoirs of Hugh L. Keenleyside.* Vol. 2, *On the Bridge of Time.* Toronto: McClelland and Stewart, 1982.

–. 'Recent Developments in the Canadian North.' *Canadian Geographical Journal* 39, 4 (October 1949): 156-76.

Kennet, Kathleen. *Self-Portrait of an Artist: From the Diaries and Memoirs of Lady Kennet (Kathleen, Lady Scott).* London: John Murray, 1949.

King, W.F. *Report upon the Title of Canada to the Islands North of the Mainland of Canada.* Ottawa: Government Printing Bureau, 1905.

Kipling, Rudyard. 'The Verdict of Equals.' In *A Book of Words,* 71-74. London: Macmillan, 1928.

Koch, Lauge. 'North of Greenland.' *Geographical Journal* 64, 1 (July 1924): 6-21.

–. 'Report on the Danish Bicentenary Jubilee Expedition North of Greenland 1920-23.' *Meddelelser om Grønland* 70 (1927).

Langton, H.H. Review of *The Friendly Arctic,* by Vilhjalmur Stefansson. *Canadian Historical Review* 3, 1 (March 1922): 86-91.

LeBourdais, Donat. 'Staking Wrangel Island.' In Vilhjalmur Stefansson, *The Adventure of Wrangel Island,* 395-410. London: Jonathan Cape, 1926.

Lee, Herbert Patrick. *Policing the Top of the World.* Toronto: McClelland and Stewart, 1928.

Low, A.P. *Report on the Dominion Government Expedition to Hudson Bay and the Arctic Islands on Board the D.G.S. Neptune, 1903-04.* Ottawa: Government Printing Bureau, 1906.

Mackenzie, George P. 'Canadian Expedition, 1925.' In J.D. Craig, F.D. Henderson, and George P. Mackenzie, *Canada's Arctic Islands,* 43-48. Ottawa: F.A. Acland, 1927.
–. 'Canadian Expedition, 1926.' In J.D. Craig, F.D. Henderson, and George P. Mackenzie, *Canada's Arctic Islands,* 49-54. Ottawa: F.A. Acland, 1927.
MacMillan, Donald B. 'Food Supply of the Smith Sound Eskimos: A Story of Primitive Life Maintained on the Natural Resources of a Frozen Arctic Land and Sea.' *American Museum of Natural History Journal* 18, 3 (1918): 161-76.
–. *Four Years in the White North.* New York: Harper, 1918.
Mathiassen, Therkel. 'Knud Rasmussen's Sledge Expeditions and the Founding of the Thule Trading Station.' *Geografisk Tidsskrift* 37, 1-2 (1934): 16-30.
–. *Report on the Expedition.* Copenhagen: Gyldendalske Boghandel, Nordisk Forlag, 1945.
McKinlay, William Laird. *Karluk: The Untold Story of Arctic Exploration.* London: Weidenfeld and Nicholson, 1976.
Nares, G.S. *Narrative of a Voyage to the Polar Sea during 1875-6 in H. M. Ships 'Alert' and 'Discovery.'* 2 vols. London: Sampson Low, Marston, Searle, and Rivington, 1878.
Oppenheim, L. *International Law: A Treatise.* 2 vols. 3rd ed. Edited by Ronald F. Roxburgh. London: Longmans, Green, 1920.
Ostermann, H., and E. Holtved. *The Alaskan Eskimos: As Described in the Posthumous Notes of Dr. Knud Rasmussen.* Report of the Fifth Thule Expedition 1921-24, Vol. 10, No. 3. Copenhagen: Gyldendalske Boghandel, Nordisk Forlag, 1952.
Pearl, Raymond. Review of *The Friendly* Arctic, by Vilhjalmur Stefansson. *Science* n.s. 55, 1421 (24 March 1922): 320-21.
Peary, Robert E. *The North Pole.* New York: Frederick A. Stokes, 1910.
–. *Northward Over the 'Great Ice': A Narrative of Life and Work along the Shores and upon the Interior Ice-Cap of Northern Greenland in the Years 1886 and 1891-1897.* 2 Vols. New York: Frederick A. Stokes, 1898.
–. 'Report of R.E. Peary, C.E., U.S.N., on Work Done in the Arctic in 1898-1902.' *Bulletin of the American Geographical Society* 35, 5 (1903): 496-534.
Permanent Court of International Justice. Series A/B (Collection of Judgments, Orders and Advisory Opinions), No. 53, *Legal Status of Eastern Greenland.* Leyden: A.W. Sijothoff, 1933.
–. Series C (Pleadings, Oral Statements and Documents), Nos. 62-67, *Legal Status of Eastern Greenland.* Leyden: A.W. Sijothoff, 1933.
Rasmussen, Knud. *Across Arctic America: Narrative of the Fifth Thule Expedition.* New York: G.P. Putnam, 1927.
–. 'The Fifth Thule Expedition, 1921-24. The Danish Ethnographical and Geographical Expedition from Greenland to the Pacific.' *Geographical Journal* 67, 2 (February 1926): 123-38.
–. *Greenland by the Polar Sea: The Story of the Thule Expedition from Melville Bay to Cape Morris Jesup.* Translated by Asta and Rowland Kenney. London: Heinemann, 1921.
–. *The Netsilik Eskimos: Social Life and Spiritual Culture.* Report of the Fifth Thule Expedition, 1921-24, Vol. 8, Nos. 1-2. Copenhagen: Gyldendalske Boghandel, Nordisk Forlag, 1931.
–. *People of the Polar North.* London: K. Paul, Trench, Trübner, 1908.
–. 'Project of a Danish Expedition to the Central Eskimo.' *Geographical Journal* 35, 3 (March 1910): 295-99.
Robertson, Douglas. *To the Arctic with the Mounties.* Toronto: Macmillan, 1934.
Scott, F.R. *The Eye of the Needle: Satire, Sorties, Sundries.* Montreal: Contact Press, 1957.
Shackleton, Ernest. *South.* London: Heinemann, 1919.

Seeman, Berthold. *Narrative of the Voyage of H.M.S. Herald during the Years 1845-51, under the Command of Captain Henry Kellett, R.N., C.B., being a Circumnavigation of the Globe, and Three Cruizes to the Arctic Regions in Search of Sir John Franklin.* 2 vols. London: Reeve, 1853.

Soper, J. Dewey. *Canadian Arctic Recollections: Baffin Island, 1923-1931.* Edited by Shirley Milligan. Saskatoon: Institute for Northern Studies, University of Saskatchewan, 1981.

Skinner, Constance L. 'Stefansson, Sentinel of Civilization.' *Canadian Forum* 7, 74 (October 1926): 398-401.

Starokadomskiy, L.M. *Charting the Russian Northern Sea Route: The Arctic Ocean Hydrographic Expedition, 1910-1915.* Translated and edited by William Barr. London/ Montreal: Arctic Institute of North America/McGill-Queen's University Press, 1976.

Steensby, H.P. 'The Polar Eskimos and the Polar Expeditions.' *Fortnightly Review,* n.s., 86, 515 (November 1909): 891-902.

Stefansson, Vilhjalmur. *The Adventure of Wrangel Island.* New York: Macmillan, 1925.

–. 'The Canadian Arctic Region.' In *Empire Club of Canada: Addresses Delivered to the Members during the Sessions, 1917-18, and May to December, 1918,* 364-79. Toronto: Warwick Bros. and Rutter, 1919.

–. *Discovery: The Autobiography of Vilhjalmur Stefansson.* New York: McGraw-Hill, 1964.

–. *The Friendly Arctic.* 2nd ed. New York: Macmillan, 1943.

–. 'The History and Importance of Wrangell Island.' *Spectator* 4934 (9 June 1923): 958-59 and 4935 (16 June 1923): 998-1000.

–. 'Letter from Mr. Stefánsson.' *Geographical Journal* 52, 4 (October 1918): 248-55.

–. 'Mr. Stefánsson's Expedition' (letter). *Geographical Journal* 46, 6 (December 1915): 457-59.

–. *The Northward Course of Empire.* New York: Harcourt, Brace, 1922.

–. 'The Region of Maximum Inaccessibility in the Arctic.' *Geographical Review* 10, 3 (September 1920): 169-72.

–. 'Wrangell Island' (letter to the editor). *Spectator* 4937 (30 June 1923): 1079.

–. *Writing on Ice: The Ethnographic Notebooks of Vilhjalmur Stefansson.* Edited by Gísli Pálsson. Hanover: University Press of New England, 2001.

Stefansson, Vilhjalmur, Burt M. McConnell, and Harold Noice. 'The Friendly Arctic' (letters). *Science* n.s. 57, 1474 (30 March 1923): 368-73.

Sverdrup, Otto. *New Land: Four Years in the Arctic Regions.* 2 vols. London: Longmans, Green, 1901.

Treaties and Other International Agreements of the United States of America, 1776-1949. Vol. 7, *Denmark-France.* Washington, DC: United States Government Printing Office, 1971.

'The Un-friendly Arctic.' *Saturday Night,* 15 September 1923, 1.

Vattel, E. de. *The Law of Nations or the Principles of Natural Law, Applied to the Conduct and to the Affairs of Nations and of Sovereigns.* Translated by Charles G. Fenwick. 3 vols. Washington, DC: Carnegie Institution of Washington, 1916 [1758].

von Wrangell, Ferdinand. *Narrative of an Expedition to the Polar Sea in the Years 1820, 1821, 1822 and 1823.* Edited by Edward Sabine. 2nd ed. London: James Madden, 1844.

Wakeham, William. *Report of the Expedition to Hudson Bay and Cumberland Gulf in the Steamship 'Diana.'* Ottawa: S.E. Dawson, 1898.

White, James. 'Place-Names – Northern Canada.' *Ninth Report of the Geographic Board of Canada,* Part IV. Ottawa: King's Printer, 1910.

Whitney, Harry. *Hunting with the Eskimos.* New York: Century, 1910.

Wild, Frank. 'The Voyage of the "Quest."' *Geographical Journal* 61, 2 (February 1923): 73-97.

Secondary Sources

Adams, John Q. 'Settlements of the Northeastern Canadian Arctic.' *Geographical Review* 31, 1 (January 1941): 112-26.

Amor, Norman. *Beyond the Arctic Circle: Materials on Arctic Explorations and Travels since 1750 in the Special Collections and University Archives Division of the University of British Columbia Library.* University of British Columbia Occasional Publication No. 1. Vancouver, 1992.

Anghie, Antony. *Imperialism, Sovereignty, and the Making of International Law.* Cambridge: Cambridge University Press, 2004.

Balch, Thomas Willing. 'The Arctic and Antarctic Regions and the Law of Nations.' *American Journal of International Law* 4, 2 (April 1910): 265-75.

Bankes, N.D. 'Forty Years of Canadian Sovereignty Assertion in the Arctic, 1947-87.' *Arctic* 40, 4 (December 1987): 285-91.

Barman, Jean. *Constance Lindsay Skinner: Writing on the Frontier.* Toronto: University of Toronto Press, 2002.

–. '"I walk my own track in life & no mere male can bump me off it": Constance Lindsay Skinner and the Work of History.' In Alison Prentice and Beverly Boutilier, eds., *Creating Historical Memory: English-Canadian Women and the Work of History,* 129-63. Vancouver: UBC Press, 1997.

Barr, William. *Back from the Brink: The Road to Muskox Conservation in the Northwest Territories.* Calgary: Arctic Institute of North America, 1991.

–. 'The Career and Disappearance of Hans K.E. Krüger, Arctic Geologist, 1886-1930.' *Polar Record* 29, 171 (October 1993): 277-304.

–. 'Eskimo Relocation: The Soviet Experience on Ostrov Vrangelya.' *Musk-Ox* 20 (1977): 9-20.

–. *Red Serge and Polar Bear Pants: The Biography of Harry Stallworthy, RCMP.* Edmonton: University of Alberta Press, 2004.

–. 'The Voyages of *Taymyr* and *Vaygach* to Ostrov Vrangelya, 1910-15.' *Polar Record* 16, 101 (May 1972): 213-34.

Barr, William, Reinhard Krause, and Peter-Michael Pawlik. 'Chukchi Sea, Southern Ocean, Kara Sea: The Polar Voyages of Captain Eduard Dallmann, Whaler, Trader, Explorer 1830-96.' *Polar Record* 40, 212 (January 2004): 1-18.

Berlin, Knud. *Denmark's Right to Greenland: A Survey of the Past and Present Status of Greenland, Iceland and the Faroe Islands in Relation to Norway and Denmark.* Translated by P.T. Federspiel. London/Copenhagen: Oxford University Press/Arnold Busck, 1932.

Billman, Christine W. 'Jack Craig and the Alaska Boundary Survey.' *Beaver* 51, 2 (Autumn 1971): 44-49.

Birket-Smith, Kai. 'The Greenlanders of the Present Day.' In M. Vahl, G.C. Amdrup, L. Bobé, and A.S. Jensen, eds., *Greenland,* vol. 2, *The Past and Present Population of Greenland,* 1-208. Copenhagen/London: C.A. Reitzel/Humphrey Milford, 1928.

Bothwell, Robert. *Loring Christie: The Failure of Bureaucratic Imperialism.* New York: Garland, 1988.

Brøsted, Jens. 'Danish Accession to the Thule District, 1937.' *Nordic Journal of International Law* 57 (1988): 259-65.

Bryant, John H., and Harold N. Cones. *Dangerous Crossings: The First Modern Polar Expedition, 1925.* Annapolis, MD: Naval Institute Press, 2000.

Castellino, Joshua, and Steve Allen. *Title to Territory in International Law: A Temporal Analysis.* Aldershot: Ashgate, 2003.

Cavell, Janice. '"As Far as 90 North": Joseph Elzéar Bernier's 1907 and 1909 Sovereignty Claims.' *Polar Record* 46, 239 (October 2010): 372-73.
–. 'Historical Evidence and the Eastern Greenland Case.' *Arctic* 61, 4 (December 2008): 433-41.
–. 'The Second Frontier: The North in English-Canadian Historical Writing.' *Canadian Historical Review* 83, 3 (September 2002): 364-89.
Cavell, Janice, and Jeff Noakes. 'Explorer without a Country: The Question of Vilhjalmur Stefansson's Citizenship.' *Polar Record* 45, 234 (July 2009): 237-41.
–. 'The Origins of Canada's First Eastern Arctic Patrol, 1919-1922.' *Polar Record* 45, 233 (April 2009): 97-112.
Coates, Ken, P. Whitney Lackenbauer, William R. Morrison, and Greg Poelzer. *Arctic Front: Defending Canada in the Far North.* Toronto: Thomas Allen, 2008.
Coates, Ken, and Bill Morrison. 'Saving our Sovereignty.' *National Post,* 23 August 2006, A13.
Connor, Michael. *The Invention of Terra Nullius: Historical and Legal Fictions on the Foundation of Australia.* Sydney: Macleay Press, 2005.
Connor, Michael. 'The Invention of Territorium Nullius.' Australian Politics. http://australian-politics.blogspot.com/2007/08/invention-of-territorium-nullius-by.html.
Copland, Dudley. *Livingstone of the Arctic.* Foreword by A.Y. Jackson. Lancaster, ON: Canadian Century Publishers, 1978 [1967].
DeMille, Dianne. 'Denmark "Goes Viking" in Canada's Arctic Islands – Strategic Resources of the High Arctic Entice the Danes.' Canadian American Strategic Review. http://www.casr.ca/id-arcticviking1.htm.
Dick, Lyle. *Muskox Land: Ellesmere Island in the Age of Contact.* Calgary: University of Calgary Press, 2001.
Dinwoodie, D.H. 'Arctic Controversy: The 1925 Byrd-MacMillan Expedition Example.' *Canadian Historical Review* 53, 1 (March 1972): 51-65.
Diubaldo, Richard. *Stefansson and the Canadian Arctic.* Montreal: McGill-Queen's University Press, 1978.
–. 'Wrangling over Wrangel Island.' *Canadian Historical Review* 48, 3 (September 1967): 201-26.
Dorion-Robitaille, Yolande. *Captain J.E. Bernier's Contribution to Canadian Sovereignty in the Arctic.* Ottawa: Department of Indian and Northern Affairs, 1978.
Esberey, Joy E. *Knight of the Holy Spirit: A Study of William Lyon Mackenzie King.* Toronto: University of Toronto Press, 1980.
Fisher, James, and Margery Fisher. *Shackleton.* London: James Barrie, 1957.
Fitzmaurice, Andrew. 'The Great Australian History Wars.' News and Events – University of Sydney. http://www.usyd.edu.au/news.84.html?newsstoryid=948.
Fogelson, Nancy. *Arctic Exploration and International Relations, 1900-1932.* Fairbanks: University of Alaska Press, 1992.
Foster, Janet. *Working for Wildlife: The Beginnings of Preservation in Canada.* 2nd ed. Toronto: University of Toronto Press, 1998.
Geller, Peter. *Northern Exposures: Photographing and Filming the Canadian North, 1920-45.* Vancouver: UBC Press, 2004.
Gilberg, Rolf. 'Inughuit, Knud Rasmussen, and Thule.' *Études/Inuit/Studies* 12, 1-2 (1988): 45-55.
–. 'Thule.' *Arctic* 29, 2 (June 1976): 83-86.
Globe and Mail. 'The U.S. Interest in Canada's Claim.' 22 August 2007, A14.
Graham, Roger. *Arthur Meighen.* Vol. 2, *And Fortune Fled.* Toronto: Clarke, Irwin, 1963.

Granatstein, J.L. 'A Fit of Absence of Mind: Canada's National Interest in the North to 1968.' In E.J. Dosman, ed., *The Arctic in Question,* 13-33. Toronto: Oxford University Press, 1976.

Grant, Shelagh. *Arctic Justice: On Trial for Murder, Pond Inlet, 1923.* Montreal/Kingston: McGill-Queen's University Press, 2002.

–. *Polar Imperative: A History of Arctic Sovereignty in North America.* Vancouver: Douglas and McIntyre, 2010.

–. *Sovereignty or Security? Government Policy in the Canadian North, 1936-1950.* Vancouver: UBC Press, 1988.

Greer, Allan, and Ian Radforth, eds. *Colonial Leviathan: State Formation in Mid-Nineteenth-Century Canada.* Toronto: University of Toronto Press, 1992.

Hall, D.J. *Clifford Sifton.* Vol. 2, *A Lonely Eminence, 1901-1929.* Vancouver: UBC Press, 1985.

Hamilton Spectator. 'Bush Agrees Arctic Is Ours, But Waterway International.' 22 August 2007, A3.

Hanson, Earl Parker. *Stefansson, Prophet of the North.* New York: Harper, 1941.

Hart, E.J. *J.B. Harkin: Father of Canada's National Parks.* Edmonton: University of Alberta Press, 2010.

Head, Ivan. 'Canadian Claims to Territorial Sovereignty in the Arctic Regions.' *McGill Law Journal* 9, 3 (1963): 200-26.

Head, Ivan, and Pierre Trudeau. *The Canadian Way: Shaping Canada's Foreign Policy, 1968-1984.* Toronto: McClelland and Stewart, 1995.

Henderson, Gordon G. 'Policy by Default: The Origin and Fate of the Prescott Letter.' *Political Science Quarterly* 79, 1 (March 1964): 76-95.

Herbert, Wally. *The Noose of Laurels: Robert E. Peary and the Race to the North Pole.* New York: Athenaeum, 1989.

Hickford, Mark. '"Vague Native Rights to Land": British Imperial Policy on Native Title and Custom in New Zealand, 1837-53.' *Journal of Imperial and Commonwealth History* 38, 2 (June 2010): 175-206.

Hilliker, John. *Canada's Department of External Affairs.* Vol. 1, *The Early Years, 1909-1946.* Montreal/Kingston: McGill-Queen's University Press, 1990.

Hillmer, Norman, and J.L. Granatstein. *Empire to Umpire: Canada and the World into the Twenty-First Century.* Toronto: Thomson Nelson, 2008.

'The Honourable Charles Stewart, 1917-21.' Legislative Assembly of Alberta. http://www.assembly.ab.ca/lao/library/PREMIERS/stewart.htm.

Huebert, Rob. 'Return of the Vikings.' *Globe and Mail,* 28 December 2002, A17.

–. 'The Return of the Vikings: New Challenges for the Control of the Canadian North.' Naval Officers' Association of Canada. http://www.noac-national.ca/article/Huebert/The_Return_of_the_Vikings.html.

Hunt, William R. *Stef: A Biography of Vilhjalmur Stefansson, Canadian Arctic Explorer.* Vancouver: UBC Press, 1986.

Hunter, Ian. 'Natural Law, Historiography, and Aboriginal Sovereignty.' *Legal History* 11, 2 (2007): 137-67.

Huntford, Roland. *Shackleton.* London: Hodder and Stoughton, 1985.

Jessup, Eric David. 'J.E. Bernier and the Assertion of Canadian Sovereignty in the Arctic.' *American Review of Canadian Studies* 38, 4 (Winter 2008): 409-27.

Johnston, V. Kenneth. 'Canada's Title to the Arctic Islands.' *Canadian Historical Review* 14, 1 (March 1933): 24-41.

Judd, David. 'Canada's Northern Policy: Retrospect and Prospect.' *Polar Record* 14, 92 (May 1969): 593-602.

Karatani, Rieko. *Defining British Citizenship: Empire, Commonwealth, and Modern Britain.* London: Frank Cass, 2003.
Kenney, Gerard. *Ships of Wood and Men of Iron: A Norwegian-Canadian Saga of Exploration in the High Arctic.* Toronto: Dundurn Press, 2005.
'Krüger Search Expeditions, 1932.' *Polar Record* 8 (July 1934): 121-29.
Lakhtine, W. 'Rights over the Arctic.' *American Journal of International Law* 24, 4 (October 1930): 703-17.
LeBourdais, D.M. 'Everything Points North.' *Canadian Forum* 25, 300 (January 1946): 240-41.
–. *Stefansson: Ambassador of the North.* Montreal: Harvest House, 1963.
Levere, Trevor H. 'Vilhjalmur Stefansson, the Continental Shelf, and a New Arctic Continent.' *British Journal for the History of Science* 12, 2 (June 1988): 233-47.
Lindley, Mark. *The Acquisition and Government of Backward Territory in International Law.* London: Longmans, Green, 1926.
Lindow, Harald. 'Trade and Administration of Greenland.' In M. Vahl, G.C. Amdrup, L. Bobé, and A.S. Jensen, eds., *Greenland,* vol. 3, *The Colonization of Greenland and Its History until 1929,* 29-77. Copenhagen/London: C.A. Reitzel/Oxford University Press, 1929.
Lloyd, Trevor. 'Knud Rasmussen and the Arctic Islands Preserve.' *Musk-Ox* 25 (1979): 85-90.
–. 'The True Story Behind the Ellesmere Island Conspiracy.' *McGill Reporter,* 11 December 1974, 3.
Long, M.H. 'The Historic Sites and Monuments Board of Canada.' Canadian Historical Association, *Report of the Annual Meeting* 33 (1954): 1-11.
Loo, Tina. *States of Nature: Conserving Canada's Wildlife in the Twentieth Century.* Vancouver: UBC Press, 2006.
Lothian, W.F. *A History of Canada's National Parks, Vol. 2.* Ottawa: Parks Canada, 1977.
Mackay, Daniel S.C. 'James White/Canada's Chief Geographer, 1899-1909.' *Cartographica* 19, 1 (Spring 1982): 51-61.
MacEachern, Alan. 'Cool Customer: The Arctic Voyage of J.E. Bernier.' *Beaver* 84, 4 (August-September 2004): 30-35.
–. *Natural Selections: National Parks in Atlantic Canada, 1935-1970.* Montreal/Kingston: McGill-Queen's University Press, 2001.
Mackinnon, C.S. 'Canada's Eastern Arctic Patrol, 1922-68.' *Polar Record* 27, 161 (April 1991): 93-101.
MacMillan, Ken. *Sovereignty and Possession in the English New World: The Legal Foundations of Empire, 1576-1640.* Cambridge: Cambridge University Press, 2006.
Malaurie, Jean. *The Last Kings of Thule.* New York: Dutton, 1982.
Maxwell, Moreau S. *Prehistory of the Eastern Arctic.* Orlando, FL: Academic Press, 1985.
McRae, Donald. 'Arctic Sovereignty? What Is at Stake?' *Behind the Headlines* 64, 1 (January 2007).
Mill, Hugh Robert. *The Life of Sir Ernest Shackleton.* London: Heinemann, 1923.
Miller, David Hunter. 'Political Rights in the Arctic.' *Foreign Affairs* 4, 1 (October 1925): 47-60.
Moore, Jason Kendall. 'Bungled Publicity: Little America, Big America, and the Rationale for Non-Claimancy, 1946-61.' *Polar Record* 40, 212 (January 2004): 19-30.
Morrison, William R. 'Canadian Sovereignty and the Inuit of the Central and Eastern Arctic.' *Études/Inuit/Studies* 10, 1-2 (1986): 245-59.

–. *Showing the Flag: The Mounted Police and Canadian Sovereignty in the North, 1894-1925.* Vancouver: UBC Press, 1985.

Niven, Jennifer. *Ada Blackjack: A True Story of Survival in the Arctic.* New York: Hyperion, 2003.

–. *The Ice Master: The Doomed 1913 Voyage of the* Karluk. New York: Hyperion, 2000.

Osborne, Season L. 'Closing the Front Door of the Arctic: Capt. Joseph E. Bernier's Role in Canadian Arctic Sovereignty.' Master's thesis, Carleton University, 2003.

Pálsson, Gísli. Introduction to *Writing on Ice: The Ethnographic Notebooks of Vilhjalmur Stefansson,* by Vilhjalmur Stefansson and edited by Gísli Pálsson, 3-78. Hanover: University Press of New England, 2001.

–. *Travelling Passions: The Hidden Life of Vilhjalmur Stefansson.* Winnipeg: University of Manitoba Press, 2003.

Pearson, L.B. 'Canada Looks "Down North."' *Foreign Affairs* 24, 4 (July 1946): 638-47.

Pope, Maurice. *Public Servant: The Memoirs of Sir Joseph Pope, Edited and Completed by Maurice Pope.* Toronto: Oxford University Press, 1960.

Rothwell, Donald. *The Polar Regions and the Development of International Law.* Cambridge: Cambridge University Press, 1996.

Saint-Pierre, Marjolaine. *Joseph-Elzéar Bernier: Capitaine et coureur des mers, 1852-1934.* Sillery, QC: Septentrion, 2004.

Sandlos, John. *Hunters at the Margin: Native People and Wildlife Conservation in the Northwest Territories.* Vancouver: UBC Press, 2007.

–. 'Where the Reindeer and Inuit Should Play: Animal Husbandry and Ecological Imperialism in Canada's North.' Paper presented at the Canadian Historical Association annual meeting, University of Saskatchewan, May 2007.

Scott, James Brown. 'Arctic Exploration and International Law.' *American Journal of International Law* 3, 4 (October 1909): 928-41.

–, ed. *Resolutions of the Institute of International Law Dealing with the Law of Nations, with an Historical Introduction and Explanatory Notes.* New York: Oxford University Press, 1916.

Seed, Patricia. 'Taking Possession and Reading Texts: Establishing the Authority of Overseas Empires.' *William and Mary Quarterly,* 3rd ser., 49, 2 (April 1992): 183-209.

Siggins, Maggie. *Bassett.* Toronto: James Lorimer, 1979.

Simsarian, James. 'The Acquisition of Legal Title to Terra Nullius.' *Political Science Quarterly* 53, 1 (March 1938): 111-28.

Skinner, Constance L. 'Stefansson, Sentinel of Civilization.' *Canadian Forum* 7, 74 (October 1926): 398-401.

Smith, Gordon W. 'Sovereignty in the North: The Canadian Aspect of an International Problem.' In R.St.J. Macdonald, ed., *The Arctic Frontier,* 194-255. Toronto: University of Toronto Press, 1966.

–. 'The Transfer of Arctic Territories from Great Britain to Canada in 1880, and Some Related Matters, as Seen in Official Correspondence.' *Arctic* 14, 1 (March 1961): 53-73.

Spence, Hugh S. 'James White, 1863-1928: A Biographical Sketch.' *Ontario History* 27 (1931): 543-44.

Steele, Harwood. *Policing the Arctic.* London: Jarrolds, 1936.

Stefansson, Vilhjalmur, ed. *Encyclopedia Arctica.* Ann Arbor, MI/Hanover, NH: University Microfilms International/Dartmouth College Library, 1974.

Tammiksaar, E. 'Wrangell or Wrangel – Which Is It?' *Polar Record* 34, 188 (January 1998): 55-56.

Tansill, Charles Callan. *The Purchase of the Danish West Indies.* Gloucester, MA: Peter Smith, 1966.

Taylor, C.J. *Negotiating the Past: The Making of Canada's National Historic Parks and Sites.* Montreal/Kingston: McGill-Queen's University Press, 1990.

Thorleifsson, Thorleif Tobias. 'Norway "Must Really Drop Their Absurd Claims Such As That to the Otto Sverdrup Islands." Bi-Polar International Diplomacy: The Sverdrup Islands Question, 1902-1930.' Master's thesis, Simon Fraser University, 2006.

Timtchenko, Leonid. 'The Russian Arctic Sectoral Concept: Past and Present.' *Arctic* 50, 1 (March 1997): 29-35.

Triggs, Gillian. *International Law and Australian Sovereignty in Antarctica.* Sydney: Legal Books, 1986.

Tynan, Thomas M. 'Canadian-American Relations in the Arctic: The Effect of Environmental Influences upon Territorial Claims.' *Review of Politics* 41, 3 (July 1979): 402-27.

Vahl, M., G.C. Amdrup, L. Bobé, and A.S. Jensen, eds. *Greenland.* Vol. 2, *The Past and Present Population of Greenland.* Copenhagen/London: C.A. Reitzel/Humphrey Milford, 1928.

–. *Greenland.* Vol. 3, *The Colonization of Greenland and Its History until 1929.* Copenhagen/London: C.A. Reitzel/Oxford University Press, 1929.

Vaughan, Richard. *Northwest Greenland: A History.* Orono: University of Maine Press, 1991.

von der Heydte, Friedrich August. 'Discovery, Symbolic Annexation and Virtual Effectiveness in International Law.' *American Journal of International Law* 29, 3 (July 1935): 448-71.

Waiser, William. 'Canada Ox, Ovibos, Woolox ... Anything But Musk-Ox.' In Kenneth Coates and William R. Morrison, eds., *For Purposes of Dominion: Essays in Honour of Morris Zaslow,* 189-200. Toronto: Captus Press, 1989.

Waldock, C.H.M. 'Disputed Sovereignty in the Falkland Islands Dependencies.' *British Yearbook of International Law* (1948): 311-53.

Webb, Melody. 'Arctic Saga: Vilhjalmur Stefansson's Attempt to Colonize Wrangel Island.' *Pacific Historical Review* 61, 2 (May 1992): 215-39.

White, Gavin. 'Henry Toke Munn (1964-1952).' *Arctic* 37, 1 (March 1984): 74.

–. 'Scottish Traders to Baffin Island, 1910-1930.' *Maritime History* 5, 1 (Spring 1977): 34-50.

Wigley, Philip. *Canada and the Transition to Commonwealth: British-Canadian Relations, 1917-1926.* Cambridge: Cambridge University Press, 1977.

Williams, M.B. *Guardians of the Wild.* London: Nelson, 1936.

Zaslow, Morris, 'Administering the Arctic Islands 1880-1940: Policemen, Missionaries, Fur Traders.' In Morris Zaslow, ed., *A Century of Canada's Arctic Islands,* 61-78. Ottawa: Royal Society of Canada, 1981.

–. *The Northward Expansion of Canada, 1914-1967.* Toronto: McClelland and Stewart, 1988.

–. *The Opening of the Canadian North, 1870-1914.* Toronto: McClelland and Stewart, 1971.

Index